Energy Fundamentals of Radio

Energy Fundamentals of Radio

Joel L. Dawson

TalkingHeads Wireless, Inc.
Boston, MA USA

WILEY

Published by John Wiley & Sons, Inc., Hoboken, New Jersey.
Published simultaneously in Canada.

Library of Congress Cataloging-in-Publication Data

Names: Dawson, Joel L. (Joel Lawrence) author | John Wiley & Sons publisher
Title: Energy fundamentals of radio / Dr. Joel Dawson.
Description: Hoboken, New Jersey : John Wiley & Sons Inc, [2026] | Includes bibliographical references and index.
Identifiers: LCCN 2025025541 | ISBN 9781119897767 (hardback) | ISBN 9781119897774 (epdf) | ISBN 9781119897781 (epub) | ISBN 9781119897798 (obook)
Subjects: LCSH: Radio—Handbooks, manuals, etc. | Radio—Transmitter-receivers | Radio—Energy consumption | Wireless communication systems
Classification: LCC TK6550 .D27 2026 | DDC 621.384—dc23/eng/20250818
LC record available at https://lccn.loc.gov/2025025541

Cover Design: Wiley
Cover Image: © Daniel Bosma/Getty Images

For my parents

Judith A. Dawson

Lawrence S. Dawson, 1937–2017

Contents

List of Figures

List of Tables

Acknowledgments

This book is the end result of years of thought and work, and has been shaped by many, many people. Some of them know the impact that they have had; some of them do not. A great pleasure of writing a book is sitting down when the manuscript is complete and thinking of all the people you are grateful to, who helped you along the way, and then letting them know.

I acknowledge my parents in the dedication. It is very moving to me, now with a son of my own, to think about how they supported me. My interest in electronics arose basically out of nowhere, after I was blown away by the Apple II computer that my elementary school bought when I was in the third grade. The following Christmas, my father bought me a copy of *Basic Electricity/Electronics Vol. 1: Basic Principles*, whose authorship is unclear, but the second edition I have was revised by Robert R. Manville. I struggled mightily to understand it. It was years before I really could, but that book is a treasured possession and is in fact on my shelf as I write these acknowledgments over 40 years later. A couple of years after that book, my mother found a McGraw-Hill correspondence course on electronics. They would send written materials, along with electronic components and solderless breadboards. Both of them continued to support me with trips to our local Radio Shack, sometimes just to look at what was on the shelves, and then through the long journey through high school, college, graduate school, academia, and ultimately as an entrepreneur. Their love and support is a debt I cannot possibly repay. I am additionally indebted to my three sisters, Lisa, Monette, and Joy, whose love and support I have depended on greatly for my whole life.

In my middle and high school years, I was greatly blessed to have mentors, informal and formal, and sometimes only through the written word. I carry an enduring sense of gratitude to Alfred Morgan for his book, *The Boys' Third Book of Radio and Electronics*. I must have checked it out a dozen times from the Prince William County Public Library. It is simply a gorgeous book, and I experience it now as a generous act of service on Morgan's part. It has become another treasured possession. Robyn Walters, the parent of a high school friend, was also an amateur radio operator and helped me get my Technician Class license, as well as my first transmitter and receiver set. And if that was not enough, my high school had an electronics lab run by two gifted, hardworking, and kind teachers who ran a brilliant program: Bill Monticone and David Bell. Last but not least in this embarrassment of riches is Dr. John Dell, who taught us physics and who knew exactly what to do with students who had a burning curiosity.

The material in this book has formed in fits and starts over the last 20 years through my teaching of a series of courses. The first such course was a graduate offering at MIT, and I benefited greatly from the strong engagement of the material with those bright students. As the material matured in my mind, I was greatly aided by engaging with working professionals. I am extremely grateful to Don Mosebrook, then of Lutron Electronics Co., Inc., and Gangadhar Burra, then of Qorvo, Inc., for sponsoring and championing course offerings at their respective companies, and for the professionals who showed up for each class with their thoughts, opinions, and experience.

When it comes to the craft of book writing, I have had the pleasure of meeting and talking to living authors whose books have had a great impact on me. My fervent hope is that they will peruse this text and see their influence. In no particular order: Tom Lee, my doctoral advisor at Stanford, from whom I learned what it means to find one's voice; Jim Roberge, a teacher and mentor at MIT who taught his students both the art and the science of design; Al Drake, who showed us that a book need not be long to be impactful; Gilbert Strang, whom I knew at a distance but whose treatment of linear algebra was very formative for me; Paul R. Gray, who signed my copy of his fantastic book when I went to visit Berkeley while deciding on graduate schools; Stephen Boyd, who writes the way he teaches, with searing brevity, clarity, and generosity; and David R. Rutledge, whose unassuming *The Electronics of Radio* is a great and rare treatment of radio links as complete systems. Finally, though I have not met them, I joyfully acknowledge all of the contributors to the 2004 edition of the *ARRL Handbook for Radio Communications*, a true labor of love and a valuable resource.

At Wiley, I have been blessed to work with Brett Kurzman, Jayashree Saishankar, Dhanapriya Velumurthy, and Ryan Couch. All one can ask of one's editors is that they treat you with professionalism and fairness. I received so much more from them: understanding, patience, support, wise feedback, and encouragement, above all, to keep going. I am extremely grateful for the opportunity to publish in a great press.

I want to thank everyone who was and is involved with Eta Devices, Inc. and Eta Wireless, Inc. The sheer breadth and depth of expertise in this group of people is pure inspiration to me. And on top of the talent, the grit and determination to push forward in the crucible of our industry gives me energy whenever I think about it, which is often. Working with this group made me grow in engineering, science, business, and as a person.

To my wife Marisol, and my son Elías: everything in life is better when one shares a home with loving people. I am glad to be on this journey with you.

1

A First Look at a Radio Link

1.1 Introduction

This book is about the magic of radio communications and how we can use our shared resources to serve more people with vital communications infrastructure. For resources, we are specifically interested in *spectrum*, always a finite resource in communications networks but a special challenge with radio waves that are not easily contained in space. We are also interested in *energy*, because Nature requires it to move bits through a network, and we are an increasingly energy-constrained world.

When it comes to spectral efficiency, the literature is vast. In the decades since the publication of Shannon's famous capacity limit in 1949, and likely before, many of the brightest minds in communications have devoted themselves to hitting the absolute maximum number of bits per second in a given slice of spectrum. That single-minded pursuit has driven almost every major trend in radio communications since the author turned to the subject in the late 1990s.

In contrast, the literature on energy efficiency, while also large, has lacked the white-hot intensity of pursuit that has been poured into spectral efficiency. The very clear priority has been maximizing data capacity. System architects, consistent with this objective, set the performance requirements for all the hardware in the signal chains of a communications system accordingly. The hardware efficiency literature very dutifully accepts these specifications as set in stone, and thoroughly explores how to minimize energy consumption in the aftermath.

That is way, way, *way* too late.

When you set targets for "performance," the physics of hardware says that you simultaneously set the minimum power consumption. It goes almost without saying you can do *worse* in your implementation: maybe you use a circuit architecture that is not the best; maybe your design is not well executed and uses more power

Energy Fundamentals of Radio, First Edition. Joel L. Dawson.

than it needs to; maybe you decided that the best electronic components are too expensive. But you will never do better than Nature's lower power limit for accomplishing what we will refer to as The Task.

There is ample reason, then, to reorient the system problem toward a more balanced treatment of spectrum and energy. But it turns out that now, in 2025, is a particularly good time to take a new look at the problem. Data traffic in communications networks around the world has been compounding annually at a rate of 20–30% for decades. If we accept that Nature requires a certain amount of energy to move a bit, then that means that Nature's energy cost for running networks has also been compounding at a similar rate. What this means is that in the early days of cellular networks, energy really did not matter from the standpoint of running a business. You have to pay for energy, sure, but with the data traffic in those days, paying employees and rent for the cell tower easily and completely overwhelmed energy as an operating expense. For all intents and purposes, energy was *free*. But exponential rises being what they are, quantities that are trivial for a long time explode out nowhere, seemingly overnight. And that is where we are as we transition to 5G deployments around the world. The big telecom companies have to reckon with large energy bills really for the first time, and so suddenly, there is an imperative from the very top of the industry to do better with energy.

Finally, there is one other major problem that this book seeks to address. Radio links touch on a very large number of disciplines: linear system theory, nonlinear system theory, information theory, circuit theory, electromagnetics, thermodynamics, digital signal processing, materials science, and numerical methods form but a partial list. The list grows longer when one looks at the governing mathematics, as it helps to have had broad exposure to algebra, geometry, calculus, multivariable calculus, differential equations, linear algebra, and complex analysis. Too often, the response to the bewildered practitioner who wants to deepen their understanding is that they should come back after they have read and digested 20 "classic" books or more, plus some untold number of journal papers.

This is not bad advice, exactly. The problem is that it actually undersells the scale of the work ahead. As big a task as it is to read and digest these books and papers, the *real* work begins afterward. The practitioner must next wade through this mountain of theory, and put it into some sensible, personalized order to enable application. This is the work of years and decades, and the truth is that some of it cannot be avoided. The author's position, though, is that *some* of this work can and should be avoided. The literature is choked, for example, with esoteric techniques that do not survive the refiner's fire of application.

This book is offered, then, in a spirit of helpfulness to the practitioner. In his own journey, the author has run the gauntlet of undergraduate and graduate courses and exams, spent years as a professor at a world-class research institution, and then endured the shock of learning what it takes to move an unwilling industry in a new direction. *Everything* about this book is informed by the triumphs and failures of that journey. At various points along the way, he has been asked by someone new to radio technology if there is "one book" that he might recommend to "learn the basics." After many years, and with the aid of many deep intellects who knowingly and unknowingly shaped his thoughts, this book is his answer.

See what you think.

1.2 A Radio Link: The Basic Parts

We'll kick things off with a cartoon of a radio or wireless link as shown in Figure 1.1. Technically, there are quite a lot of ways to communicate "wirelessly." We could use the optical spectrum by modulating lasers; we could use acoustic waves, as we do when we simply talk; we could use smoke signals, or flag semaphor. While all known communication methods are bound by a specific information-theoretic law called the Shannon limit, we will focus here on communications using electronic machines as depicted in Figure 1.1. The electromagnetic spectrum that contains the lion's share of users worldwide starts at the bottom end with the amplitude modulation (AM) spectrum at 540 kHz, and tops out in polite company somewhere around 60 GHz for millimeter wave transmissions. In just over a century since Marconi's triumphant transatlantic transmission on December 12, 1901, wireless communications have gone from a total astonishment to a vital part of the global infrastructure on par with highways, water, and electricity distribution. Who could have predicted?

A beginning of understanding radio is to look at the underlying physical laws that we use in service of our information-sharing goal. Let's examine.

1.3 On Using Nature's Laws to Build Machines

The scientist discovers for us the Laws of Nature, and it falls to the engineer to put these laws to use by building machines that serve humanity. It is extremely helpful if these Laws can be expressed mathematically. The engineer is then able to use them with a great deal of precision.

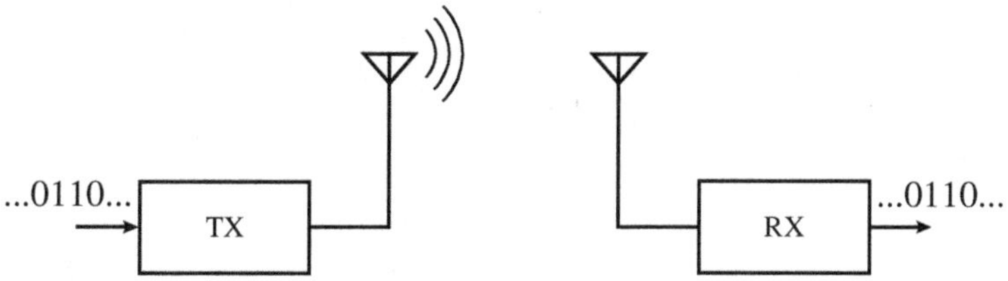

Figure 1.1 A cartoon radio link.

A simple mechanical balance is actually a powerful example. Suppose that you are presented with the balance and set of weights as shown in Figure 1.2, and suppose further that we are able to slide the balance to vary l_1 and l_2. We could operate this simple machine as a mechanical adder as follows:

1. Set $l_1 = l_2$ by placing the fulcrum exactly in the middle.
2. Pick our operands and stack them, say, on the left side.
3. Perform a systematic search of blocks to place on the right side, searching until balance is achieved.
4. The block on the right side gives us the answer that we sought, which is the sum of the operands that we stacked on the left side.

See Figure 1.3.

It's worth playing around with this on your own. What you'll find is that subtraction, multiplication, and division are all reachable for this mechanical calculator.[1]

The overall point is that we are exploiting the principle of balance to build a mechanical calculator. Or as our physicist friends might say, "when the net torque on a rigid body is zero, so too is its angular acceleration." The engineering lies in taking advantage of this principle to build a four-function mechanical calculator.

And so it goes in all of engineering. The physical principles that we make use of, and the ends that we achieve, vary widely. But the underlying spirit is the same. In the next section, we take a look at the physical principles at the very heart of radio communications.

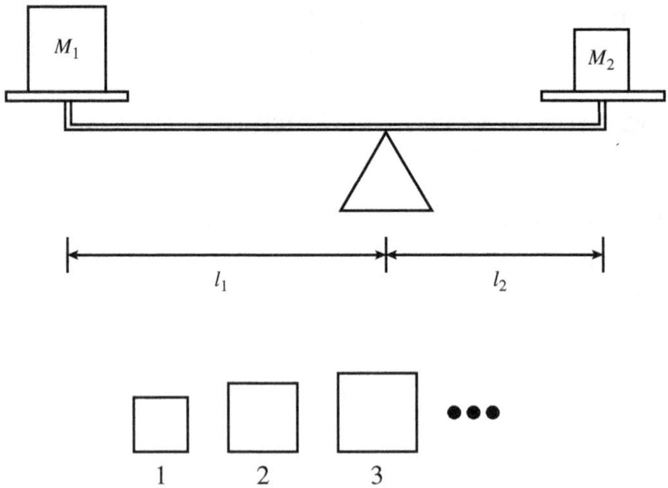

Figure 1.2 Setup for a mechanical calculator.

1 Multiplication and division will require you to be flexible about where you place the fulcrum.

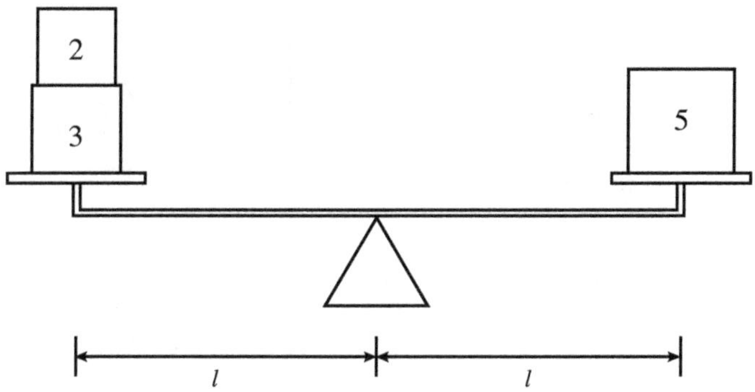

Figure 1.3 What is 2+3? It is 5, evidently.

Figure 1.4 Force on test charge q_{test} due to Q_0.

1.4 The Key Natural Law for Radio Communications

Understanding the key natural laws for radio communications can start by examining the situation in Figure 1.4. Here, we have a charge of Q_0 coulombs frozen at the top of, say, a nonconducting wand. How do we know that the charge is Q_0? We know because when we bring a test charge q_{test} in the vicinity of Q_0, that test charge experiences a force that is *exactly*

$$\vec{F} = \frac{Q_0 q_{\text{test}}}{4\pi\epsilon_0 r^2}\hat{r}. \tag{1.1}$$

We are in the habit of describing Q_0 as creating an "electric field" of magnitude $|\vec{E}| = \frac{Q_0}{4\pi\epsilon_0 r^2}$. And any point charge q immersed in that field will feel a force of magnitude $q|E|$.

Notice that we could stop right here and get busy constructing a communications machine based on this law. We could, for example, encode our information as

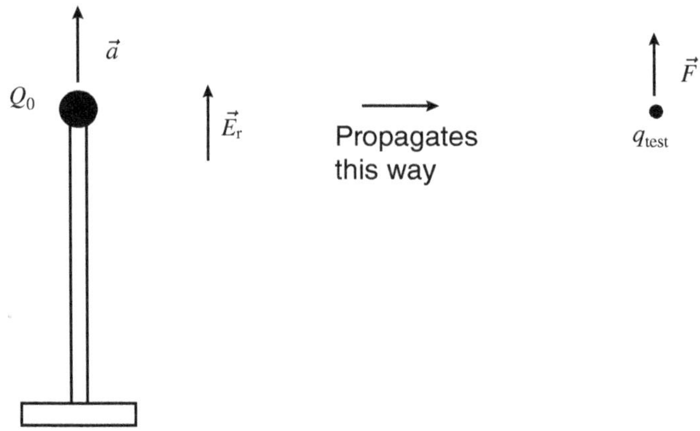

Figure 1.5 Accelerating charges is where the action begins.

spatial positions of Q_0. In this scenario, we keep q_{test} constant, and simply record the force required to keep it still. The range over which this technique would be effective would depend on how sensitive our force measurement sensor is. In any event, the strength of the electric field would fall with distance as $1/r^2$. We will come to appreciate this as a very rapid drop off.

Now at this point in our development, if we pay enough attention and are careful with our measurements, we might discover a strange and wonderful gift that Nature gives us when we *move* Q_0. The gift is this: when Q_0 accelerates, Nature launches a radiative electric field that points in the direction of acceleration, radiates perpendicularly to the direction of acceleration, and *falls off according to* $1/r$ *instead of* $1/r^2$! See Figure 1.5. That beautifully slow $1/r$ decay with distance is the key Natural Law that makes shockingly long-distance communications possible. It is what enables us to communicate with the Mars Curiosity Rover, for example, which is 170 million miles from Earth.

We will see in Chapter 7 that an E field decay of $1/r$ corresponds to a power flux density decay of $1/r^2$. For most people, even those with deep backgrounds in science and engineering, it will be difficult to muster the appropriate level of shock and gratitude for this slow decay. The best way to appreciate the gift of radiating E fields is via comparison with the gold standard of high-throughput communications, the transatlantic optical fiber.

We love optical fiber. The MAREA cable, operational since 2018, spans 6644 undersea kilometers beneath the Atlantic Ocean between Virginia Beach in the US and the city of Bilbao in Spain. It combines eight fiber pairs, and boasts a capacity of 28 terabytes per second per fiber pair. The beauty of optical fiber is that there is no issue of fiber pairs placed near one another causing mutual interference.

One could, in principle, just keep laying parallel cables and expand this information super highway.

But.

Cables under the ocean don't last forever. They are expected to last somewhere on the order of 25 years, which is not that long when it comes to infrastructure. In today's dollars, the cost to install that cable was on the order of \$300 M. And as amazing as optical fibers are, they exhibit a signal power loss on the order of 0.15 dB/km.

What's that, you say? 0.15 dB/km is not so bad? You're right, in part, in that it is state-of-the-art and amazing. On the other hand, it is still *exponential decay.* Let's look as an example at the signal attenuation that occurs between 5000 and 6000 km. For our wireless link, the attenuation is

$$\text{Att.} = 10\log_{10}\left(\frac{5000}{6000}\right)^2 = -1.6\ \text{dB}. \tag{1.2}$$

For the cable, that attenuation comes to

$$\text{Att.} = -1000\ \text{km} \cdot 0.15\ \text{dB/km} = -150\ \text{dB}. \tag{1.3}$$

Fifteen orders of magnitude is massive. We are comparing multiplying by 10^{-15} to multiplying by a factor of 0.69.

In practice, what this means is that we do not span the Atlantic with uninterrupted cables. Instead, we place repeaters every 70–100 km.[2] And these repeaters are not mere amplifiers. They must demodulate the signal and detect the underlying digital symbols, and then *retransmit* our signal with a reset signal-to-noise ratio. And the power source for each of these repeaters is not a local, submarine diesel generator (with all the problems associated with keeping it fueled). The power to run each repeater must be routed separately along the cable from one of the two ends.

By contrast, amateur radio operators who specialize in "QRP" operations routinely muster communications links of thousands of kilometers on a transmit power of 5 W or less. To be sure, the data capacities are nowhere near a terabit per second. But the truly long-distance possibilities of radio links as we currently build them are a fundamental advantage that is easy to lose track of.

Both wireless and cabled telecommunications will always have their place. With wireless links, because electromagnetic waves travel everywhere in space so easily, spectrum can become a scarce resource in a way that it does not for cabled links.

2 Actually, the MAREA cable uses the unusually short repeater spacing of 56 km.

1.5 A Simple Radio Transmitter and Receiver

Newly enamored of our Key Natural Law, we devote ourselves to designing machines that exploit that law in order to realize a communications link. It turns out that in order to launch one of these special, radiating electromagnetic waves, we do not have to mechanically move a charged object. It is enough if we accelerate charges by exciting time-varying electrical currents. Figure 1.6 shows how we might establish a crude communications link using on-off keying. It is simple enough, and in the case of the venerable crystal receiver shown, we do not even need to provide a power supply(!). Generations of us got our start in electronics by connecting headphones to such receivers and listening to AM broadcast stations.

As with our mechanical calculator from Section 1.3, mental clarity about Nature's laws leads very quickly to a machine that carries out our desired functions. After that, pencils down, right? What more is there to do?

1.6 The Most Important Question in Engineering

After the first flush of triumph from getting our communications system to work, questions arise. What if we want to have two transmit/receive pairs in close proximity, but have independent, noninterfering exchanges of information? Well, it turns out that the variable capacitor and inductor in Figure 1.6 allow us to *tune our transmit/receive pairs to different frequencies.* We can thereby arrange for multiple TX/RX pairs, with each one assigned to use a different carrier frequency ω.

How far can we pursue this trick? We will find that the range of carrier frequencies for which radios are convenient to build is not infinite. And with that, we

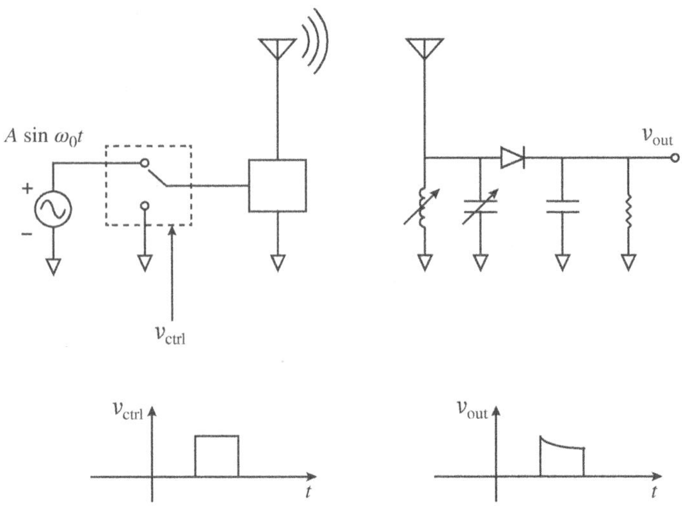

Figure 1.6 A very simple radio link.

run headfirst into our first resource constraint problem. Spectrum, or our range of possible ωs, is finite. How can we know if we are putting this finite resource to the best possible use? How can we, for example, maximize our rate of information exchange given a fixed amount of spectrum? How can we maximize the number of users we can accommodate? These questions have driven communications theory for most of the 20th century and for all of the 21st century up to the time of writing. Understanding the maximal use of spectrum is a major theme of this book.

Very good. Suppose, now, that we want to make our transmitter and receiver portable, so we can communicate on the go. Doing so requires that we carry a source of energy around with us (at least for the transmitter). The less energy we must carry, the smaller and lighter our energy storage will be, which translates to a smaller and lighter communications machine. Following this line of thought leads to a question that is of towering importance, but that has not received the same attention as the spectral efficiency problem: what is the minimum energy we require to send a bit, or equivalently, what is the minimum power we must provide to support a given bit rate?

We thus arrive at two versions of the most important and interesting question in any engineering project: *are we using the minimum amount of resources to achieve the required level of performance*? In the particular case of radio links, "performance" is usually measured in bits per second of information communicated without error. Our resources are spectrum, and energy. But the list of examples more broadly in engineering is endless once you start to look through this lens. Race cars carry a finite amount of fuel, and engineers are strongly motivated to maximize energy efficiency because less fuel equals less weight and more speed. In various types of manufacturing, scarcity of materials manifests as the cost of manufacturing. What is the best product we can build, at scale, while minimizing cost?

Returning to radio links, some readers may be quietly unmoved, especially when it comes to energy. Their objection might go something like this: "Saving energy is nice, but at the end of the day what I care about is performance." If this is you, hopefully just seeing this thought written explicitly is enough to give you pause. There is simply no way to separate energy usage and performance. If you push hard enough for performance, what you find is that you are *always* ultimately energy constrained. In a resource-constrained environment, if you are not maximizing efficiency, you are not maximizing performance. Period.

1.7 Overview of This Book

And with that, we set the table for the 11 chapters to follow. The system to come back to is the generic transceiver shown in Figure 1.7. We will look at the various pieces of Figure 1.7 in some detail, but use this overall diagram to keep yourself oriented on the journey.

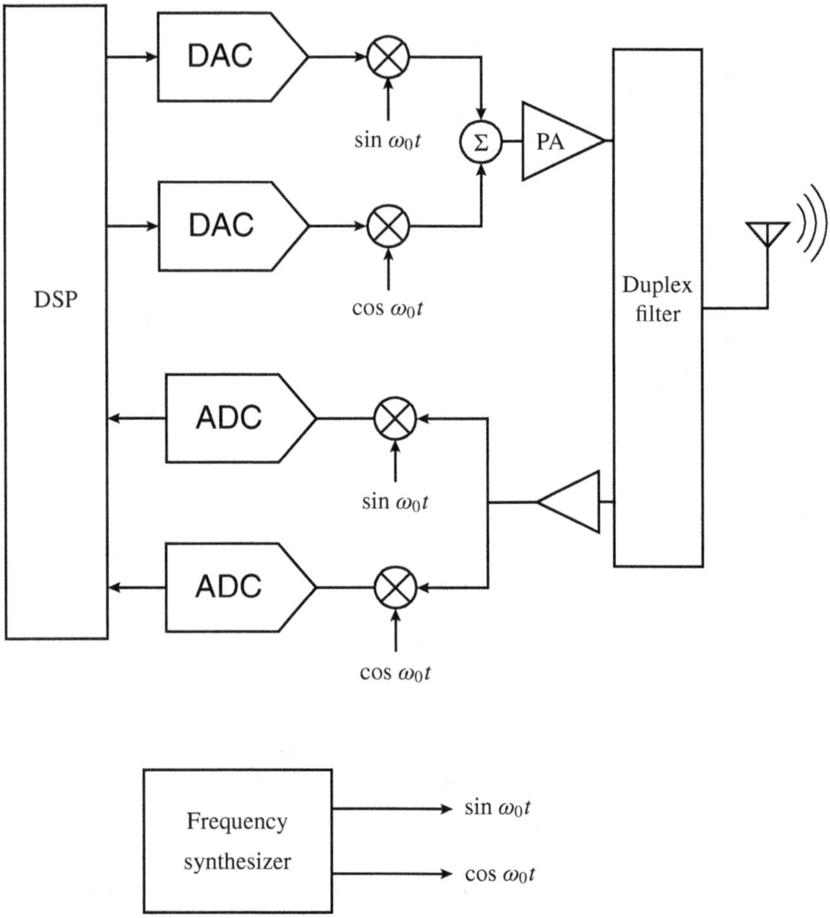

Figure 1.7 Overview of a radio transceiver.

Chapter 2 is a condensed treatment of the linear system theory and signal processing concepts that are foundational to modern radio links. Chapter 3 covers some fundamental concepts in analog electronics. It does so with a focus on energy that frees the reader to think of system implications without getting bogged down in the intricacies of analog circuit design.[3] Chapter 4 focuses on the physics and mathematics of noise, particularly thermal noise. Chapters 2 and 4 are important background for Chapter 5, which discusses the mighty Shannon capacity limit. It is no overstatement to say that this singular result has driven just about everything

3 The author is a former analog circuit designer. If you are too, *do* give this chapter an extra careful look!

in communications for the last few decades. And then in Chapter 6, we take the first step beyond this huge theoretical result and look at modulating signals for communications in the real world.

A recurring theme of this book is that a maniacal focus on data capacity has been the default, while the energy cost of getting bits through a link has been secondary. And it's not that the Shannon limit leaves power out completely: signal *power* is in the numerator of the signal-to-noise ratio, after all. It just turns out that the physics of radio links allows us to get a much more sophisticated understanding of energy costs if we diligently apply all that we know.

So, we move on to Chapter 7, reacquainting ourselves with the fundamental physics of electromagnetic wave propagation. Chapter 8 attempts to tame the vast, wild jungle that is the antenna literature. Chapters 9 and 10 are a primer for practitioners on what we can do with multiple antennas. And then Chapter 11, a big one, is intended to introduce readers with a wide variety of backgrounds to the implementation energy costs of high-capacity radio links. It is hard to compete with the austere brevity of something like the Shannon limit, but over many years the author has found thinking this way about implementation has yielded a clarity that is comparable and...thrilling.

We tie things together with Chapter 12. It is not so long or difficult, and comes together rather quickly. A reasonable way to approach this book, in fact, is to read Chapter 12 *first*, so that you know where we are headed. Then you can either read the rest of the book in order, or pick and choose results in Chapter 12 to follow-up on and free yourself to simply follow your curiosity. Even if your goal is to reckon with the whole book, this apparently meandering approach *will* get you there with time.

Finally, a word on the intensely mathematical nature of some parts of this book. The derivations in particular are directed to graduate students and those of a more academic bent, and should not be taken as a barrier to entry to this wonderful subject. The author has thought much about this recently, as recent work has put him in conversation with field engineers who install and configure cell towers, and whose job it is to guarantee coverage to thousands of customers. Not all of them can recite to you the mathematical form of the Shannon limit or the Friis equation, but the careful ones who learn by experience have gained deep physical insight. If it has been a long time since you studied complicated mathematics, or if derivations bore you to tears, know that the author has made every attempt to *prioritize insight in the text.* Trust yourself to gravitate to the level of mathematical expression that suits you. And know that you can always come back to the derivations later if you like.

2

Highlights from Linear System Theory

A towering ambition for this book is that it be useful to the self-learner as a standalone text. We will treat topics in a chapter, or in a section of a chapter, to which many long, thorough, well-written books have been devoted. It remains true that to be a specialist in a particular subject, a mere chapter of reading will not do. However, if your goal is to understand wireless communications links, which draw on many deep subdisciplines, the good news is that you can assemble a powerful, coherent, *actionable* picture of the whole system if you know, deeply, the key fundamentals from select subdisciplines.

With this chapter, you will see a first example of the form. Linear system theory is normally and deservedly treated in a multiple-term course sequence at top engineering schools. What is presented here is a highly curated conceptual core that has served the author well in many academic and industrial contexts.

2.1 What Are "Signals," Come to Think of It?

We are all familiar with the concept of measurement, and with a nearly endless list of things to measure: temperature, voltage, current, mechanical pressure, sound intensity, light intensity, acceleration, velocity, weight, height, tension, and so on. We transition from thinking about individual measurements to thinking about signals when the quantity that we measure *varies with time*. When that is true, the particular sensor output may in fact have a story to tell. It may be important to take many measurements, and pay attention to the order in which the measurements were taken, as well as to the time interval between measurements. As it happens,

Energy Fundamentals of Radio, First Edition. Joel L. Dawson.

a reasonable working definition of a signal is that it is a *time-ordered sequence of measurements, for which the interval between measurements is known.*[1]

We care about ordered sequences of measurements because ordered sequences of measurements carry information. For example, one can think of the human eardrum as an exquisite sensor of air pressure. If we ignore all the other miracles in the design of the ear, we may simplify the act of hearing to be the processing of a continuous stream of ear pressure readings at each eardrum.

Now, close your eyes and imagine listening to music that you really enjoy. (Even better, put down this book for a few minutes and listen to some music that you really enjoy right now! *Do remember to close your eyes!*) How's that for some powerful signal processing? Now go further, and imagine the last time you went to a great outdoor event like a parade, or a concert, or a sporting event. Your eyes take in a constant stream of measurements of the light that falls on them, your eardrums doggedly continue their air pressure measurements, your nose samples the chemicals in the air...and your mind takes all of this time-sequenced data, and more, and processes it into an experience that you never forget.[2]

Bringing it all back to communications systems, what we wish to do is transmit signals, in all of their richness, from one location to another. We focus in this book on a particular mode of wireless transmission, which involves electromagnetic waves. By way of preparation for thinking about signals, let's consider a low-stakes problem: the signal output from the simple kinematic sensors in a car.

Suppose that we have a car with a two-dimensional (2D) accelerometer. That is, it can measure the instantaneous acceleration in both the direction that the car is going, and in the direction orthogonal to that as shown in Figure 2.1. For the first part of the problem, we are interested in using the on-board accelerometer signal to measure the stopping distance in the case of a sudden stop. Suppose the data that we get from accelerometer is as shown in Figure 2.2. Let's suppose further that we have speedometer data that tells us the speed as a function of time, and that its output looks like that shown in Figure 2.3. Your challenge is this: using

---

1 If you want to be more general, you might not be so rigid about the measurements being spaced out in time, but rather allow for spatial variation as well. For example, if we are representing a pixelated photograph as a stored signal, then our measurements might be of red, blue, and green (RBG) intensities as a function of the (x, y) position in the frame.

2 It goes *almost* without saying that the signal processing carried out by the human brain in this example is of an entirely different character than what goes on in digital processing machines at the time of this writing. On top of these sensory inputs, we package with the experience a memory of who we were with, the nature of those interactions, what our emotions were like, any gap between our expectations and reality, and so much more. All we mean to highlight here is the profound experiential gap between the snapshot of a single measurement and the richness of time-ordered data.

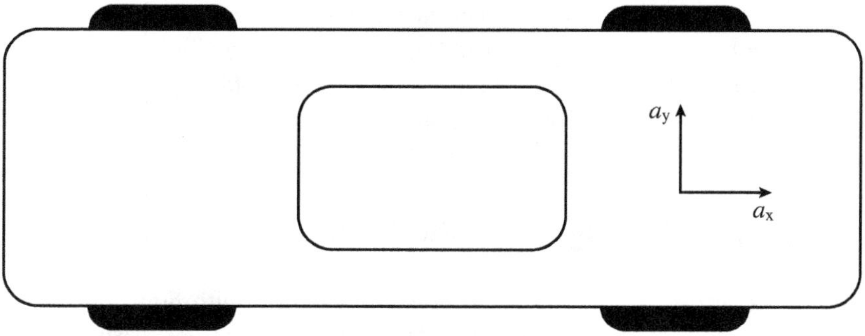

Figure 2.1 A car equipped with a 2D accelerometer.

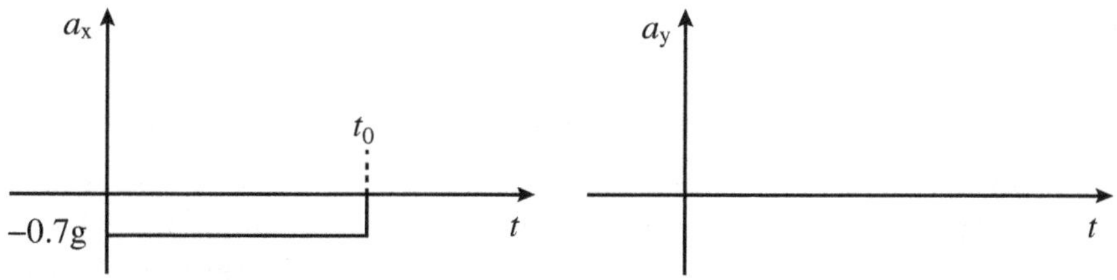

Figure 2.2 Example accelerometer data.

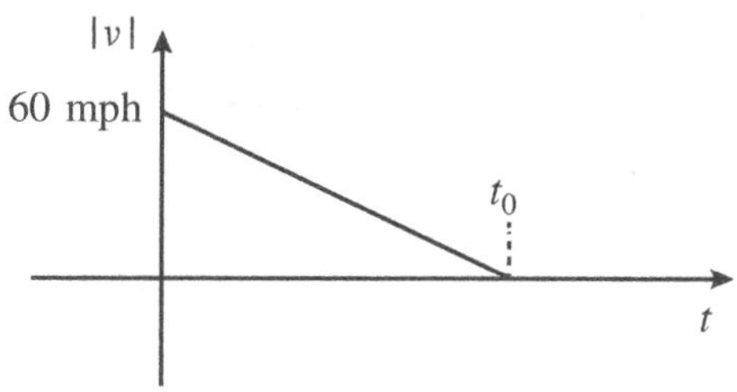

Figure 2.3 Example speedometer data.

the accelerometer and the speedometer data, determine the distance this vehicle required in order to come to a full stop.

Those of you from a physics background may laugh at this "simple" kinematic problem and just solve it. But if this is *not* you, do not despair if you have forgotten (or never learned) how to deal with such calculations. Think carefully and focus on the fact that *knowing the sensor data as a function of time is key.* If we just know the speedometer readout at a particular time, that's all we have. But looking at the time-ordered data points, a dynamic picture emerges of the car coming to a stop under uniform acceleration. Even if you cannot get the final quantitative answer, devote some time to seeing what qualitative features of the solution you can deduce. How would you expect the stopping distance to be affected, for example, by a higher initial speed? And for those of you who are able to simply write down a quantitative answer to this problem, it is doubly important for *you* to spend some time thinking about the qualitative aspects. A deep understanding of a physical situation always has both quantitative and qualitative dimensions.

2.2 Representing Signals in the Time Domain

In wireless communications, we are of course concerned with how our signals evolve over time. It will turn out, however, that there will be ample reason to look at "frequency-domain" representations of our signals as well. Before we get to that in Section 2.3, let's slow down and look in more detail at what we mean by a time-domain representation.

2.2.1 Continuous-time Representations

Suppose that we have an electronic amplifier whose output is a voltage, $v(t)$, as shown in Figure 2.4. What we mean when we write $v(t)$ is that for *any* value of time instant t, no matter how close it is to another time instant, we have a distinct value for $v(t)$. We could choose a vector of time values according to

$$t \in [0 \text{ sec}, 1 \text{ sec}, 2 \text{ sec}, 3 \text{ sec}] \tag{2.1}$$

or according to

$$t \in [0 \text{ sec}, 1 \times 10^{-15} \text{ sec}, 2 \times 10^{-15} \text{ sec}, 3 \times 10^{-15} \text{ sec}] \tag{2.2}$$

or according to times scattered over extremely irregular intervals[3]

$$t \in [0 \text{ sec}, h \text{ sec}, e \text{ sec}, \pi \text{ sec}, c \text{ sec}]. \tag{2.3}$$

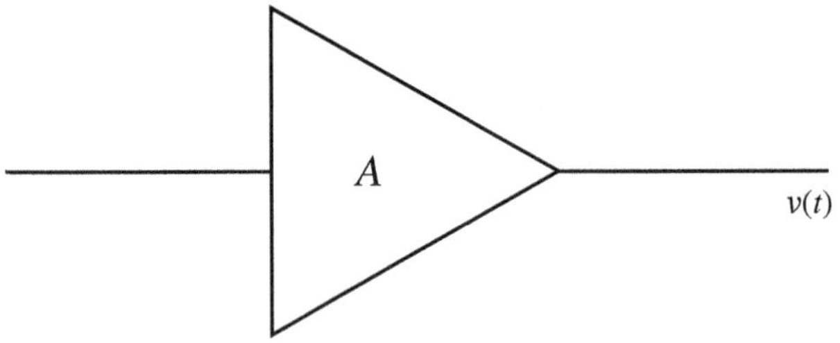

Figure 2.4 A voltage amplifier.

3 $h \approx 6.6261 \times 10^{-34}$; $e \approx 2.7182\ldots$; $\pi \approx 3.1415\ldots$; $c \approx 2.9979 \times 10^{8}$.

It does not matter. For the purposes of representation in our analyses, the output $v(t)$ has a defined value for any t we can think of.

Depending on your background, this may just seem normal and unremarkable. If our signals then happen to be expressible in closed forms, as in the examples of Eq. 2.4,

$$\begin{aligned} v_1(t) &= k \cdot t \\ v_2(t) &= A \cdot \sin \omega t \\ v_3(t) &= (1 - e^{-\alpha t})u(t), \end{aligned} \tag{2.4}$$

then, in our graphical representations, we draw these functions as smooth, unbroken lines as shown in Figure 2.5.

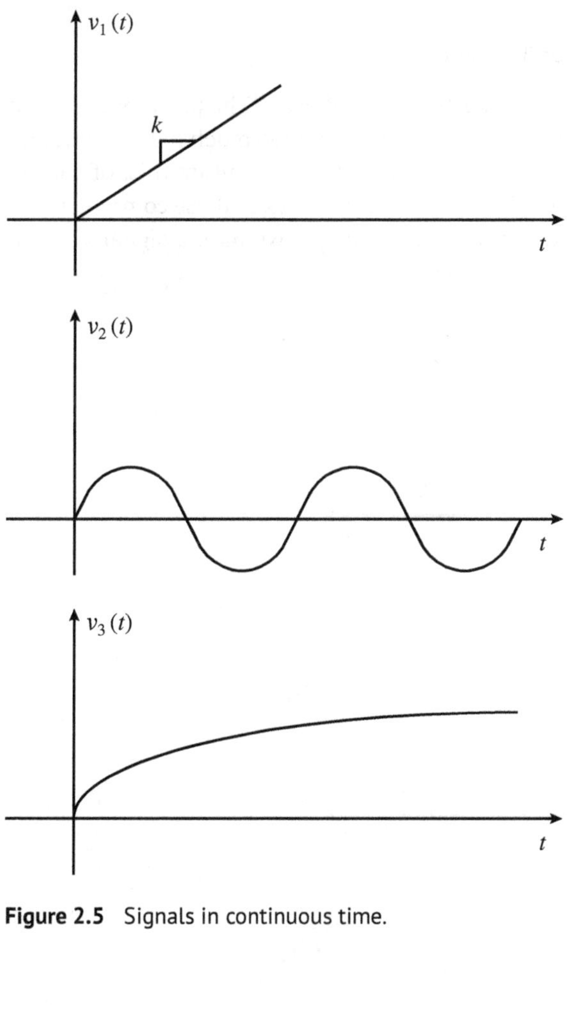

Figure 2.5 Signals in continuous time.

2.2.2 Discrete-time Representations

By contrast with its continuous-time (CT) cousin, the signal $x[n]$ is only defined for integer values of n. That is to say

$$n \in [\ldots, -3, -2, -1, 0, 1, 2, 3, \ldots]. \tag{2.5}$$

For noninteger values of n, $x[n]$ is not defined. We capture this graphically in Figure 2.6. It is actually quite important to fix in your mind that, in discrete-time (DT) representations, these integer-indexed signal instants are *all* we have by way of information about the signal. If a DT signal was derived by sampling a CT signal, for example, it does no good in general to make assumptions about what may have happened between the samples. Unless...unless...certain conditions are met. And what might those conditions be?

2.2.3 Sampling and Hazards Thereof

A decent qualitative answer to this question is as follows: "The primary condition is that the CT signal we are sampling not move around too much in between sampling instants." There are formal, mathematical ways to capture this, of course, and we will get to those. But first, it is helpful to step back and use common sense to see where we might get into trouble. Suppose that we have a signal $x(t)$ and

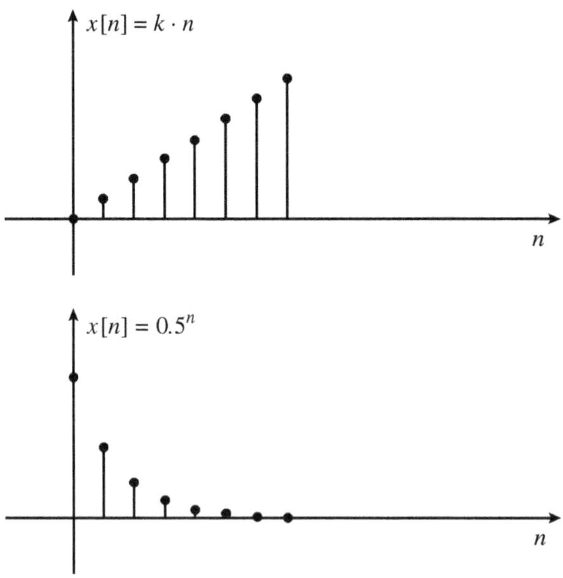

Figure 2.6 Signals in DT.

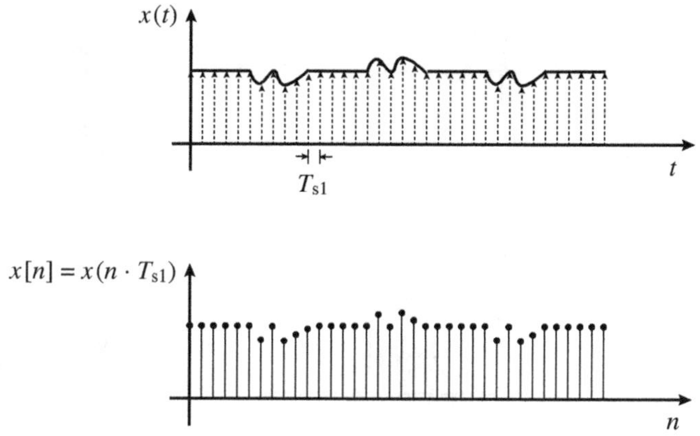

Figure 2.7 Sampling "often enough" to represent a signal well in DT.

we decided to make measurements every T_s seconds of its instantaneous value. We could construct a DT signal based on these samples according to

$$x_d[n] = x(nT_s). \tag{2.6}$$

Let's now look graphically at what happens when we sample at two different sample periods, T_{s1} and T_{s2}. First, T_{s1} is shown in Figure 2.7. Now if you had to guess, just based on looking, you would probably say that we have taken enough samples to "roughly capture what is going on." You would not bet your life on it, because you are worldly enough to have heard of things like "aliasing" and the "Nyquist sampling rate," which sound very mathematical, and you haven't seen any math yet, right? But in your heart, you would be surprised if a sampling period of T_{s1} failed to do the trick. And it turns out that your intuition would be correct.

Now let's contrast this by sampling with a longer sampling period T_{s2}, shown in Figure 2.8. This time, using only the eye test, we can see that our sampling process woefully misrepresents the underlying CT signal. There is way too much wiggling in between the samples for us to say that the samples capture the signal in a meaningful way.

It bears emphasis that you should give conceptual insight at least as much weight as you give to analytical developments such as the one to follow. Especially, look for instances when your conceptual understanding *clashes with* or *contradicts* the implications of your mathematical analysis. Learn to prize these clashes and to resolve them. Resolving them requires work and persistence, but unfailingly leads to great steps forward in understanding. Sometimes such efforts even lead to great discoveries. Be persistent, and put in the work.

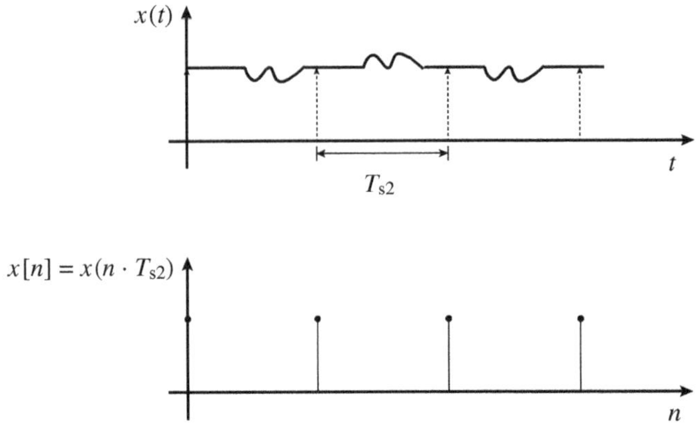

Figure 2.8 An example of extreme undersampling.

Now, we proceed to analysis. Again, we collect our samples according to

$$x_{\mathrm{d}}[n] = x(nT_{\mathrm{s}}). \tag{2.7}$$

Consider next two possible choices for $x(t)$, which we will call $x_1(t)$ and $x_2(t)$. Define these two signals as:

$$x_1(t) = \sin(2\pi ft) \tag{2.8}$$

and

$$x_2(t) = \sin\left(2\pi\left(f + \frac{k}{T_{\mathrm{s}}}\right)t\right) \tag{2.9}$$

where k is an integer. What we can now show is that these two signals x_1 and x_2 will produce exactly the same samples. First, we take samples of the first signal:

$$x_{\mathrm{d1}}[n] = \sin(2\pi fnT_{\mathrm{s}}). \tag{2.10}$$

Straightforward enough. Now for samples of the second signal:

$$x_{\mathrm{d2}}[n] = \sin\left(2\pi\left(f + \frac{k}{T_{\mathrm{s}}}\right)nT_{\mathrm{s}}\right). \tag{2.11}$$

Let's do some manipulations in the argument of the sin:

$$\begin{aligned} x_{\mathrm{d2}}[n] &= \sin\left(2\pi fnT_{\mathrm{s}} + 2\pi\frac{knT_{\mathrm{s}}}{T_{\mathrm{s}}}\right) \\ &= \sin\left(2\pi fnT_{\mathrm{s}} + 2\pi kn\right). \end{aligned} \tag{2.12}$$

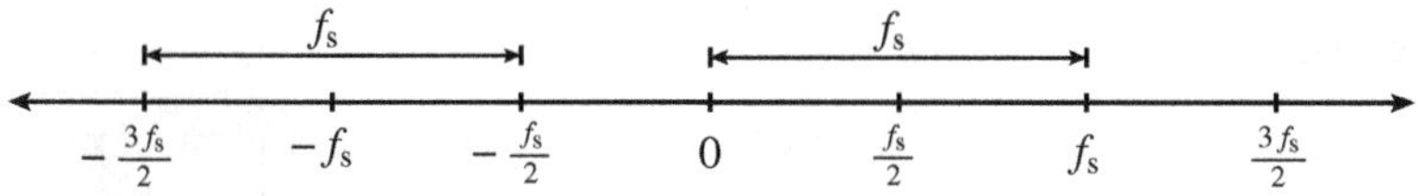

Figure 2.9 A frequency number line. For a sampling rate of f_s, there will be no ambiguity among frequencies in a band that is f_s Hz wide.

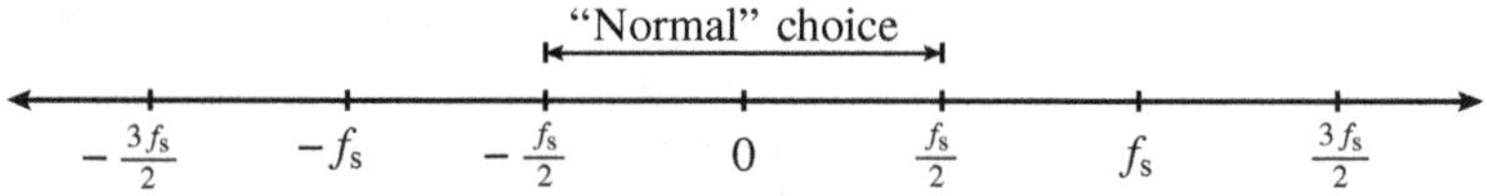

Figure 2.10 We normally enforce the assumption that the input to a sampler is restricted to frequencies of magnitude less than $f_s/2$. An important exception is subsampling receiving architectures, which seek to downconvert frequency bands that are f_s wide but centered at high carrier frequencies.

Remembering that both k and n are integers, we have that

$$\begin{aligned} x_{d2}[n] &= \sin(2\pi fnT_s)\cos(2\pi kn) + \cos(2\pi fnT_s)\sin(2\pi kn) \\ &= \sin(2\pi fnT_s). \end{aligned} \tag{2.13}$$

That is, as advertised, both $x_1(t)$ and $x_2(t)$ result in exactly the same samples.

This essential ambiguity when it comes to sinusoids and samplers is formally referred to as *aliasing*. If we were to draw a number line for frequencies, we can say that any two sinusoids separated by $f_s = \left(\frac{1}{T_s}\right)$, or an integer multiple thereof, produce exactly the same samples, as illustrated in Figure 2.9. To avoid ambiguity, then, we arrive at the condition that allows us to make assumptions about what happens in between the samples that we alluded to at the end of Section 2.2.2. The key condition is that we make sure ahead of time that the input to our sampler is bandlimited such that it occupies a swath of spectrum no wider than f_s Hz. A typical choice for that golden zone is between $-\frac{f_s}{2}$ and $+\frac{f_s}{2}$, as shown in Figure 2.10. And this, friends, is the origin of the so-called Nyquist sampling rate. The default for that golden zone is, in fact, normally between $-\frac{f_s}{2}$ and $+\frac{f_s}{2}$,[4] and with this choice we may say that *if we keep the highest frequency component of the sampled signal at less than half the sampling rate*, we can proceed with the rest of our signal processing without worrying about aliasing.

4 You can go with other choices as hinted at in the number line figures, as long as you keep the swath of frequencies no wider than f_s Hz. This is the animating insight behind subsampling receivers.

2.2.4 Discretizing Amplitude and the Concept of Bits

It is a fact of modern engineering that our fastest, most efficient, and most powerful computation engines are built around digital microprocessors. At the very foundation of such computation engines is the electronic flip-flop, which has the following remarkable properties:

- It has two distinct, stable states. By "stable," we mean that it persists in one of these states indefinitely; when perturbed with a perturbation up to a certain strength, a restoring force strongly pushes it back to its unperturbed state.
- It can be made to switch between these states extremely quickly.
- It can be made very small, such that millions of them can be placed in an area of only a few square millimeters.[5]
- The states are "far enough apart" in the signal space such that the state can be determined with a vanishingly small probability of error.
- When it is in one of these states, it consumes almost no power. When switching between these states, there is some power consumed, but as of now (here early in the 21st century) this dynamic switching power had been halving roughly every 18 months for an astonishing, decades-long run.

It is worth pretending for a moment that you know nothing about computers, and that you are tasked with designing a computation engine around this remarkable little widget.

Your first concern might be how to represent numbers. At first, wedded to base 10 arithmetic since learning to count on your fingers as a child, this seems hopeless. But then you open up a little, and a primary insight from which almost all others would follow is that you could do computation with binary representations, or base 2. That is, one of the stable states of the flip-flop is designated as a "1," and the other as a "0." In our machine, we could represent numbers by parallel sets of flip-flops. Consider the bank of flip-flops shown in Figure 2.11.

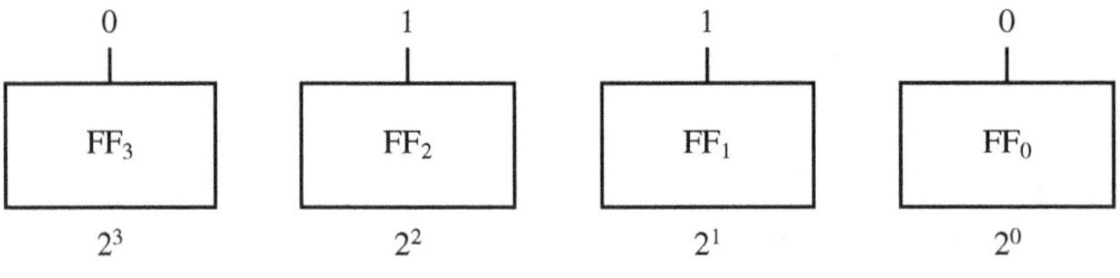

Figure 2.11 A four-bit digital register composed of four flip-flops.

5 With the reasonable expectation that all of them will work. Think about *that*, and marvel.

In the figure, the number we have stored is, in binary, "0110." To convert it back to our native decimal, we calculate according to

$$0 \times 2^3 + 1 \times 2^2 + 1 \times 2^1 + 0 \times 2^0 = 5. \tag{2.14}$$

In all, if we have m flip-flops, it allows us to represent a total of 2^m digital codes. Remember this result.

It's a long way from here to a digital computer, of course, but we can take another step forward by recognizing that once we have a way to represent and store numbers, the next thing we could use is a set of logical operators that take one or more binary digits as inputs and produce one or more binary digits as outputs according to a well-defined set of rules. It turns out that this particular problem was tackled by none other than Claude Shannon, of Shannon limit fame (getting to that in Chapter 5), as his master's thesis for the Massachusetts Institute of Technology, submitted on August 10, 1937, which he worked out while at the famous Bell Telephone Laboratories (then in New York City). What we find is that the same electronic circuit elements that give us flip-flops also give us the three foundational binary logic functions of Boolean algebra, as shown in Figure 2.12. What may not be obvious, but nevertheless is miraculously true, is that if we combine enough of these three basic "logic gates" intelligently, we can realize computation engines of arbitrary complexity. Combined with the storage ability of the previously described flip-flop, what we have here is the foundation of modern digital computing.

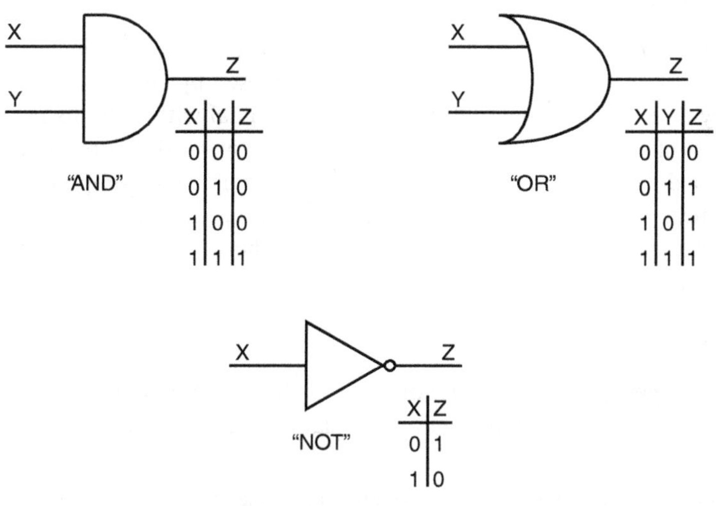

Figure 2.12 The three basic logic functions at the foundation of Boolean algebra.

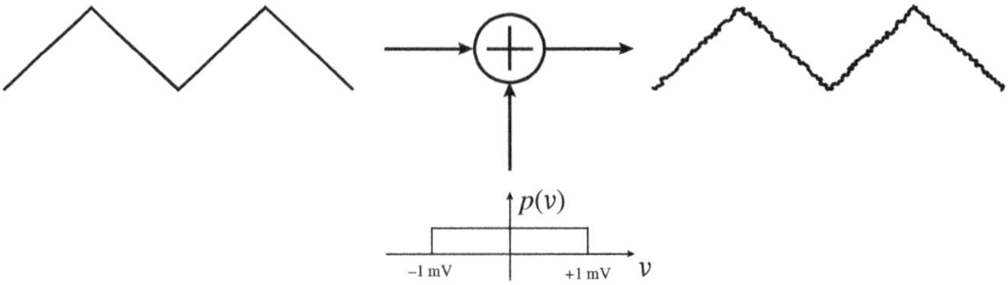

Figure 2.13 Introducing additive noise to a voltage.

It is impossible to overstate the impact that the modern computer and modern computational structures have had on how we think about information in communications systems. In anticipation of our signal being processed by a digital computer, for example, it is very common and natural to look at an analog voltage and determine how many reliably distinguishable levels there are *given a certain amount of stochastic noise in the system.* We will discuss this critical topic in depth in Chapter 4 and dwell on its implications for communications in Chapter 5. But for the moment, imagine having a voltage signal that could range anywhere from −1 to +1 V, and suppose further that there was an additive noise source that could range anywhere from −1 to +1 mV at any given sample instant with uniform distribution[6] as shown in Figure 2.13. In this case, convince yourself that we could, without any chance of error, distinguish a number of levels given by

$$l = \frac{1\,\text{V} - (-1\,\text{V})}{1\,\text{mV} - (-1\,\text{mV})} = 1000. \tag{2.15}$$

A figure may help with the detailed accounting; see Figure 2.14. So, taking into account the limits of our signal, and the limitations imposed by our noisy environment, we say that we can meaningfully encode our signal with 1000 distinct codes with no loss of information. Given this number, how many flip-flops might we need to store a digital representation of our signal internal to our computer?

As before, if there are m flip-flops, that corresponds to representing 2^m levels. Inverting this relation, if we want to represent l levels, the number of flip-flops required is given by

$$m = \log_2(l). \tag{2.16}$$

Now, m has a very special name: it is the number of "bits," or binary digits, that we need to represent the signal. Notice that as the signal range expands, while the

6 We have to add here that the sample spacing is far enough apart that the *autocorrelation* function with offset of one sample spacing or greater is zero. We get slightly ahead of ourselves, here: see Chapter 4 for a discussion on autocorrelation, and then come back and try to see why we are being a stickler for such a seemingly minor point here.

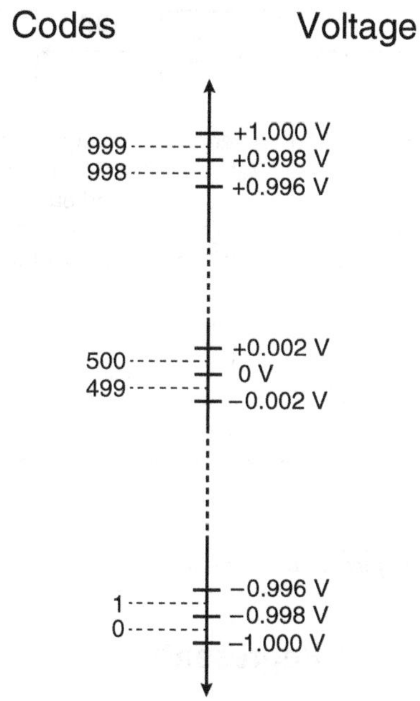

Figure 2.14 Detailed accounting for the number of distinguishable levels of a signal contaminated by noise.

range of the noise values stays constant, the number of bits goes up (logarithmically, as it happens, with the range). Alternatively, note that as the noise spread goes down, the number of distinguishable levels goes up, and therefore the number of bits also goes up. Do not be alarmed that this formula does not restrict the number of bits to be an integer. Although we started with the concept of a digital computer firmly in mind, information theorists move very quickly beyond implementation details to m being a quantitative measure of the amount of information. In the example shown in Figure 2.14, the number of bits m would be 9.965784.[7]

So to recap, we talk about bits and distinguishable levels for at least two reasons in electronic systems. One reason is that our most efficient computational engines at the time of this writing are digital computers that use binary arithmetic. It follows that it is very, very common to discretize our signals not just in time, but also in amplitude, envisioning a system as shown in Figure 2.15. That is the practical, implementation-minded reason. Another reason we talk about bits and distinguishable levels is that it gives a language for expressing *how much information is extractable from a signal that is contaminated by noise.* We scratch the surface here; you will see more in Chapters 4 and 5. But if you can get this initial example, you will have gained some very important grounding in information theory that will serve you well in your study of communications systems.

7 Had we arranged for 1024 levels, we would have had exactly 10 bits.

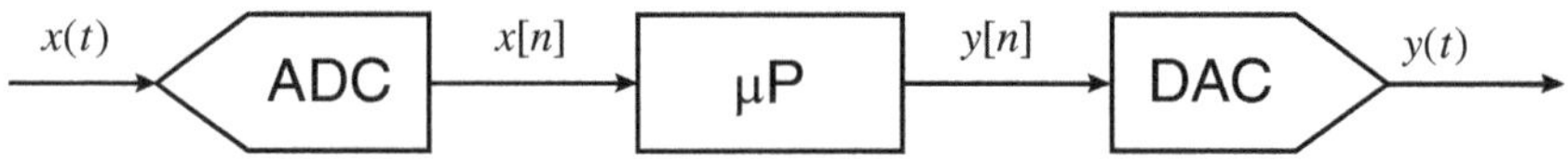

Figure 2.15 A very common signal processing lineup: a CT signal $x(t)$ is the input to an analog-to-digital converter (ADC), which produces discretized samples $x[n]$; any number of processing steps are carried out by a microprocessor (μP) producing discretized samples $y[n]$; a digital-to-analog converter (DAC) (plus, usually, some sort of analog filter) then turns discretized samples $y[n]$ into CT signal $y(t)$.

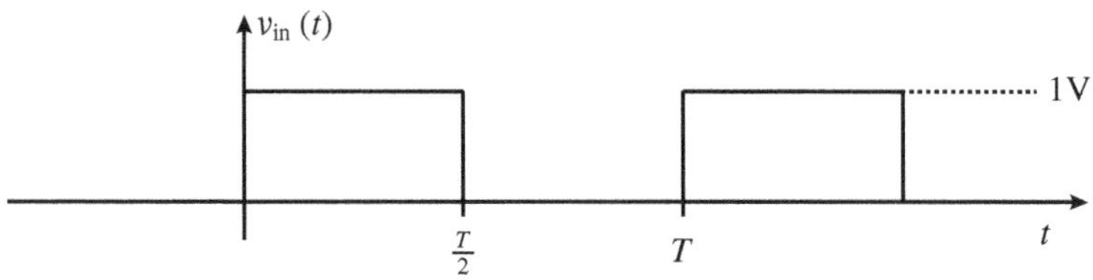

Figure 2.16 A square wave.

2.3 Representing Signals in the Frequency Domain

We are going to kick this section off by putting a perfect square wave at the input of a particular electronic circuit, examining its output, and reproducing a verbal analysis of the output that will be familiar to nearly any practicing engineer. First, the input square wave is shown in Figure 2.16. We call this a square wave with period T, because it repeats every T seconds. At time $t = 0, T, 2T, \ldots$ the signal rises instantaneously from 0 to 1 V, and then falls again for $t = \frac{T}{2}, \frac{3T}{2}, \frac{5T}{2}, \ldots$ and so on. This is our input.

As promised, we feed this square wave input into a "mystery circuit" as shown in Figure 2.17, and look at the output. Now suppose that we look at the output on an imaginary oscilloscope as shown in Figure 2.18. An experienced engineer might describe what the circuit has done to the input to produce the output in the following way:

> Well, you can see right away that the circuit has some sort of DC block. The zero-frequency component, or DC offset, has been removed, which is why the output perfectly straddles the time axis whereas in the input the voltage was always greater than or equal to zero. The next thing that is clear is that the circuit applies some sort of low-pass filtering. I can tell, because you need those super-high-frequency components to make the sharp transitions that you can see at the input. The gentler roll up to the +0.5 V and back down to the −0.5 V is a dead giveaway that high-frequency components have been removed.

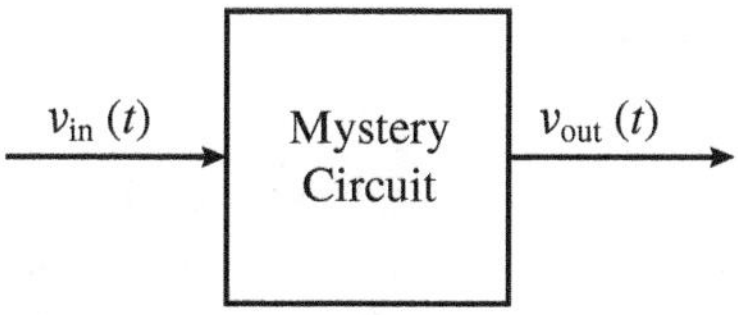

Figure 2.17 We pass our square wave into the input of a mystery circuit...

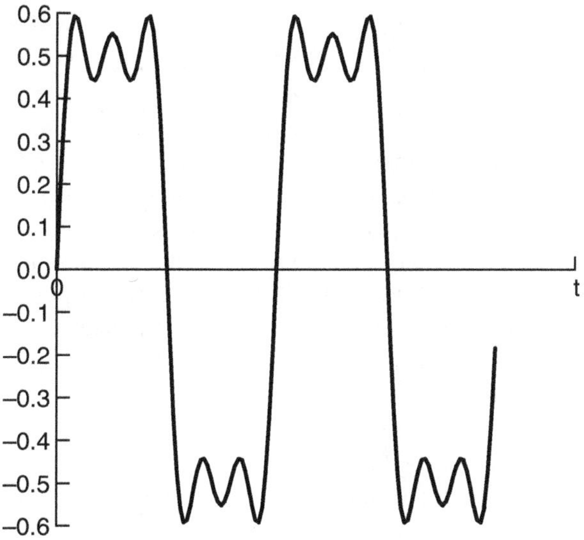

Figure 2.18 ...and see this on the display of our oscilloscope.

Even if this way of talking is deeply familiar to you...in fact, especially if this way of talking is familiar to you, pause and really look at the passage with fresh eyes. The speaker is viewing the signal in a very different way than we have up to this point, which is as a voltage that we chart along a steadily moving time index. Instead, the speaker talks as though this signal can be expressed as a sum of signals of a specific type called "frequency components." We are apparently free to subtract individual frequency components at will, without affecting the other constituent components.[8]

These frequency components are called *sinusoids*, and looking at signals as sums of sinusoidal frequency components is called "looking at signals in the frequency domain," as opposed to before, where we looked at signals "in the time domain." Let's get into this.

---

8 Remarkable when you think about it, really.

2.3.1 On the Centrality of Exponentials and Sinusoids in Linear System Theory

If you are a longtime practitioner, the centrality of sinusoids and exponentials in discussing systems is so far beyond taken for granted that you may well have forgotten a time when you did *not* think this way. If you are new to the field, said centrality is still very much a mystery. The purpose of this section is to put thinkers in both camps on solid conceptual footing.

In brief, we care about exponentials and sinusoids because: (1) they are treated with special deference in the equations that describe linear and time-invariant (LTI) systems; (2) it turns out that we can easily express almost any practical signal as a sum of sinusoids; (3) we find that when we do break a signal down into its sinusoidal components, *noise* energy tends to be spread uniformly throughout the frequency space, whereas our *signal* energy is restricted to certain predetermined bands. This last aspect helps to explain why "filtering" is an obsession among designers of communications systems: you can increase the ratio of signal power to noise power by filtering to block frequencies that carry only noise power while passing frequencies that contain signal power. Keep an eye on this idea, and return to it when you study Chapters 4 and 5.

Exponentials and Sinusoids are Eigenfunctions of Linear, Time-Invariant (LTI) Systems

What do we mean by "eigenfunctions"? For now, we'll define exponentials and sinusoids separately according to

$$x(t) = e^{at} \tag{2.17}$$

for $-\infty < t < \infty$, for an exponential, and for a sinusoid

$$\begin{aligned} x(t) &= C_1 \sin \omega t + C_2 \cos \omega t \\ &= A \sin(\omega t + \phi_s) \\ &= A \cos(\omega t + \phi_c), \end{aligned} \tag{2.18}$$

where again $-\infty < t < \infty$. For Eq. 2.18, we maintain equivalence among the forms if we require

$$\begin{aligned} C_1 &= A \cos \phi_s = -A \sin \phi_c \\ C_2 &= A \sin \phi_s = A \cos \phi_c. \end{aligned} \tag{2.19}$$

Now, it is important to note that we are absolutely serious when we specify $-\infty < t < \infty$, at least for the purposes of our mathematical analysis. When we speak of pure sinusoids, there is no time at which the sinusoid just turned on...it has always been on, and will always be on. *That* is a pure sinusoid.

Great. With these definitions in place, we move on to a remarkable fact of Nature, which is that many, many practical systems are well modeled by linear differential equations with constant coefficients. Amplifiers, filters, antennas, attenuators, sensors of all types, you name it. Even better, for many system elements whose physics are manifestly *not* captured by linear differential equations with constant coefficients...under a wide variety of circumstances, we can *approximate* their behavior with these famously well-behaved equations.

If a system is in fact governed or well approximated by linear differential equations with constant coefficients, we say that it is LTI.[9] When such a system is stimulated by an exponential or a sinusoid, we get the response illustrated in Figure 2.19. That is to say, the output is the same as the input, only scaled by a multiplicative constant and in the case of a sinusoid possibly phase shifted. This property of passing through essentially unmolested is why we call exponentials and sinusoids *eigenfunctions* of LTI systems.

The concept of an eigenfunction is related to a similar concept from linear algebra, or the algebra of matrices. In that context, when the matrix A operates on

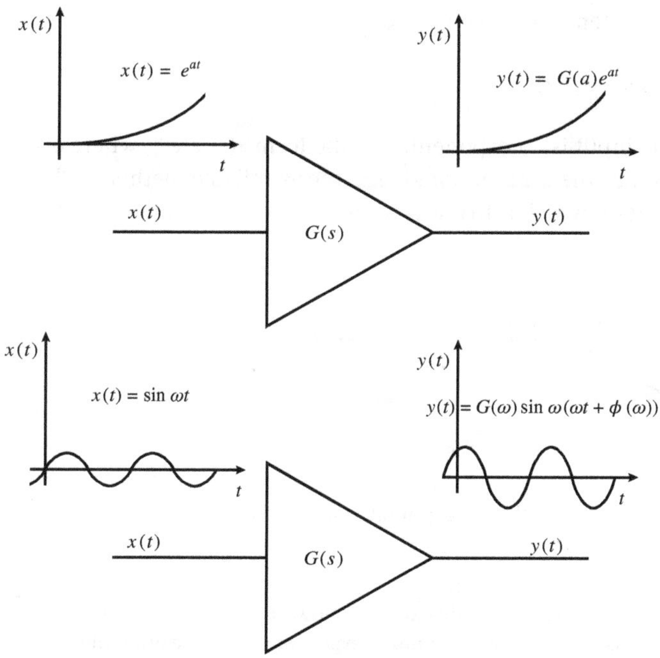

Figure 2.19 Exponentials and sinusoids as eigenfunctions of LTI systems.

9 Siebert. *Circuits, Signals, and Systems*, Cambridge, MA: MIT Press, 1986.

one of its special eigenvectors $\vec{x}_e$, the result is that the direction of the vector $\vec{x}_e$ is unchanged, while its length is scaled by a simple number λ which is the eigenvalue of that eigenvector. That is,

$$A \cdot \vec{x}_e = \lambda \vec{x}_e. \tag{2.20}$$

It is normally true that when the matrix A operates on vector $\vec{x}$, the result is a new vector $\vec{y}$ whose direction and length differ from $\vec{x}$. But the eigenvectors of A are special, in that A only scales their length but does not alter their direction.

Anyway, we have opened this section by asserting that exponentials and sinusoids are eigenfunctions of LTI systems. How can we show this?

We start by restating the interchangeability of describing a system as LTI with saying that its behavior is governed by a linear differential equation with constant coefficients. Then, we stop talking in such generalities and let an example do the heavy lifting. To wit, consider the simple circuit shown in Figure 2.20. We show an ideal amplifier, which takes a voltage $x(t)$ as its input and produces the scaled voltage $Ax(t)$ at its output.[10] This output is then fed into the simple RC network that is more fully described in Chapter 3. It turns out that the differential equation governing this simple system is given by

$$\frac{d}{dt}y(t) + \frac{1}{RC}y(t) = \frac{A}{RC}x(t). \tag{2.21}$$

Now, suppose that the input is an exponential of the form $x(t) = e^{\sigma t}$, where σ is a real constant,[11] and recall the time-honored and utterly reliable method of "solving" differential equations, which is to know the answer ahead of time. If we think

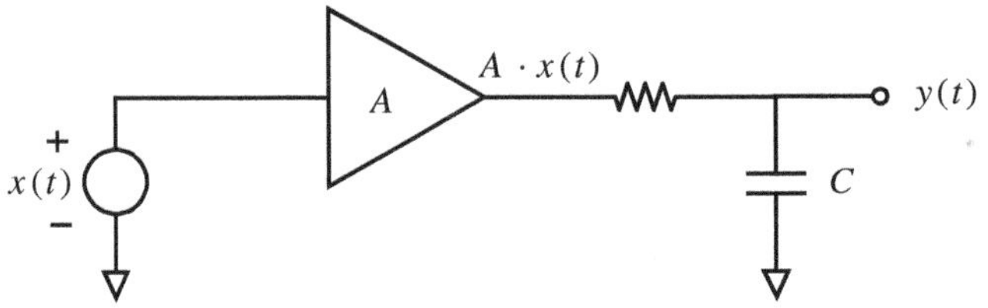

Figure 2.20 A simple circuit for exploring eigenfunctions of LTI systems.

---

10 *All* signals are eigenfunctions of ideal amplifiers.

11 Don't let it bother you that for nonzero σ, this function blows up at either $t = \infty$ or $t = -\infty$. Going forward, we will ultimately find that it is complex exponentials that are primarily of interest, for which the σ values are purely imaginary. Such complex exponentials are oscillatory and do not go off to infinity, and we will get to this when we unveil Euler's relation later on in this subsection. For now, note that (1), an exponential input with purely real σ does blow up to infinity at one time or another, and (2) the differential equation does not care...the exponential is *still* an eigenfunction no matter what anyone else says.

that $x(t) = e^{\sigma t}$ is indeed an eigenfunction of this system, then it must be true that $y(t) = ke^{\sigma t}$. Plugging in for $x(t)$ and $y(t)$, and doing some algebra, ought to tell us what k has to be.

$$\frac{d}{dt}[ke^{\sigma t}] + \frac{1}{RC} \cdot ke^{\sigma t} = \frac{A}{RC}e^{\sigma t}$$

$$k\sigma e^{\sigma t} + \frac{1}{RC} \cdot ke^{\sigma t} = \frac{A}{RC}e^{\sigma t}$$

$$k = \frac{A}{RC\sigma + 1}. \tag{2.22}$$

We take the fact that we can solve successfully for k as proof of our eigenfunction thesis, and confidently write down that for stimulus $x(t) = e^{\sigma t}$, the output $y(t)$ is given by

$$y(t) = \frac{A}{RC\sigma + 1} \cdot e^{\sigma t}. \tag{2.23}$$

There are a few items left on our to-do list, which we spell out here by way of staying oriented over a lengthy development:

1. We will show that sinusoids (sin ωt and cos ωt) are also eigenfunctions of this simple system.
2. We will show this eigenfunction property for a general linear differential equation with constant coefficients.
3. We will marvel at how awkward it truly is to work with sin ωt and cos ωt, and look for a way out.
4. We will introduce Euler's relation, one of the most remarkable results in all of mathematics, to show how we can deal with sin ωt and cos ωt with the same ease with which we dealt with $e^{\sigma t}$.

Starting with item no. 1 will lead to a development that is unfamiliar to many of you. If you have seen anything like it, it would have been in an undergraduate mathematics class devoted to differential equations. And even there, they would have moved quickly past this to the ease of working with complex exponentials like $e^{j\omega t}$. But if you are reading this book, chances are high that you have either a basic, healthy skepticism about the methods that you learned, or alternatively, a curiosity about why accepted methods became so accepted. What follows is a powerful demonstration of why you do not remember or were never shown how to work with differential equations using purely real sinusoids.

Recall that our working assumption is that sin ωt and cos ωt are eigenfunctions of LTI systems. But when we test this out, our early efforts are at first disappointing. Suppose, for example, that our input to the same circuit of Figure 2.20 was $x(t) =$ sin ωt. We might follow the steps that we had before, and guess that $y(t) = k$ sin ωt.

We plug in and rush through, expecting to be able to solve for k:

$$\frac{d}{dt}[k \sin \omega t] + \frac{1}{RC} \cdot k \sin \omega t = \frac{A}{RC} \sin \omega t$$

$$k\omega \cos \omega t + \frac{1}{RC} \cdot k \sin \omega t = \frac{A}{RC} \sin \omega t. \tag{2.24}$$

Stuck! We hit a wall. Whatever the solution, it does not appear to have the neat, clean "eigen" sort of outcome that we had for the exponential in Eq. 2.22.

It turns out that giving up now would be giving up too soon. Consider the question: if you put $x(t) = \sin \omega t$ through the system, and at the output you saw $k \sin(\omega t + \phi)$, would you still say that $\sin \omega t$ is an eigenfunction of the system? Why yes, yes you would. You put in a sine wave, and what came out was a sine wave identical in all important respects, only amplitude scaled and phase shifted. So how can we make the math behave?

We will see that our escape lies in recognizing that the sum

$$k_1 \sin \omega t + k_2 \cos \omega t \tag{2.25}$$

is really just a phase-shifted $\sin \omega t$. We can see this if we use a common trigonometric identity:

$$\begin{aligned} k \sin(\omega t + \phi) &= k \sin \omega t \cos \phi + k \cos \omega t \sin \phi \\ &= (k \cos \phi) \sin \omega t + (k \sin \phi) \cos \omega t. \end{aligned} \tag{2.26}$$

We can ensure equality of

$$k_1 \sin \omega t + k_2 \cos \omega t = k \sin(\omega t + \phi) \tag{2.27}$$

if we enforce[12]

$$\phi = \tan^{-1} \frac{k_2}{k_1} \tag{2.28}$$

and

$$k = \sqrt{k_1^2 + k_2^2}. \tag{2.29}$$

We can now go back and show the reason we got stuck in Eq. 2.24. With $\sin \omega t$ as our input, as it happens, we were too restrictive in the form of our output by committing to $k \sin \omega t$. Setting the problem up this way is fine if we get away with a clean amplitude scaling, but we have to be more creative if we want to accommodate a phase shift.

---

12 If you choose to show this for yourself, you will find another trigonometric identity handy: $\sin^2\phi + \cos^2\phi = 1$.

The path forward is that for an input $x(t) = \sin \omega t$, we should assume a form for the output as follows:

$$y(t) = k_1 \sin \omega t + k_2 \cos \omega t. \tag{2.30}$$

Let's try it. Remember the main rule of the game, which is that we plug in our guess for $y(t)$ and try to solve for k_1 and k_2. If we are able to do this successfully, we have our solution indeed. Plugging into our original differential equation, we start off with:

$$\frac{d}{dt}[k_1 \sin \omega t + k_2 \cos \omega t] + \frac{1}{RC}[k_1 \sin \omega t + k_2 \cos \omega t] = \frac{A}{RC} \sin \omega t. \tag{2.31}$$

Differentiating and collecting terms, we have

$$\left[\frac{k_1}{RC} - k_2\omega\right] \sin \omega t + \left[k_1\omega + \frac{k_2}{RC}\right] \cos \omega t = \frac{A}{RC} \sin \omega t. \tag{2.32}$$

Matching sines with sines and cosines with cosines, it looks like we require

$$k_1\omega + \frac{k_2}{RC} = 0 \tag{2.33}$$

and

$$\frac{k_1}{RC} - k_2\omega = \frac{A}{RC}. \tag{2.34}$$

Two equations in our two unknowns k_1 and k_2 are readily handled. What we find is that

$$k_1 = \frac{A}{1 + \omega^2 R^2 C^2} \tag{2.35}$$

and

$$k_2 = -\frac{A\omega RC}{1 + \omega^2 R^2 C^2}. \tag{2.36}$$

Since we are able to solve for k_1 and k_2, we can exhale and see that $y(t)$ is in fact a sinusoid at the same frequency ω as the sinusoid at the input, simply scaled in amplitude and shifted in phase. We have for $y(t)$

$$\begin{aligned} y(t) &= \frac{A}{1 + \omega^2 R^2 C^2} \sin \omega t - \frac{A\omega RC}{1 + \omega^2 R^2 C^2} \cos \omega t \\ &= k \sin(\omega t + \phi), \end{aligned} \tag{2.37}$$

where after some algebra, we can now express ϕ and k as

$$\phi = \tan^{-1}(-\omega RC) \tag{2.38}$$

and

$$k = \frac{A}{\sqrt{1 + \omega^2 R^2 C^2}}. \tag{2.39}$$

Well! That was...terrible. Which is to say, that was a fair amount of work for a humble first-order differential equation. There is no reason to expect that it will get better for more complicated, higher order differential equations, and there is maybe even some question in your mind as to whether the method generalizes at all. We will soon enough breathe the free air of complex exponentials. But let us not, in our haste, skip item no. 2.

This happens to be not as difficult as it might first appear. We may write down the general form of a linear differential equation with constant coefficients as

$$\left(\sum_{i=0}^{n} A_i \frac{d^i}{dt^i}\right) y(t) = \left(\sum_{k=0}^{m} B_k \frac{d^k}{dt^k}\right) x(t). \tag{2.40}$$

If we put in $x(t) = e^{\sigma t}$ as our input and follow steps similar to those that led to Eq. 2.22, we wind up with a solution for $y(t)$ as:

$$y(t) = \frac{\sum_{k=0}^{m} B_k \sigma^k}{\sum_{i=0}^{n} A_i \sigma^i} e^{\sigma t}. \tag{2.41}$$

Complicated looking, but not complicated. Check against Eq. 2.22, as it should, of course, work for the first-order case.

Putting in a real sinusoid makes us work much harder on the book keeping. If $x(t) = \sin \omega t$, our eigenfunction thesis demands that $y(t) = k_1 \sin \omega t + k_2 \cos \omega t$ and we have

$$\left(\sum_{i=0}^{n} A_i \frac{d^i}{dt^i}\right) [k_1 \sin \omega t + k_2 \cos \omega t] = \left(\sum_{k=0}^{m} B_k \frac{d^k}{dt^k}\right) \sin \omega t. \tag{2.42}$$

In the following romp with indices, understand that when a summation limit goes to $n/2$ or $m/2$, we increment the summation index up to exactly that when n or m is even, and up to $(n-1)/2$ or $(m-1)/2$ when n or m is odd. That said, we hold on tight and rewrite Eq. 2.42 as:

$$\begin{aligned}
&\sin \omega t \left[k_1 \cdot \sum_{i=0}^{n/2} A_{2i} \omega^{2i} (-1)^i + k_2 \cdot \sum_{i=0}^{n/2} A_{2i+1} \omega^{2i+1} (-1)^{i+1} \right] \\
&+ \cos \omega t \left[k_1 \cdot \sum_{i=0}^{n/2} A_{2i+1} \omega^{2i+1} (-1)^i + k_2 \cdot \sum_{i=0}^{n/2} A_{2i} \omega^{2i} (-1)^i \right] \\
&= \sin \omega t \left[\sum_{k=0}^{m/2} B_{2k} \omega^{2k} (-1)^k \right] + \cos \omega t \left[\sum_{k=0}^{m/2} B_{2k+1} \omega^{2k+1} (-1)^k \right].
\end{aligned} \tag{2.43}$$

As complicated as this looks, equating coefficients of the sin ωt terms and the cos ωt terms gives two equations in two unknowns k_1 and k_2. Equating coefficients of the sin ωt terms gives

$$k_1 \cdot \sum_{i=0}^{n/2} A_{2i}\omega^{2i}(-1)^i + k_2 \cdot \sum_{i=0}^{n/2} A_{2i+1}\omega^{2i+1}(-1)^{i+1} = \sum_{k=0}^{m/2} B_{2k}\omega^{2k}(-1)^k, \tag{2.44}$$

while equating coefficients of the cos ωt terms gives

$$k_1 \cdot \sum_{i=0}^{n/2} A_{2i+1}\omega^{2i+1}(-1)^i + k_2 \cdot \sum_{i=0}^{n/2} A_{2i}\omega^{2i}(-1)^i = \sum_{k=0}^{m/2} B_{2k+1}\omega^{2k+1}(-1)^k. \tag{2.45}$$

Messy, but here we are. And it *is* just a matter of bookkeeping. A good first check of this generalization is to use the example of Eq. 2.21, where $B_0 = A/RC$ and all other B_is are 0; and $A_0 = 1/RC$, $A_1 = 1$, and all other A_is are 0. If you are careful with your algebra, and patient, you might put this generalization through its paces with a second- or even third-order example for which you already have the answer. Just don't lose track of the overall point, which is that we derive this ungainly generalization *by way of showing the need for a better method.* Sufficiently motivated at last, we check item no. 3 off our list and turn gratefully to the complex exponential.

Our salvation from this index-riddled fever dream lies in one of the most mysterious and remarkable results in all mathematics, called Euler's formula:

$$e^{j\omega t} = \cos \omega t + j \sin \omega t. \tag{2.46}$$

Recall that j itself is borne of algebraic necessity. Without it, we lack a solution to the simple equation[13]

$$x^2 = -1. \tag{2.47}$$

The trigonometric functions sin and cos, on the other hand, come to us from geometry as illustrated in Figure 2.21. This connection between algebra and geometry is mysterious and beautiful.

As long as we keep our focus and treat our new complex exponential the same way we treated the real exponential, it is straightforward to show what happens to a general LTI system whose input is $e^{j\omega t}$. The output will be $Ce^{j\omega t}$, where C is a complex constant given by

$$C = \frac{\sum_{k=0}^{m} B_k(j\omega)^k}{\sum_{i=0}^{n} A_i(j\omega)^i}. \tag{2.48}$$

13 The author acknowledges, with gratitude, Feynman's development on complex numbers, which can be found in the *Feynman Lectures on Physics,* Vol. 1, Chapter 22, Addison-Wesley, 1963.

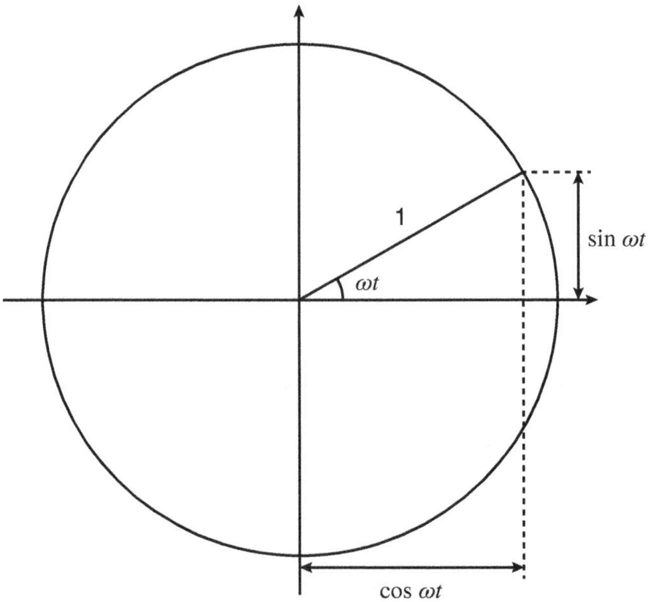

Figure 2.21 The unit circle.

Compare this with Eq. 2.41.

Given, now, this amazing Euler's formula, how can we use it to solve our original problem of determining the response of our systems to a real sinusoidal input? We'll close out this rather long subsection with two possibilities, both of which rely on the *linearity* of ordinary differential equations.

What we mean by "linearity" is as follows. Suppose that the response of our system to input $x_1(t)$ is $y_1(t)$, and the response to input $x_2(t)$ is $y_2(t)$. Then, it turns out that the response to $\alpha x_1(t) + \beta x_2(t)$ is just $\alpha y_1(t) + \beta y_2(t)$.[14]

This insight leads to one possible alternative to the way we have worked with purely real sinusoids up to this point. If we want to know the response of our system to $\cos \omega t$, then we could proceed by posing $x(t) = e^{j\omega t}$, and then read the response $y(t)$ as simply the real part of the total output $y(t)$! Similarly:

- If we want to know the response to $\sin \omega t$, we put again as our input $x(t) = e^{j\omega t}$ but this time take the *imaginary part* of the total complex response $y(t)$ as our answer.
- If we want the response to $\cos(\omega t + \phi)$, then we use as in input $x(t) = e^{j\phi}e^{j\omega t}$, and then as usual take the real part of the result.

14 Maybe give a try at showing this using Eq. 2.40.

Another common way forward is to make use of the following equalities:

$$\cos \omega t = \frac{1}{2}e^{j\omega t} + \frac{1}{2}e^{-j\omega t} \tag{2.49}$$

and

$$\sin \omega t = -j \cdot \left(\frac{1}{2}e^{j\omega t} - \frac{1}{2}e^{-j\omega t}\right). \tag{2.50}$$

By expressing our basic, real sinusoids this way, we again make our differential equations easy to work with.

At the end of the day, then, we have the following understanding that is important to carry with you as you study communications systems. Start with the fact that an astonishing portion of the physical world is reasonably modeled by linear differential equations with constant coefficients. Given this fact, we find that exponentials and sinusoids are eigenfunctions of these LTI systems. Through Euler's formula, we find that exponentials and sinusoids are but two sides of the same coin. We therefore *want* to privilege complex exponentials in the study of LTI systems. We find all the justification we need in the next subsection.

Almost All Signals Can Be Expressed as a Sum of Complex Exponentials

It's true: if we wish, we can express almost any signal $x(t)$ as a sum of complex exponentials. This remarkable fact finds its most elegant expression in the so-called synthesis/analysis equations of the Fourier transform:

$$\begin{aligned} x(t) &= \frac{1}{2\pi}\int_{-\infty}^{\infty} X(j\omega)e^{j\omega t}d\omega \\ X(j\omega) &= \int_{-\infty}^{\infty} x(t)e^{-j\omega t}dt. \end{aligned} \tag{2.51}$$

We'll delve into the origins of the Fourier transform shortly, but for now it is more important to focus on getting meaning out of the math.

Start with the first equation, which we call the *synthesis* equation. We call it that because it expresses synthesizing the signal $x(t)$ by summing a whole bunch of complex exponentials. If it has been a while since you studied calculus, you may have forgotten than an integral is simply a fancy way of writing a sum. In calculus, you would call the underlying sum a Riemann sum, and let's regress back to that for a moment for the synthesis equation:

$$\begin{aligned} x(t) &= \lim_{\Delta\omega \to 0} \frac{1}{2\pi} \sum_{k=-\infty}^{\infty} X(j(k \cdot \Delta\omega)) \cdot e^{j(k \cdot \Delta\omega)t} \Delta\omega \\ &= \lim_{\Delta\omega \to 0} \frac{\Delta\omega}{2\pi} \sum_{k=-\infty}^{\infty} X(j(k \cdot \Delta\omega)) \cdot e^{j(k \cdot \Delta\omega)t}. \end{aligned} \tag{2.52}$$

Save worrying about how we got here for Section 2.3.3, and focus for now on putting words to the math. What Eq. 2.52 tells us is that if we determine $X(j\omega)$ properly, then it tells us how to construct our signal $x(t)$ out of a weighted sum of sinusoids.

This synthesis gives us at least the option of looking at a signal *either* in the "normal" time domain $x(t)$ form, *or* in the "frequency domain." In the frequency-domain view, we consider a signal according to the complex sinusoids, or frequency components, that comprise it.

A couple of examples are helpful here. We noted that with the help of Euler's formula (Eq. 2.46), we can write $\cos \omega t$ as:

$$\cos \omega t = \frac{1}{2}e^{j\omega t} + \frac{1}{2}e^{-j\omega t}. \tag{2.53}$$

So a frequency-domain picture of $\cos \omega t$ might look like Figure 2.22.

Another example is the square wave that we opened Section 2.3 with. It turns out that you can form this square wave by stacking a bunch of real sinusoids on top of each other according to

$$x_{sq} = \frac{4}{\pi}\left(\sin \omega_0 t + \frac{1}{3}\sin 3\omega_0 t + \frac{1}{5}\sin 5\omega_0 t + \cdots\right) \tag{2.54}$$

where $\omega_0 = \frac{2\pi}{T_0}$, and T_0 is the fundamental period of the square wave. Calling it a "square wave" is a very time-domain thing to do. In the frequency domain, it looks as depicted in Figure 2.23.

The overall point is that sinusoids are also special because, through them, we can look at our signals in a so-called frequency domain, analyzing them according to their constituent frequency components. We *can* do this. But *should* we?

The Frequency-domain View Is Natural in the Presence of Noise

Yes, we should!

The easiest way to understand this is to get some understanding of noise in communications systems. There are many types of noise in Nature. But one type of

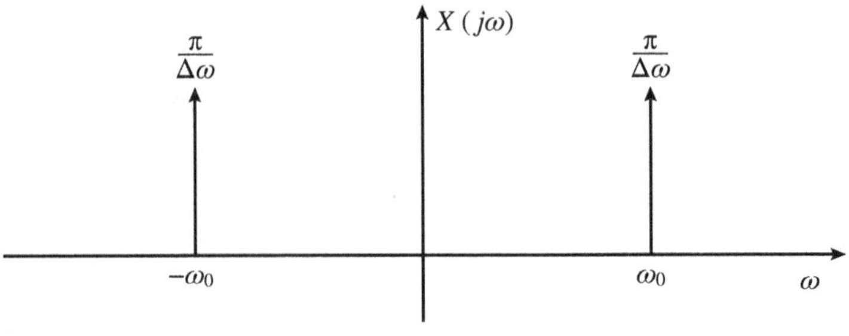

Figure 2.22 $\cos \omega_0 t$ in the frequency domain.

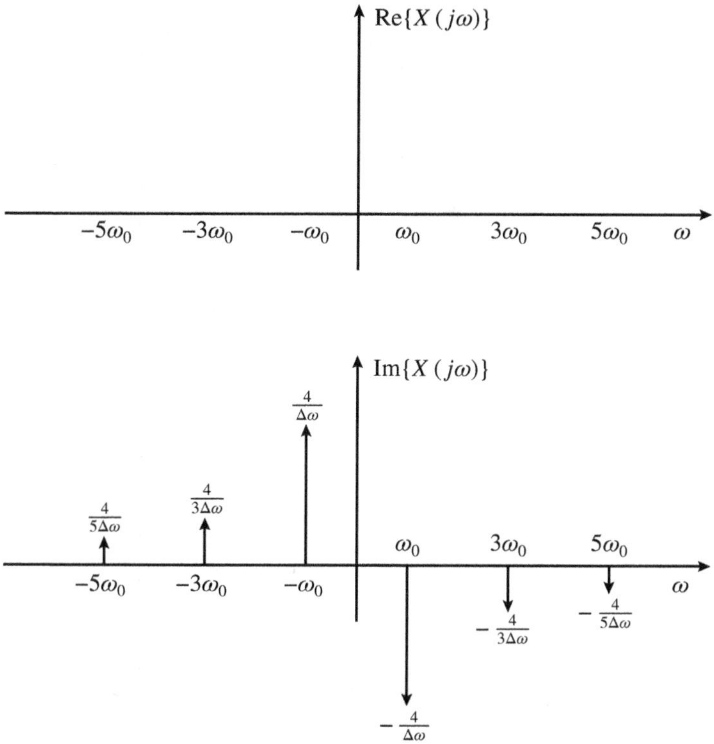

Figure 2.23 A square wave in the frequency domain. If you want to check the math, use Eqs. 2.50, 2.52, and 2.54.

noise that is everywhere is thermal noise, which in electronic circuits owes to the thermal agitation of charge carriers.[15] We have already stated that almost all signals can be represented as a sum of sinusoids. This statement is spectacularly true of thermal noise, which shows up in our systems as signal energy spread evenly at all frequencies that we commonly measure.

A visualization is helpful here. Suppose that we had a perfect sinusoidal oscillator, and we connected its output to a laboratory spectrum analyzer. In the absence of thermal noise, what we would see on the spectrum analyzer display is shown in Figure 2.24. But in the presence of thermal noise, we would see something like Figure 2.25. It is customary, and useful, to speak of noise in terms of "power spectral density," which is how much noise power is measured per Δf of bandwidth.

---

15 If you have seen this turn of phrase so often as to render it meaningless, first notice and appreciate the high level of self-awareness in you as a learner. Then, you can either patiently wait for Chapter 4 or jump to that chapter and scratch this particularly deserving itch.

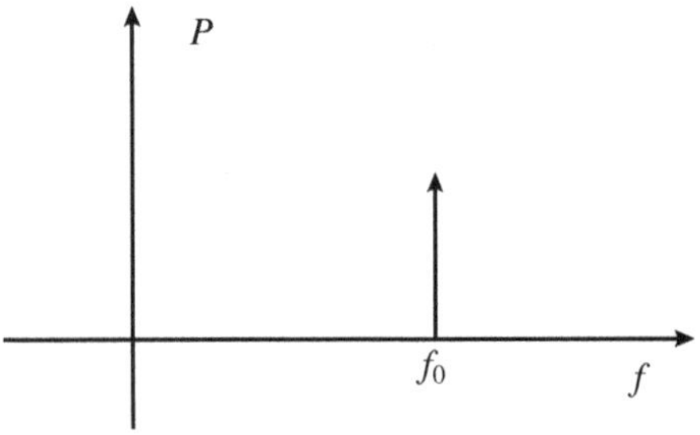

Figure 2.24 A pure sinusoid in a noiseless environment, as measured on a spectrum analyzer.

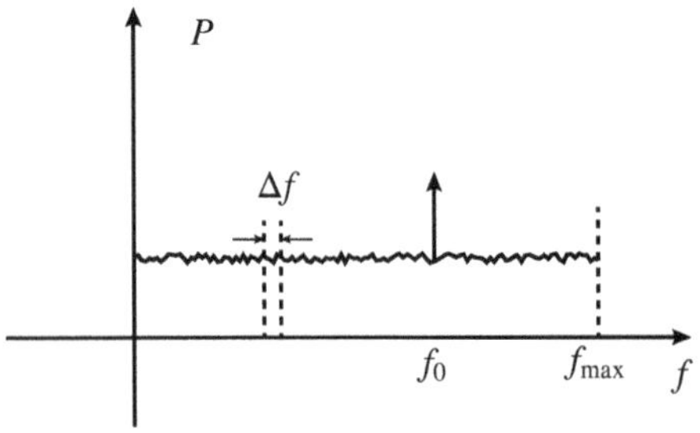

Figure 2.25 A pure sinusoid in a noisy environment, as measured on a spectrum analyzer.

We will find in coming chapters that when it comes to communications, the ratio of signal power to *total* noise power, or signal-to-noise ratio (SNR), is critical in determining how quickly we can send information. In this case, suppose our sinusoid had a signal power of P_{sig}, that the thermal noise in the system had a power spectral density of N_D W/Hz, and that the spectrum analyzer had an upper measurement limit of f_{max}. We would say that the SNR in this measurement scenario is given by

$$\mathrm{SNR} = \frac{P_{\mathrm{sig}}}{N_{\mathrm{D}} \cdot (f_{\mathrm{max}} - 0)}. \tag{2.55}$$

Now suppose that we put a filter between the oscillator and the spectrum analyzer. The idea of the filter is that it passes, with unity gain, all signal energy between f_{low} and f_{high}, and blocks energy at all other frequencies. Our spectrum analyzer display would look very different, as shown in Figure 2.26. If we redo our SNR measurement, we will find that our SNR has improved according to

$$\mathrm{SNR} = \frac{P_{\mathrm{sig}}}{N_{\mathrm{D}} \cdot (f_{\mathrm{high}} - f_{\mathrm{low}})} > \frac{P_{\mathrm{sig}}}{N_{\mathrm{D}} \cdot (f_{\mathrm{max}})}. \tag{2.56}$$

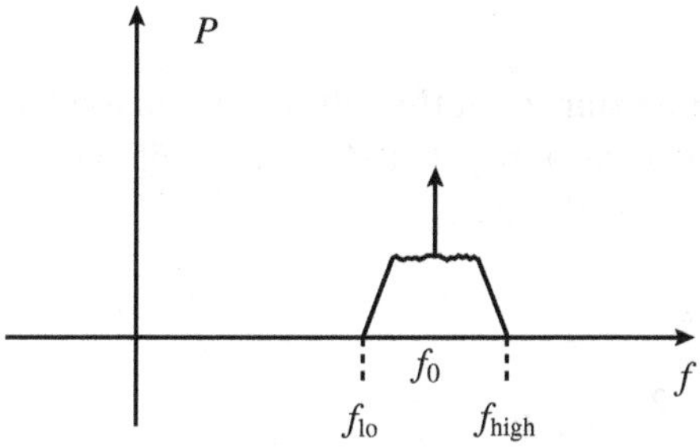

Figure 2.26 A pure sinusoid in a noisy environment, as measured on a spectrum analyzer, but this time with a bandpass filter preceding the spectrum analyzer.

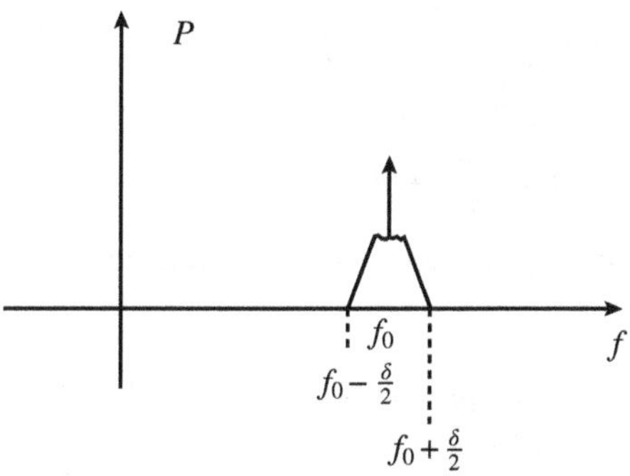

Figure 2.27 A pure sinusoid in a noisy environment, as measured on a spectrum analyzer, again with a bandpass filter. This time, we have tightened the bandpass filter to *really* raise the SNR.

Taking on faith for the moment that better SNR is a good thing, the physics of the situation suggests going as far with this filtering trick as we can, according to Figure 2.27. Just following our analytical nose:

$$\frac{P_{\text{sig}}}{N_{\text{D}} \cdot \delta} > \frac{P_{\text{sig}}}{N_{\text{D}} \cdot (f_{\text{high}} - f_{\text{low}})} > \frac{P_{\text{sig}}}{N_{\text{D}} \cdot (f_{\text{max}})}. \tag{2.57}$$

Take a moment to reflect on Eq. 2.57. This equation alone explains why there is a huge, rich literature devoted to the art and science of frequency-selective filters: zooming in to the part of the frequency spectrum where the signal is known to reside raises the SNR of the communications channel. For reasons that we will explore heavily in Chapter 5, a high SNR is key to communicating at a high number of bits per second. Very early in the history of electronic communications systems, then, frequency-selective filtering evolved from an interest, to a preoccupation, to an obsession. In this context, frequency-domain thinking is inevitable and natural.

2.3.2 The Fourier Series

To understand the Fourier series, one must start with the following remarkable fact: if a signal $x(t)$ is periodic with period T, meaning that $x(t) = x(t + T)$, then it is true that such a signal can be represented as the sum:

$$\begin{aligned} x(t) &= C_0 + A_1 \cos\left(\frac{2\pi}{T}t\right) + A_2 \cos\left(2 \cdot \frac{2\pi}{T}t\right) + A_3 \cos\left(3 \cdot \frac{2\pi}{T}t\right) + \cdots \\ &+B_1 \sin\left(\frac{2\pi}{T}t\right) + B_2 \sin\left(2 \cdot \frac{2\pi}{T}t\right) + B_3 \sin\left(3 \cdot \frac{2\pi}{T}t\right) + \cdots \end{aligned} \tag{2.58}$$

Through the magic of trigonometric identities, Eq. 2.58 can also be written as:[16]

$$x(t) = A_0 + A_1 \cos\left(\frac{2\pi}{T}t + \phi_1\right) + A_2 \cos\left(2 \cdot \frac{2\pi}{T}t + \phi_2\right) + \cdots \tag{2.59}$$

or, of course,

$$x(t) = B_0 + B_1 \sin\left(\frac{2\pi}{T}t + \theta_1\right) + B_2 \sin\left(2 \cdot \frac{2\pi}{T}t + \theta_2\right) + \cdots. \tag{2.60}$$

Convinced by now of the utility of complex exponentials, we can jump right into a rewrite of Eq. 2.58 as

$$\begin{aligned} x(t) &= \cdots + C_{-3}e^{-j3\cdot\frac{2\pi}{T}t} + C_{-2}e^{-j2\cdot\frac{2\pi}{T}t} + C_{-1}e^{-j\cdot\frac{2\pi}{T}t} + C_0 + C_1 e^{j\cdot\frac{2\pi}{T}t} \\ &+ C_2 e^{j2\cdot\frac{2\pi}{T}t} + C_3 e^{j3\cdot\frac{2\pi}{T}t} + \cdots. \end{aligned} \tag{2.61}$$

There are two reorganizations of Eq. 2.61 that will prove useful going forward. First, we group terms according to

$$\begin{aligned} x(t) = C_0 &+ \left(C_1 e^{j\cdot\frac{2\pi}{T}t} + C_{-1}e^{-j\cdot\frac{2\pi}{T}t}\right) + \left(C_2 e^{j2\cdot\frac{2\pi}{T}t} + C_{-2}e^{-j2\cdot\frac{2\pi}{T}t}\right) \\ &+ \left(C_3 e^{j3\cdot\frac{2\pi}{T}t} + C_{-3}e^{-j3\cdot\frac{2\pi}{T}t}\right) + \cdots. \end{aligned} \tag{2.62}$$

Second, we collapse Eq. 2.61 into the compact

$$x(t) = \sum_{k=-\infty}^{\infty} C_k e^{j\cdot k\frac{2\pi}{T}t}. \tag{2.63}$$

If we define $\omega_0 = \frac{2\pi}{T}$, we have

$$x(t) = \sum_{k=-\infty}^{\infty} C_k e^{j\cdot k\omega_0 t}. \tag{2.64}$$

16 It is useful as an exercise to work out for yourself how you would go back and forth between these various forms. When it comes to jumping in with complex exponentials, the key identities are again Eqs. 2.49 and 2.50. To make it all work, remember that the C_is will have real and imaginary parts.

At least two issues come up, now:

- If we accept that $x(t)$ is in fact a sum of sinusoids, how exactly do we determine the C_ks?
- If $x(t)$ is purely real, then that must imply *something* about the C_ks to make sure that no rogue imaginary component emerges out of the Fourier sum. Details, please?

We take these items one at a time. In order to determine the C_ks, we make note of the following:

$$\int_{t_0}^{t_0+T} e^{jk\omega_0 t} e^{-jl\omega_0 t} dt = \int_{t_0}^{t_0+T} e^{j(k-l)\omega_0 t} dt$$
$$= \begin{cases} \frac{1}{j(k-l)\omega_0} [e^{j(k-l)\omega_0 t}]_{t_0}^{t_0+T} & k \neq l \\ T & k = l \end{cases} \tag{2.65}$$
$$= \begin{cases} 0 & k \neq l \\ T & k = l. \end{cases}$$

So if...if...we accept that a periodic signal $x(t)$ can be represented as the Fourier series of Eq. 2.64, we can determine the C_ks according to

$$\int_{t_0}^{t_0+T} x(t)e^{-jl\omega_0 t} dt = \int_{t_0}^{t_0+T} \left(\sum_{k=-\infty}^{\infty} C_k e^{jk\omega_0 t} \right) e^{-jl\omega_0 t} dt$$
$$= TC_l. \tag{2.66}$$

From here it follows that

$$C_l = \frac{1}{T} \int_{t_0}^{t_0+T} x(t)e^{-jl\omega_0 t} dt. \tag{2.67}$$

At this point, it is helpful and instructive to make a couple of observations. The first is that the C_ls will themselves be complex, since $x(t)$ is purely real and $e^{-jl\omega_0 t}$ is a complex quantity. But notice something important, which is the second observation:

$$C_l = \frac{1}{T} \int_{t_0}^{t_0+T} x(t)e^{-jl\omega_0 t} dt$$
$$C_{-l} = \frac{1}{T} \int_{t_0}^{t_0+T} x(t)e^{+jl\omega_0 t} dt. \tag{2.68}$$

That is, we have that C_l and C_{-l} are *complex conjugates.* If $C_l = C_{l\mathrm{R}} + jC_{l\mathrm{I}}$, then it is true that $C_{-l} = C_{l\mathrm{R}} - jC_{l\mathrm{I}}$. Go back now to the suggestive form Eq. 2.62,

$$\begin{aligned} x(t) = C_0 &+ \left(C_1 e^{j\cdot\omega_0 t} + C_{-1} e^{-j\cdot\omega_0 t}\right) + \left(C_2 e^{j2\cdot\omega_0 t} + C_{-2} e^{-j2\cdot\omega_0 t}\right) \\ &+ \left(C_3 e^{j3\cdot\omega_0 t} + C_{-3} e^{-j3\cdot\omega_0 t}\right) + \cdots . \end{aligned} \tag{2.69}$$

We find something interesting and important if we examine, say, the ω_0 term. Writing $C_1 = C_{1\mathrm{R}} + jC_{1\mathrm{I}}$ and $C_{-1} = C_{1\mathrm{R}} - jC_{1\mathrm{I}}$ and doing the substitution, we wind up with the equality

$$C_1 e^{j\cdot\omega_0 t} + C_{-1} e^{-j\cdot\omega_0 t} = 2C_{1R} \cos \omega_0 t - 2C_{1I} \sin \omega_0 t. \tag{2.70}$$

Outstanding. Not only is this purely real, as it must be, but it allows us to tie our complex exponential formulation back to Eq. 2.58. Evidently, if we go with complex exponentials, we have the equivalence between this formulation and that of Eq. 2.58 if

$$C_{i\mathrm{R}} = \frac{1}{2} A_i \tag{2.71}$$

and

$$C_{i\mathrm{I}} = -\frac{1}{2} B_i. \tag{2.72}$$

Nice.

All you need now is proof that periodic signals can, in fact, be represented by a sum of sinusoids, and then a nice bow to wrap things up. For the proof, well...it's complicated, and we'll suggest two possible routes. To begin exploring the rather rich mathematical literature, see the discussion by Siebert as a start.[17] If the reader is interested in exploring the thought process for themselves, a great beginning can be had by pondering the DT problem along the lines of the discussion in Section 2.3.4.

If you are satisfied that the foundations are sound, the neat bow on top is the synthesis/analysis equation pair. For synthesis:

$$x(t) = \sum_{k=-\infty}^{\infty} C_k e^{j\cdot k\omega_0 t}. \tag{2.73}$$

And for analysis:

$$C_l = \frac{1}{T} \int_{t_0}^{t_0+T} x(t) e^{-jl\omega_0 t} dt. \tag{2.74}$$

17 Siebert. *Circuits, Signals, and Systems,* Cambridge, MA: MIT Press, 1986. The relevant section is 12.3 A Special Case—Fourier Series.

Finally, if you are new to this, it is helpful to work out an example for yourself. Using the waveform of Figure 2.16 and the Fourier series analysis Eq. 2.76, calculate for yourself the first few coefficients C_l in the Fourier sum. Your answers should be consistent with Figure 2.23 and Eq. 2.54.

2.3.3 The Fourier Transform

If you have a good understanding of the Fourier series, you have an excellent foundation for understanding the frequency-domain view of the world. We just have one bit of untidiness to clean up, which is that the Fourier series deals with periodic signals...yet in the real world, most of our signals are not periodic. How do we deal with these signals analytically?

We start back with the Fourier series defining equations from before:

$$x(t) = \sum_{k=-\infty}^{\infty} C_k e^{j\cdot k\omega_0 t}. \tag{2.75}$$

$$C_l = \frac{1}{T} \int_{t_0}^{t_0+T} x(t) e^{-jl\omega_0 t} dt. \tag{2.76}$$

Now, rather than treat the aperiodic waveform as an entirely separate problem, we'll instead approach it as a *limit* of the Fourier series case, wherein the period T gets very "large" (see Figure 2.28). One thing we notice right away is that as the fundamental period T gets large, the frequency steps in our Fourier series get smaller. We write

$$\Delta\omega = \omega_0 = \frac{2\pi}{T}. \tag{2.77}$$

Our journey to the Fourier transform begins by writing our synthesis equation in terms of $\Delta\omega$:

$$x(t) = \sum_{k=-\infty}^{\infty} C_k e^{j\cdot k\Delta\omega t}. \tag{2.78}$$

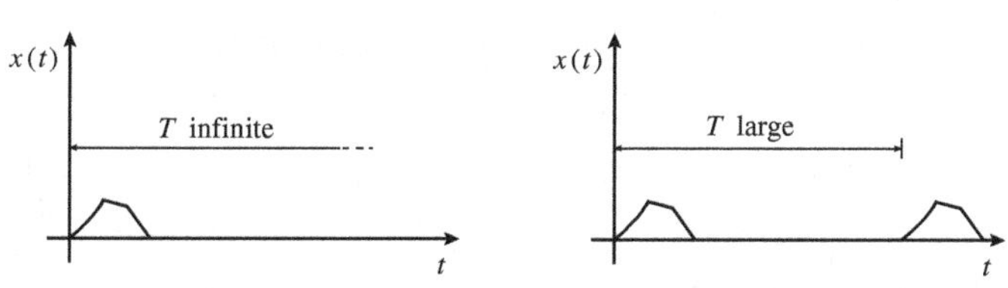

Figure 2.28 We approach the Fourier transform as a limit of the Fourier series, stretching the fundamental period T so that it is large compared to the time span of our signal.

To get to the Fourier transform integral, we need to get Eq. 2.78 to look like a Riemann sum. A tricky way to do this is to redefine C_l from Eq. 2.76 without the $\frac{1}{T}$ out front, according to

$$C_l = \int_{t_0}^{t_0+T} x(t)e^{-jl\omega_0 t}dt. \tag{2.79}$$

With this definition for the Fourier coefficients, the synthesis equation now becomes

$$x(t) = \frac{1}{T}\sum_{k=-\infty}^{\infty} C_k e^{j\cdot k\Delta\omega t}. \tag{2.80}$$

But we have $\frac{1}{T} = \frac{\Delta\omega}{2\pi}$, so we can get the desired Riemann sum form as follows:

$$x(t) = \frac{1}{2\pi}\sum_{k=-\infty}^{\infty} C_k e^{j\cdot k\Delta\omega t}\Delta\omega. \tag{2.81}$$

We then slightly modify Eq. 2.79 to be

$$C_l = \int_{-\frac{T}{2}}^{\frac{T}{2}} x(t)e^{-jl\omega_0 t}dt, \tag{2.82}$$

where we have made a nice, symmetrical choice for the limits of integration for the C_ls that nevertheless covers a full period T.

Now, finally, we examine our equations in the limit of $T \to \infty$. We have for $x(t)$

$$\begin{aligned} x(t) &= \lim_{T\to\infty} \frac{1}{2\pi}\sum_{k=-\infty}^{\infty} C_k\, e^{j\cdot k\Delta\omega t}\,\Delta\omega \\ &= \lim_{\Delta\omega\to 0} \frac{1}{2\pi}\sum_{k=-\infty}^{\infty} C_k\, e^{j\cdot k\Delta\omega t}\,\Delta\omega. \end{aligned} \tag{2.83}$$

We have our Riemann sum, which is the fundamental definition of an integral. If it has been a while since you studied calculus,[18] this summation form may be unfamiliar and you are more accustomed to this limit being formally written as:

$$x(t) = \frac{1}{2\pi}\int_{-\infty}^{\infty} C(\omega)e^{j\omega t}d\omega. \tag{2.84}$$

---

18 If you come at this topic from a completely different, self-taught direction, and have never formally studied calculus, ignore these references to what you “may remember” altogether. A *major* goal of this book is to be helpful to people like you.

We call $C(\omega)$ the *Fourier transform* of $x(t)$, and we will now rewrite this expression to follow convention in a couple of ways: (1) it is normal to write C as a function of $j\omega$, not ω; (2) convention is also to use the X in place of C. If we follow these conventions, the synthesis/analysis equation pair becomes

$$x(t) = \frac{1}{2\pi} \int_{-\infty}^{\infty} X(j\omega)e^{j\omega t}d\omega$$

$$X(j\omega) = \int_{-\infty}^{\infty} x(t)e^{-j\omega t}dt. \tag{2.85}$$

Take all the time you need to digest the foregoing treatment, culminating in the Fourier transform. Frequency-domain thinking saturates the field of wireless communications. A deep understanding of the Fourier transform will greatly increase your comfort in discussions and your understanding of the literature.

2.3.4 In Praise of Discretized Representations: The Discrete Fourier Transform (or Fast Fourier Transform)

We have just gone through a treatment of the Fourier transform that will be familiar to those whose education took place in what we call "traditional" engineering college programs. This is also how it tends to be introduced in math and physics departments, or at least it was when the author was an undergraduate in the mid-1990s.

But there are also other readers whose primary experience of frequency-domain thinking comes from digital signal processing. For you, the material up to this point may be unfamiliar, and even seem a little old-fashioned. And if you *do* feel it is old-fashioned...you have a point! Modern communications systems deal almost exclusively in discrete-time signals, and you may never have seen a Fourier transform that was not discretized (the discrete Fourier transform, or DFT) and computed using the fancy computational structures now known as the FFT, or fast Fourier transform.

We need to bridge the gap between continuous and discretized Fourier transform representations. We will do so in two steps. First, we examine the real-world implications of discretizing the Fourier transform. Then, we'll go all the way through the looking glass and dive into frequency-domain representation of DT signals.

All of this is worth the effort. Digital computation is a modern miracle that demands discretized representations. We need analytical tools to match.

Discretizing the Fourier Transform

Having carefully built our way up to the synthesis/analysis equations in Eq. 2.85, we now make progress by going backward. We start, again, with

$$x(t) = \frac{1}{2\pi} \int_{-\infty}^{\infty} X(j\omega)e^{j\omega t} d\omega \tag{2.86}$$

$$X(j\omega) = \int_{-\infty}^{\infty} x(t)e^{-j\omega t} dt.$$

With the synthesis equation, now, undo the fancy integral and return to the Reimann sum:

$$\begin{aligned} x(t) &= \lim_{\Delta\omega \to 0} \frac{1}{2\pi} \sum_k X(jk \cdot \Delta\omega) e^{jk \cdot \Delta\omega t} \Delta\omega \\ &= \lim_{T \to \infty} \frac{1}{T} \sum_k X\left(jk \cdot \frac{2\pi}{T}\right) e^{jk \cdot \frac{2\pi}{T} t}. \end{aligned} \tag{2.87}$$

Now we will do something quite interesting. Note that $X(j\omega)$ remains rigorously defined for all ω as a continuous variable. The interesting thing we will do is *no longer insist that T go to infinity*, but leave the equation otherwise unchanged. The right side of Eq. 2.87 relaxes to

$$\frac{1}{T} \sum_k X\left(jk \cdot \frac{2\pi}{T}\right) e^{jk \cdot \frac{2\pi}{T} t}. \tag{2.88}$$

That is, it is a Fourier series, and the coefficients C_k are just uniformly spaced samples of the Fourier transform $X(j\omega)$! It must follow that the proper left side of Eq. 2.88 is as follows:

$$\tilde{x}(t) = \sum_n x(t - nT) = \frac{1}{T} \sum_k X\left(jk \cdot \frac{2\pi}{T}\right) e^{jk \cdot \frac{2\pi}{T} t}. \tag{2.89}$$

This is so much better in pictures; see Figure 2.29. This is another moment to take your time and absorb. If you are a digital signal processor, the foregoing is extremely good news. It turns out that in our frequency-domain representations, we can capture $x(t)$ with a finite number of Fourier coefficients $X(jk \cdot \frac{2\pi}{T})$ provided that we meet two conditions:

1. When we invert the sampled Fourier transform according to $\tilde{x}(t) = \sum_n x(t - nT)$, we can discard all values of $\tilde{x}(t)$ for which $n \neq 0$, because those time replicas are artifacts of discretizing ω in the Fourier transform.
2. We can only take our frequency-domain samples so far apart before our replicas in time overlap, invalidating our discretized representation in the

frequency domain. In particular, we can see from Figure 2.29 that T has a minimum value $T_{\min}$. This means that

$$\frac{2\pi}{T} \leq \frac{2\pi}{T_{\min}}. \tag{2.90}$$

If we designate $\Delta\omega$ as the increment between frequency samples $\frac{2\pi}{T}$, then we have that

$$\Delta\omega \leq \frac{2\pi}{T_{\min}}. \tag{2.91}$$

Note that this maximum spacing for samples in the frequency domain is analogous to the spacing of time samples we required in Section 2.2.3 to avoid aliasing.

Frequency-domain Representations of DT Signals

At last, it is time to come to grips with how signals are kept track of in modern, digital communications systems. It is almost always true that we deliver our information in packets of DT samples. We bring them in in a burst, store them in a data buffer, and process them when and how it is convenient. Pictorially, we might represent our stored data buffer as in Figure 2.30, with M samples.

As we think about a frequency-domain representation, we'll take a cue from how we approached the issue in CT and suppose that we have a periodic signal

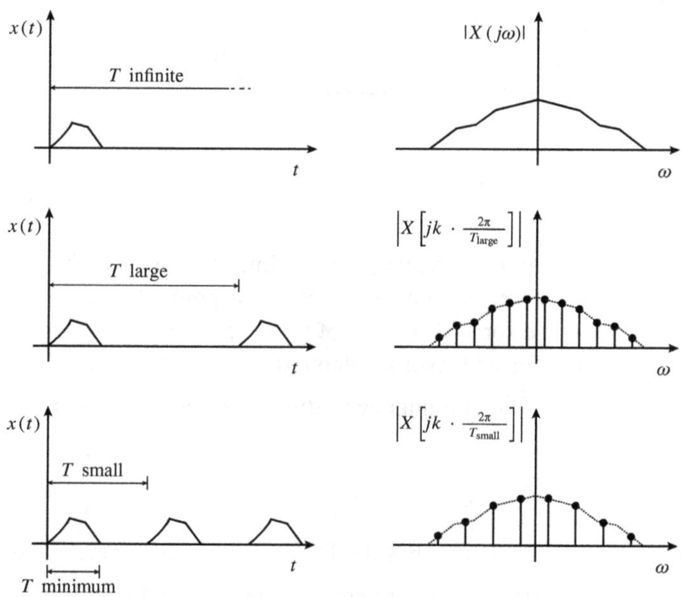

Figure 2.29 Discretizing the Fourier transform.

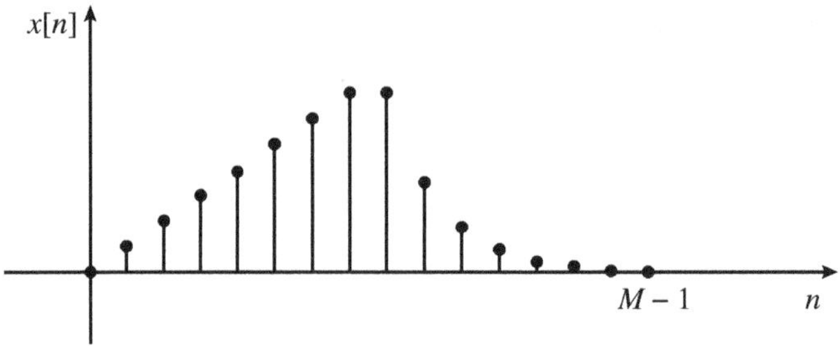

Figure 2.30 A short, DT data buffer.

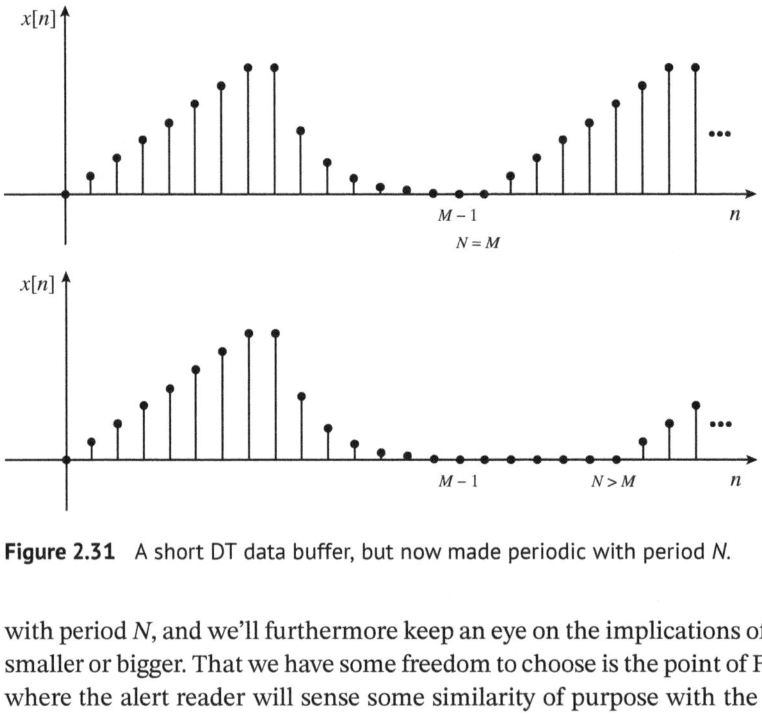

Figure 2.31 A short DT data buffer, but now made periodic with period *N*.

with period *N*, and we'll furthermore keep an eye on the implications of making *N* smaller or bigger. That we have some freedom to choose is the point of Figure 2.31, where the alert reader will sense some similarity of purpose with the discussion surrounding Figure 2.29. Now, for whatever fundamental period *N* we choose, it corresponds to a DT sinusoid $e^{j\left(\frac{2\pi}{N}\right)n}$. Frequency multiples of this sinusoid are

$$e^{j\left(\frac{2\pi}{N}\cdot 0\right)n},\ e^{j\left(\frac{2\pi}{N}\cdot 1\right)n},\ e^{j\left(\frac{2\pi}{N}\cdot 2\right)n},\ \ldots,\ e^{j\left(\frac{2\pi}{N}\cdot (N-1)\right)n}. \tag{2.92}$$

...and we encounter a surprise, which is that the list of distinct sinusoids in Eq. 2.92 is, in fact, exhaustive. It's true: for a complex sinusoid defined by $e^{j\left(\frac{2\pi}{N}\cdot k\right)n}$, where *k* and *n* are integers, any *N* consecutive values of *k* give us all the distinct sinusoids. To see this, imagine choosing a value for k, and then looking at the sinusoid corresponding to $k + N$. We find that they are the same:

$$\begin{aligned} e^{j\left(\frac{2\pi}{N}\cdot(k+N)\right)n} &= e^{j\left(\frac{2\pi}{N}k+2\pi\right)n} \\ &= e^{j\left(\frac{2\pi}{N}\cdot k\right)n}e^{j2\pi n} \\ &= e^{j\left(\frac{2\pi}{N}\cdot k\right)n}. \end{aligned} \tag{2.93}$$

This is a little uncomfortable. When we first discussed the Fourier series in CT, we did not *prove* that any periodic signal could be represented as a sum of sinusoids. The true mathematicians among you stomped off into the wilderness that is the Fourier analysis literature, while the rest of us took comfort in the fact that the CT Fourier series is an *infinite* sum. Surely there's a lot one can do with an infinite number of degrees of freedom, right? But here, where there's only a finite number of sinusoids wherewith to synthesize our time sequence, doubt can creep in.

Probably the clearest path to understanding here lies in the tools of linear algebra, or the mathematics of matrices. The vocabulary and concepts of linear algebra are important enough to wireless communications that we devote a chapter to the bare essentials (Chapter 9). If linear algebra is unfamiliar to you, or if you have forgotten what you once knew, a valid course of action is to skip ahead to Chapter 9 to get your bearings, and then come back here. An *equally* valid course is to focus here on the application *first*, and get what insight you can. You will then be able to bring some helpful context to your later study of the principles of linear algebra. Give it some careful consideration, then follow your instincts.

If you're still here, let's start with a clear statement of our problem. We have a time sequence $x[n]$, defined for $0 \leq n \leq N-1$, and we suppose that we can reproduce it in this range of ns with the sum

$$x[n] = \sum_{k=0}^{N-1} C_k e^{j\left(\frac{2\pi}{N}\cdot k\right)n}. \tag{2.94}$$

We can beautifully cast this as a system of linear equations, N equations with N unknowns (the unknowns in this case are the C_ks). Let's do just that, and write the system of equations as follows:

$$\begin{aligned} C_0 + C_1 + \cdots + C_{N-1} &= x[0] \\ C_0 e^{j\left(\frac{2\pi}{N}\cdot 0\right)\cdot 1} + C_1 e^{j\left(\frac{2\pi}{N}\cdot 1\right)\cdot 1} + \cdots + C_{N-1} e^{j\left(\frac{2\pi}{N}\cdot (N-1)\right)\cdot 1} &= x[1] \\ C_0 e^{j\left(\frac{2\pi}{N}\cdot 0\right)\cdot 2} + C_1 e^{j\left(\frac{2\pi}{N}\cdot 1\right)\cdot 2} + \cdots + C_{N-1} e^{j\left(\frac{2\pi}{N}\cdot (N-1)\right)\cdot 2} &= x[2] \\ &\vdots \\ C_0 e^{j\left(\frac{2\pi}{N}\cdot 0\right)\cdot (N-1)} + C_1 e^{j\left(\frac{2\pi}{N}\cdot 1\right)\cdot (N-1)} + \cdots + C_{N-1} e^{j\left(\frac{2\pi}{N}\cdot (N-1)\right)\cdot (N-1)} &= x[N-1]. \end{aligned} \tag{2.95}$$

The equivalent form in the grammar of matrices is as follows:

$$\begin{bmatrix} 1 & 1 & \cdots & 1 \\ 1 & e^{j\left(\frac{2\pi}{N}\cdot 1\right)\cdot 1} & \cdots & e^{j\left(\frac{2\pi}{N}\cdot (N-1)\right)\cdot 1} \\ \vdots & \vdots & \vdots & \vdots \\ 1 & e^{j\left(\frac{2\pi}{N}\cdot 1\right)\cdot (N-1)} & \cdots & e^{j\left(\frac{2\pi}{N}\cdot (N-1)\right)(N-1)} \end{bmatrix} \begin{bmatrix} C_0 \\ C_1 \\ \vdots \\ C_{N-1} \end{bmatrix} = \begin{bmatrix} x[0] \\ x[1] \\ \vdots \\ x[N-1] \end{bmatrix} \tag{2.96}$$

Now here is the key. Our matrix of complex exponentials has N columns, and each of those columns has N rows. It is a truth of linear algebra that if we can establish that the columns of this matrix are linearly independent (that is, no column can be reproduced with a weighted sum of the other columns), then we can produce *any* column vector $[x[0]\ x[1]\ \ldots\ x[N-1]]$ with a suitable choice of C_ks.

Astonishing!

And it gets better. It turns out that the columns of our matrix in Eq. 2.96 have the absolute cleanest form of linear independence possible, which is that they are *orthogonal*. So nice. We say that a set of complex vectors $\vec{v}_i$ is orthogonal if, for the inner product (or dot product) $\vec{v}_k \cdot \vec{v}_l^*$, we have[19]

$$\vec{v}_k \cdot \vec{v}_l^* = \begin{cases} 0 & k \neq l \\ P & k = l, \end{cases} \tag{2.97}$$

where P is simply a nonzero constant (for tidiness, it is common to *normalize*, or scale, the vectors so that the inner product $\vec{v}_k \cdot \vec{v}_k^*$ has the same real value P for all k). To see that orthogonality is what we have, consider the inner product of two columns of our matrix of exponentials:

$$\sum_{n=0}^{N-1} e^{j\left(\frac{2\pi}{N}\cdot k\right)n} e^{-j\left(\frac{2\pi}{N}\cdot l\right)n} = \sum_{n=0}^{N-1} e^{j\left(\frac{2\pi}{N}\cdot (k-l)\right)n}. \tag{2.98}$$

The case of $k = l$ is easy:

$$\sum_{n=0}^{N-1} e^{j\left(\frac{2\pi}{N}\cdot 0\right)n} = N. \tag{2.99}$$

19 The vector $\vec{v}_l^*$ is the *complex conjugate* of $\vec{v}_l$. The complex conjugate of a complex number is simply a new complex number where the sign of the imaginary part is reversed. If complex number $x = a + jb$, then the complex conjugate x^* is given by $x^* = a - jb$.

For $k \neq l$, it is a little harder. The math fact we need is the closed-form formula for a finite-term geometric series. A geometric series is a sum of terms with a fixed ratio r between each successive term:

$$a + ar + ar^2 + ar^3 + \cdots + ar^{N-1} = \sum_{n=0}^{N-1} ar^n. \tag{2.100}$$

It turns out that the well-known result is

$$\sum_{n=0}^{N-1} ar^n = a\left(\frac{1-r^n}{1-r}\right). \tag{2.101}$$

In our case, $a = 1$, and the common ratio r is $e^{j\left(\frac{2\pi}{N}\cdot(k-l)\right)}$. We have, then

$$\begin{aligned}\sum_{n=0}^{N-1} e^{j\left(\frac{2\pi}{N}\cdot(k-l)\right)n} &= \left(\frac{1-e^{j\left(\frac{2\pi}{N}\cdot(k-l)\right)N}}{1-e^{j\left(\frac{2\pi}{N}\cdot(k-l)\right)}}\right) \\ &= \frac{1-e^{j(2\pi\cdot(k-l))}}{1-e^{j\left(\frac{2\pi}{N}\cdot(k-l)\right)}} \\ &= \frac{1-1}{1-e^{j\left(\frac{2\pi}{N}\cdot(k-l)\right)}} \\ &= 0.\end{aligned} \tag{2.102}$$

We have at last

$$\sum_{n=0}^{N-1} e^{j\left(\frac{2\pi}{N}\cdot(k-l)\right)n} = \begin{cases} 0 & k \neq l \\ N & k = l. \end{cases} \tag{2.103}$$

We have established orthogonality, and therefore linear independence, of the columns of our matrix of exponentials in Eq. 2.96. In linear algebra, we say that these columns "are an orthogonal basis that spans the column space of vectors of length N."

We are ready to complete our mission of determining the synthesis/analysis Fourier pair for DT signals like we had in Eq. 2.85 for CT signals. Using the foregoing results, we have

$$\begin{aligned} x[n] &= \sum_{k=0}^{N-1} C_k e^{j\left(\frac{2\pi}{N}\cdot k\right)n} \\ C_k &= \frac{1}{N}\sum_{n=0}^{N-1} x[n]\cdot e^{-j\left(\frac{2\pi}{N}\cdot k\right)n}. \end{aligned} \tag{2.104}$$

A couple more notes on the oddities of DT signals. Suppose we have a complex sinusoid in DT given by

$$x[n] = e^{j\Omega n}. \tag{2.105}$$

We saw with Eq. 2.93 that the range of Ω that gives a unique set of sinusoids is only 2π wide. In Eq. 2.104, you can see that the particular 2π swath that we chose is for Ω in the range between 0 and 2π. This is a fine choice of frequency range, all the math works out, but for those coming from a CT background, we have the bizarre artifact of the "fastest" sinusoid being at the center of this range, at $\Omega = \pi$ (see Figure 2.32). Meantime, sinusoids at $e^{j(0+\epsilon)}$ and $e^{j(2\pi-\epsilon)}$, where $\epsilon \ll 1$, correspond to sinusoids that vary much more slowly. Again, there is nothing wrong with this mathematically. But for some, it is a bit of an eyesore in that it lacks the pleasing symmetry of positive and negative frequencies that we had in the CT case.

We can again use the following math fact, true for all integers m and n:

$$e^{j\Omega n} = e^{j(\Omega+2\pi m)n}. \tag{2.106}$$

We use this fact as follows. In our synthesis equation, we will leave it untouched for all values of k for which $0 \leq \frac{2\pi}{N}k \leq \pi$. For those sinusoids for which $\pi < \frac{2\pi}{N}k < 2\pi$, we will replace these sinusoids with their equivalents

$$\frac{2\pi}{N}k \rightarrow \frac{2\pi}{N}k - 2\pi = \frac{2\pi}{N}(k - N). \tag{2.107}$$

If we are careful with our accounting, our synthesis/analysis pair takes one of two forms, depending on whether N is even or odd. For N EVEN:

$$x[n] = \sum_{k=\left(-\frac{N}{2}+1\right)}^{\frac{N}{2}} C_k e^{j\left(\frac{2\pi}{N}\cdot k\right)n} \tag{2.108}$$

$$C_k = \frac{1}{N}\sum_{n=0}^{N-1} x[n] \cdot e^{-j\left(\frac{2\pi}{N}\cdot k\right)n}.$$

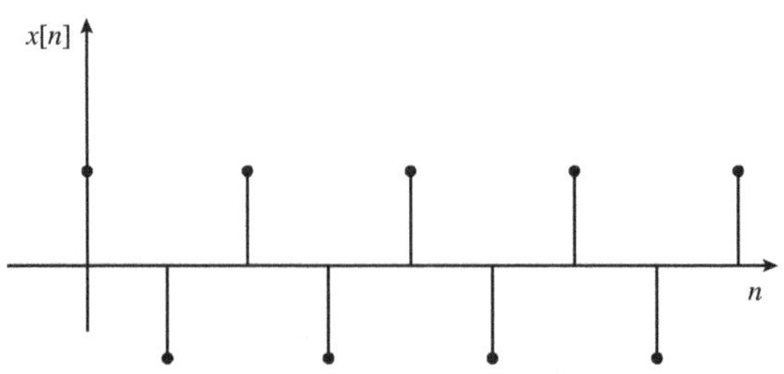

Figure 2.32 The highest frequency sinusoid in DT, with $\Omega = \pi$.

For N ODD:

$$x[n] = \sum_{k=-\frac{(N-1)}{2}}^{\frac{(N-1)}{2}} C_k e^{j\left(\frac{2\pi}{N}\cdot k\right)n} \tag{2.109}$$

$$C_k = \frac{1}{N}\sum_{n=0}^{N-1} x[n]\cdot e^{-j\left(\frac{2\pi}{N}\cdot k\right)n}.$$

Some readers will carefully pick these equations apart to make sure the math is correct. For you, a few guideposts:

1. The number of sinusoids in each synthesis sum is N, as it should be.
2. In a couple of special cases, the sinusoids that we pick from for synthesis are purely real, as when $\Omega = 0$ or $\Omega = \pi$. For all other sinusoids, for every complex $e^{j\left(\frac{2\pi}{N}\cdot k\right)}$, we have a matching $e^{-j\left(\frac{2\pi}{N}\cdot k\right)}$ term. We need this for the (normal) cases when $x[n]$ is purely real. This in turn requires the conjugate symmetry that we first saw in Eq. 2.68. That is, $C_k = C^*_{-k}$.
3. One has to admit that having one set of equations for N odd and the other for N even is a little untidy. But notice that as N gets large compared to 1 (think hundreds, thousands, millions) the little $\pm$1s get less important and we get closer to nice, symmetric $\pm\frac{N}{2}$. Note also that if it is your job to implement discrete signal algorithms on an actual computation processor, these index headaches are *absolutely* yours to master!

Unifying CT and DT Frequency-domain Methods

It is natural for most students of signal processing to have a church-and-state, hard separation between CT and DT methods. But by this point, there is clear carryover in terms of the concepts. Can we show that the mathematics is also, in every way that is important, the same?

We can. Start this time with the synthesis/analysis pair for, say, the N-even version of the DFT:

$$x[n] = \sum_{k=\left(-\frac{N}{2}+1\right)}^{\frac{N}{2}} C_k e^{j\left(\frac{2\pi}{N}\cdot k\right)n} \tag{2.110}$$

$$C_k = \frac{1}{N}\sum_{n=0}^{N-1} x[n]\cdot e^{-j\left(\frac{2\pi}{N}\cdot k\right)n}.$$

We'll compare this to a regression to the Riemann sum versions of *both* of the synthesis/analysis pair from Eq. 2.85:

$$x(n\cdot\Delta t)=\lim_{T\to\infty}\frac{1}{T}\sum_{k=-\infty}^{\infty}X(jk\cdot\frac{2\pi}{T})e^{j\left(k\cdot\frac{2\pi}{T}\right)n\cdot\Delta t} \tag{2.111}$$

$$X\left(jk\cdot\frac{2\pi}{T}\right)=\lim_{\Delta t\to 0}\sum_{n=-\infty}^{\infty}x(n\cdot\Delta t)e^{-j\left(k\cdot\frac{2\pi}{T}\right)n\cdot\Delta t}\Delta t.$$

Now in this Riemann sum regression, we are not going to take the whole limits business of $T\to\infty$ and $\Delta t\to 0$ so literally. Instead, we are going to make, in our heads, T *large enough* and Δt *small enough* that there is no practical difference between a discretized representation and a continuous representation, even though the math will be explicitly discretized.[20] With $\frac{T}{\Delta t}=N$ time intervals, our synthesis/analysis pair becomes, for N even:

$$x(n\cdot\Delta t)=\frac{1}{T}\sum_{k=\left(-\frac{N}{2}+1\right)}^{\frac{N}{2}}X\left(jk\cdot\frac{2\pi}{T}\right)e^{j\left(k\cdot\frac{2\pi}{T}\right)n\cdot\Delta t} \tag{2.112}$$

$$X\left(jk\cdot\frac{2\pi}{T}\right)=\sum_{n=\left(-\frac{N}{2}+1\right)}^{\frac{N}{2}}x(n\cdot\Delta t)e^{-j\left(k\cdot\frac{2\pi}{T}\right)n\cdot\Delta t}\Delta t.$$

Look an awful lot like the synthesis/analysis pair of Eq. 2.110, don't they? One of the biggest differences is in the summation limits of the analysis equation, but we remember that we've constructed things so that $x(n\cdot\Delta t)$ has a period of N. Thus, this summation is unchanged as long as we do the sum over N consecutive samples. We are free to make the bottom limit 0 and the top limit $N-1$ if we like, or leave the limits as they are. Also, we're going to redefine $X\left(jk\cdot\frac{2\pi}{T}\right)$ to move $\frac{1}{T}$ to a more convenient place:

$$X\left(jk\cdot\frac{2\pi}{T}\right)=\frac{1}{T}\sum_{n=\left(-\frac{N}{2}+1\right)}^{\frac{N}{2}}x(n\cdot\Delta t)e^{-j\left(k\cdot\frac{2\pi}{T}\right)n\cdot\Delta t}\Delta t. \tag{2.113}$$

To be consistent, we must now take that factor of $\frac{1}{T}$ out of the synthesis equation:

$$x(n\cdot\Delta t)=\sum_{k=\left(-\frac{N}{2}+1\right)}^{\frac{N}{2}}X(jk\cdot\frac{2\pi}{T})e^{j\left(k\cdot\frac{2\pi}{T}\right)n\cdot\Delta t}. \tag{2.114}$$

We'll end with one more manipulation: in our construction, each of the $\frac{\Delta t}{T}$s that appear in these equations (don't forget about the exponential) is equal to $\frac{1}{N}$. This

20 For example, perhaps we make T, the fundamental period, a million years, and Δt, our time increment, 0.01 fs.

means that our result for the Riemann sum regression of the CT Fourier transform can be written as, for N even:

$$\begin{aligned} x(n\cdot\Delta t) &= \sum_{k=\left(-\frac{N}{2}+1\right)}^{\frac{N}{2}} X\left(jk\cdot\frac{2\pi}{T}\right)e^{j\left(k\cdot\frac{2\pi}{N}\right)n} \\ X\left(jk\cdot\frac{2\pi}{T}\right) &= \frac{1}{N}\sum_{n=\left(-\frac{N}{2}+1\right)}^{\frac{N}{2}} x(n\cdot\Delta t)e^{-j\left(k\cdot\frac{2\pi}{N}\right)n}. \end{aligned} \tag{2.115}$$

And there we have it! Comparing Eqs. 2.115 and 2.110, we see that they are the same.

So when you get down to it, CT and DT Fourier methods are different neither conceptually nor analytically. The notation of CT can obscure this sometimes, what with the limit of this and that going to zero or infinity, and the sums replaced by integrals. But if you are trying to stay oriented, it can be helpful in the math of continuous variables to remember that if you take small enough steps in your variables, there is no practical difference between a truly continuous representation and a discretized one. Once you realize this, it is not much work at all to show, as we have done, that the methods are identical. Understanding this foundational concept is what enables us to make accurate computer simulations of continuous systems that occur in Nature.

2.4 Parseval's Theorem and Signal Energy

In this section, we introduce an important result to have firmly in your mind when jumping back and forth between the time and frequency domains. It is a relation called Parseval's Theorem, and is stated as follows: For a signal $x(t)$ and its Fourier transform $X(j\omega)$, we have the equality:

$$\int_{-\infty}^{\infty} |x(t)|^2 dt = \frac{1}{2\pi}\int_{-\infty}^{\infty} |X(j\omega)|^2 d\omega. \tag{2.116}$$

For DT signals, where we have

$$\begin{aligned} x[n] &= \sum_{k=0}^{N-1} C_k e^{j\left(\frac{2\pi}{N}\cdot k\right)n} \\ C_k &= \frac{1}{N}\sum_{n=0}^{N-1} x[n]\cdot e^{-j\left(\frac{2\pi}{N}\cdot k\right)n}, \end{aligned} \tag{2.117}$$

we will first bow to convention and use the more common notation

$$x[n] = \sum_{k=0}^{N-1} X[k] e^{j\left(\frac{2\pi}{N}\cdot k\right)n}$$

$$X[k] = \frac{1}{N}\sum_{n=0}^{N-1} x[n] \cdot e^{-j\left(\frac{2\pi}{N}\cdot k\right)n}. \tag{2.118}$$

It turns out that Parseval's theorem for DT is written

$$\sum_{n=0}^{N-1} |x[n]|^2 = \frac{1}{N}\sum_{k=0}^{N-1} |X[k]|^2. \tag{2.119}$$

As we will see in Section 2.5, this handy little relation helps us get to the famous Uncertainty Principle for Fourier transform pairs. In addition, a good understanding of Parseval's theorem will help you to understand the connections between noise power spectral density, noise bandwidth, and mean squared noise power in Chapter 4.

2.4.1 Signal Energy and Power for Time-domain Signals

Suppose we had a voltage source driving a resistor with a voltage waveform $v(t)$ as shown in Figure 2.33. When we build real-world physical machines, including computational engines, we are *always* interested in how much energy is involved. We want to know where the energy is coming from, where it is going, and, especially, how much of it is used for the intended task vs. wasted in some other way. In the case of the humble circuit of Figure 2.33, all of the energy flow is from the voltage source and out into the environment, dissipated as heat by the resistor. The instantaneous rate at which energy dissipates from the resistor is given by

$$P = \frac{v^2(t)}{R}, \tag{2.120}$$

which has units of watts (joules (of energy) per second). The total energy that has left the supply, and dissipated into the atmosphere (never to be recovered!) over an interval T is given by

$$E_{\text{tot}} = \frac{1}{R}\int_T v^2(t)dt. \tag{2.121}$$

We would properly say that the *average*, as opposed to instantaneous, power provided by the voltage supply in an interval T is given by

$$\begin{aligned} P_{\text{avg}} &= \frac{E_{\text{tot}}}{T} \\ &= \frac{1}{T}\cdot\frac{1}{R}\int_T v^2(t)dt. \end{aligned} \tag{2.122}$$

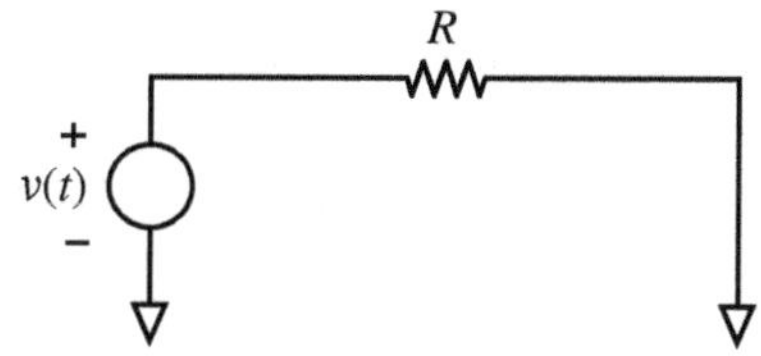

Figure 2.33 A simple circuit, dissipating energy at a rate of $\frac{v^2(t)}{R}$ J/sec.

It is conventional to call the integral here the "mean squared voltage," $\overline{v^2}$:

$$\overline{v^2} = \int_T v^2(t)dt. \tag{2.123}$$

Following along, joyfully putting words to the mathematics, we say that the root-mean-square voltage is

$$v_{\text{rms}} = \sqrt{\overline{v^2}} = \sqrt{\int_T v^2(t)dt.} \tag{2.124}$$

From here, in signal analysis, it is conventional and useful when discussing the signal $x(t)$ to refer to $x^2(t)$ as the "signal power"...units be damned. Some authors will even get a little sloppy and use signal "energy" and "power" interchangeably. This is all okay, and does little harm in practice. But if this little nit has confused, disturbed, or frustrated you in the past, the purpose of this subsection is to lay it out clearly so you can choose for yourself how to organize things in your mind.

2.4.2 Deriving Parseval's Theorem

If we have the voltage waveform $v(t)$, we can write down immediately the total energy delivered into a 1Ω load over an interval T as

$$\int_T v^2(t)dt = E_{\text{delivered}}. \tag{2.125}$$

Now we're going to write $v(t)$ in terms of its Fourier transform, and do what has become our favorite trick of regressing to the Reimann sum:

$$\begin{aligned} v(t) &= \frac{1}{2\pi} \int_{-\infty}^{\infty} V(j\omega)e^{j\omega t}d\omega \\ &= \lim_{T\to\infty} \frac{1}{T} \sum_k V(jk \cdot \Delta\omega)e^{j(k\cdot\Delta\omega)t}, \end{aligned} \tag{2.126}$$

where $\Delta\omega = \frac{2\pi}{T}$. Now using the Riemann sum, we can write (something proportional to) the instantaneous power as:

$$|v(t)|^2 = v(t)v^*(t) = \lim_{T\to\infty} \frac{1}{T^2} \sum_k V(jk \cdot \Delta\omega) e^{j(k\cdot\Delta\omega)t} \sum_l V^*(jl \cdot \Delta\omega) e^{-j(l\cdot\Delta\omega)t}. \tag{2.127}$$

We can write this as the compact

$$|v(t)|^2 = v(t)v^*(t) = \lim_{T\to\infty} \frac{1}{T^2} \sum_k \sum_l V(jk \cdot \Delta\omega) V^*(jl \cdot \Delta\omega) e^{j((k-l)\cdot\Delta\omega)t}. \tag{2.128}$$

Now integrate over the interval T:

$$\int_T |v(t)|^2 dt = \lim_{T\to\infty} \frac{1}{T^2} \sum_k \sum_l V(jk \cdot \Delta\omega) V^*(jl \cdot \Delta\omega) \int_T e^{j((k-l)\cdot\Delta\omega)t} dt. \tag{2.129}$$

Recall that k and l are integers, and that $\Delta\omega = \frac{2\pi}{T}$. We have that

$$\int_T e^{j((k-l)\cdot\Delta\omega)t} dt = \begin{cases} 0 & k \neq l \\ T & k = l. \end{cases} \tag{2.130}$$

The fearsome double summation that we started with collapses in a particularly satisfying way:

$$\int_T |v(t)|^2 dt = \lim_{T\to\infty} \frac{1}{T} \sum_k |V(jk \cdot \Delta\omega)|^2. \tag{2.131}$$

Retaking the limit, we ease into Parseval's theorem

$$\int_T v(t)|^2 dt = \frac{1}{2\pi} \int_{-\infty}^{\infty} |V(j\omega)|^2 d\omega. \tag{2.132}$$

There we have it!

2.4.3 The Concept of Power Spectral Density

Take another look at Parseval's theorem, now, and let's continue to interpret what we see:

$$\int_T |v(t)|^2 dt = \frac{1}{2\pi} \int_{-\infty}^{\infty} |V(j\omega)|^2 d\omega. \tag{2.133}$$

We agreed that $|v(t)|^2$ is proportional to the instantaneous power delivered into a resistive load. Therefore, $\int_T |v(t)|^2 dt$ is proportional to the total energy delivered into that load during a time interval T. It logically follows that in the integral on the right, we can interpret $|V(j\omega)|^2$ as an *energy spectral density*. That is to say, $|V(j\omega_0)|^2 \cdot \Delta\omega$ tells us how much energy the sinusoids in the $\Delta\omega$-wide region centered at ω_0 contributed to the energy that wound up in the resistive load.

There are two more things to say about this. In this text, we have defaulted to ω, radians per second, as the frequency variable, rather than f, which is hertz or cycles per second. There's no right or wrong here, and for the most part, it comes down to personal preference. For whatever reason, though, there is a pretty strong leaning in the literature toward using f when discussing energy spectral density. We, therefore, rewrite Eq. 2.133 with the substitution $d\omega = 2\pi\ df$, and have

$$\int_T |v(t)|^2 dt = \int_{-\infty}^{\infty} |V(f)|^2 df. \tag{2.134}$$

The second thing is that you will often see $|V(j\omega)|^2$ or $|V(f)|^2$ being referred to as a *power* spectral density rather than an *energy* spectral density. This is, strictly speaking, a minor abuse. But to see that it does no harm, observe that if the total energy delivered to the load is proportional to $\int_{-\infty}^{\infty} |V(f)|^2 df$, then the average power delivered is given by something like

$$P_{\text{avg}} = \frac{1}{T}\int_{-\infty}^{\infty} |V(f)|^2 df. \tag{2.135}$$

Once we stopped carrying around explicitly the resistance of the load and began speaking in terms of proportionalities, one more step doesn't hurt. You can call $|V(f)|^2$ the power or energy spectral density as you see fit.

2.4.4 Noise and Energy Spectral Density

For deterministic signals, talking about energy/power spectral density may seem less impressive than knowing exactly $x(t)$ and $X(j\omega)$ for any values of t and ω. But for noisy signals $n(t)$, we don't know ahead of time what, say, $n(t_0)$ will be. We can only meaningfully characterize such signals through their statistical moments, the most useful of which tend to be the average value $\overline{n(t)}$ and the mean squared departure from the mean $\overline{n^2(t)}$. Well, it turns out that it is also useful to characterize noisy signals by their energy spectral density $|N(f)|^2$. A handy fact to carry around is that Parseval's theorem holds for noisy signals too, so that

$$\overline{n^2(t)} = \int_{-\infty}^{\infty} |N(f)|^2 df. \tag{2.136}$$

Now $n(t)$ is normally real, which means that $N(f)$ is conjugate symmetric. This in turn means that $|N(f)|^2 = |N(-f)|^2$, and so

$$\overline{n^2(t)} = 2 \cdot \int_0^{\infty} |N(f)|^2 df. \tag{2.137}$$

This explains the convention of dealing only with positive frequencies when treating spectral analysis for noisy signals. (Note also that keeping track of this factor of 2 is not always strictly enforced. Proportionalities wind up being the most important thing usually, especially with hand analysis. You only have to track down all of your proportionality constants when you need numerical results.)

With probabilistic variables, we identify $\overline{n^2(t)}$ as the *variance* of a zero-mean signal, which is roughly speaking the width of the statistical distribution. Compare, then, the energy spectral densities in Figure 2.34. Based on Eq. 2.137, we certainly have that

$$\overline{n_1^2(t)} < \overline{n_2^2(t)}. \tag{2.138}$$

This little result illustrates a vital truth about communications systems in general. It is almost always true that we will send our signals in the presence of noise, usually thermal noise, and *a characteristic of thermal noise is that it has a flat energy spectral density*. The wider the bandwidth of our communications, the greater the noise variance $\overline{n^2(t)}$ and, therefore, the lower the overall SNR if the signal power is left constant.

We will have much more to say about this in coming chapters. But with a concept this important, repetition never hurts. Fix it in your mind: thermal noise is everywhere, so the more broadband your communications channel, the more noise energy you let in.

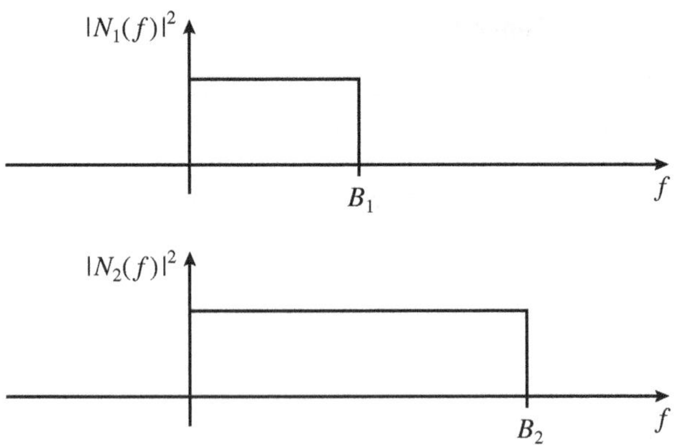

Figure 2.34 Contrasting thermal noise spectra with identical heights but differing bandwidths.

2.5 Bandwidth Equals Speed: The Uncertainty Principle

Once you accept the idea that time-domain signals can be expressed as the sum of sinusoids, it will probably not surprise you that signals that move faster in time require higher frequency components in their decomposition in order to do so. It turns out that we can formalize this bit of mathematical intuition in a way that is very satisfying (if a fair amount of work). The end result is as follows. Suppose we have a pulse in time $x(t)$ and its Fourier transform $X(j\omega)$ as shown in Figure 2.35. Suppose further that we set down a sensible mathematical definition of ΔT and $\Delta\Omega$ as indicated in the figures. Then it will turn out that the following inequality holds:

$$\Delta T \cdot \frac{\Delta\Omega}{2\pi} \geq \frac{1}{\pi}$$
$$\rightarrow \Delta T \cdot \Delta\Omega \geq 2. \tag{2.139}$$

The big news here is less the value of the constant on the right side of the inequality, and more that it is a constant. *We observe that small widths in the time domain imply large widths in the frequency domain, and vice versa.* You will see that this relationship holds between any two variables that happen to be a Fourier transform pair.

This inequality has many important implications for communications systems. For example, it is nearly universal at the time of this writing to tightly restrict the amount of electromagnetic spectrum that an individual operator may occupy. A limit, in other words, on $\Delta\Omega$. This, in turn, says that if the operator is encoding his/her data as pulses, there is a limit to how narrow those pulses can be in time. Therefore, there is an upper bound on how many such pulses can fit into a given time window T. Therefore, the regulatory bandwidth limit implies a limit on the transmittable bit rate! We are getting our first glimpse, as it happens, of the Shannon limit, with which we will occupy ourselves throughout Chapter 5.

Section 2.5.1 is devoted to deriving Eq. 2.139, showing that this form follows from the fact that $x(t)$ and $X(j\omega)$ are Fourier transform pairs. For many readers, a sensible course of action is to skip over this rather involved development and let

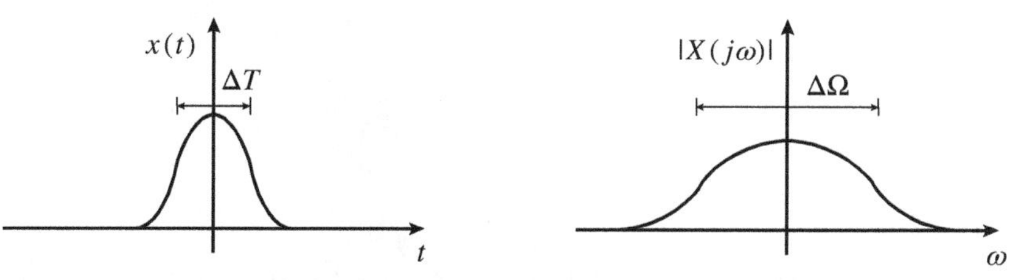

Figure 2.35 A pulse Fourier transform pair.

the fact of the uncertainty principle and its implications (outlined in Section 2.5.2) settle in your mind. When it is comfortable, and when you are curious about how the underlying mathematical machinery works, the author's goal is for Section 2.5.1 to be a satisfying read. For those of you who have known about the uncertainty principle forever, whether through communications systems or in its more famous incarnation in quantum mechanics, and wondered where it all comes from, this next section is for you.

2.5.1 Deriving the Uncertainty Principle

Okay. The first order of business is to set down a mathematical definition for the width of waveforms as shown in Figure 2.35.

It is actually worth puzzling over this problem a little on your own. Using your eye and your intuition, you probably do a decent, if numerically imprecise, job of estimating the width of a distribution. Can you now come up with an impersonal, automatic, analytical crank that takes $x(t)$ as an input and spits out a justifiable and *useful* width? Whatever you come up with should not be confounded by any of the examples shown in Figure 2.36. Spend as much time as you wish coming up with a solution. Consider returning to this exercise if you find yourself surprised, frustrated, skeptical, or outraged by any of the steps the author takes to get to the final answer. This is a well-worn path to learning and, sometimes, to true innovation. Can you do better?

A natural first effort might be to determine a box whose height is the peak value of the distribution, and whose width is chosen such that the area of the box

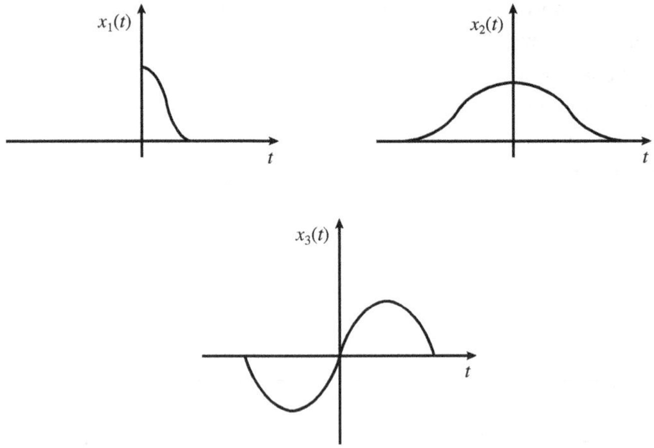

Figure 2.36 At minimum, our analytical definition of width should deal successfully with all three of these examples.

matches the area under the curve of the real function. Then, the width of that box becomes the "width" of the curve. This method works well for the first two examples, as shown in Figure 2.37. However, this method fails in the case of the third example from Figure 2.36: in the case of true odd symmetry about the vertical axis, the integration keeps coming out to zero. Interestingly, your eye is not fooled by the odd symmetry in terms of estimating the width of the distribution. It just turns out that our dumb crank is a tad *too* dumb to get the job done.

Onward. We could modify our method quite naturally by insisting that when we determine the width of $x(t)$, we work with the absolute value $|x(t)|$ instead. Using this trick leaves the width of the first two examples in Figure 2.36 unchanged, and helps us to successfully escape the difficulties posed by the third in that we now work with the distribution in Figure 2.38. This is all good, sound reasoning, and one could, within reason, stop here when it comes to defining the width of a distribution. We don't stop here, though, for reasons that are equal parts aesthetic and historic. The *main* reason is that it is fairly common for $\int_{-\infty}^{\infty} |x(t)| dt$ to lack a closed-form solution. Consider, for example,

$$x(t) = \begin{cases} e^{-\sigma t} \cos \omega t & t \geq 0 \\ 0 & t < 0. \end{cases} \tag{2.140}$$

With modern computational aids, of course, we can quite easily tackle $|e^{-\sigma t} \cos \omega t| dt$, but with pencil and paper we are a little stuck.

So instead of using $| \cdot |$, we're going to do something that seems a little strange but is actually okay. Rather than deal directly with the width of $x(t)$...we're going to look at the width of $x^2(t)$!

Can we do this? Well, let's look graphically at our three examples, and compare $|x(t)|$ to $x^2(t)$ in Figure 2.39. Now, no matter what definition we selected for the analytical width, you can perhaps agree that: (1) $|x(t)|$ and $x^2(t)$ would not yield *identical* answers for the width; (2) while not identical, they would be *close*.

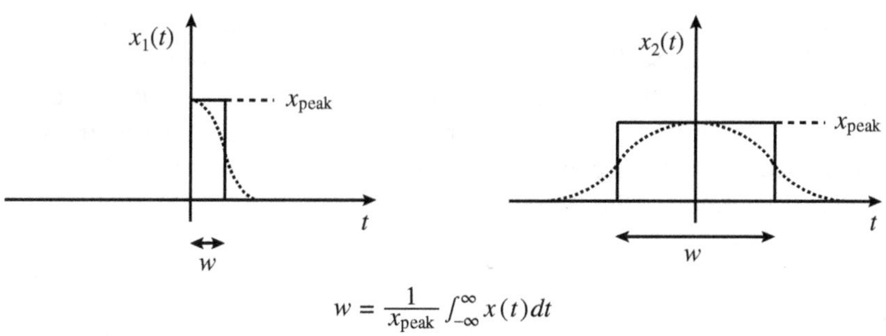

Figure 2.37 A try at defining the width of a distribution $x(t)$.

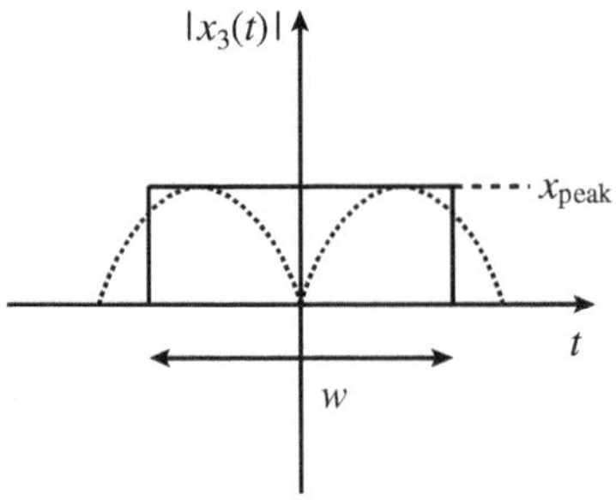

$$w = \frac{1}{x_{\text{peak}}} \int_{-\infty}^{\infty} x(t)dt$$

Figure 2.38 Improving our definition of w by employing the absolute value $|x(t)|$.

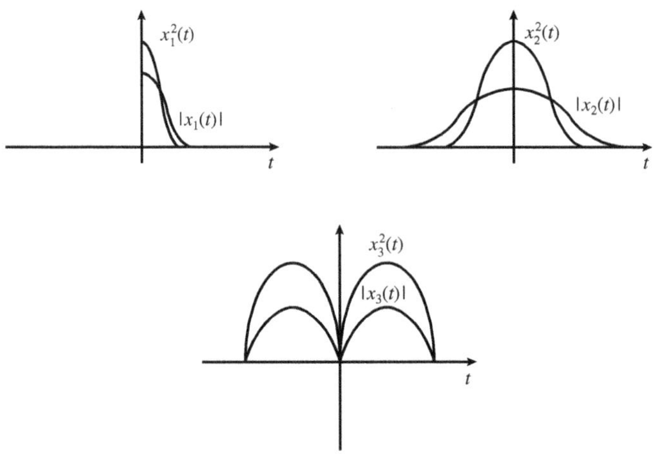

Figure 2.39 Using the square $x^2(t)$ vs. the absolute value $|x(t)|$.

And in exchange for a little nonlinear warping, many functions whose integrals get awkward with $|x(t)|$ come happily back into the pencil-and-paper fold as $x^2(t)$.

Off we go now to the really challenging part, which is coming up with an analytical definition of the "width" of $x^2(t)$. It turns out that with some patience, an extremely clever, wonderful, and imaginative answer comes to us from the realm of rotational mechanics. Consider a point mass free to rotate around a central axis as shown in Figure 2.40. In response to a torque $\tau = Fl_1$ (force vector into the page), the angular acceleration of the structure is given by

$$\frac{d\omega}{dt} = \frac{Fl_1}{m_1 l_1^2}. \tag{2.141}$$

The quantity $m_1 l_1^2$ is called the "moment of inertia," or the second moment of the mass.

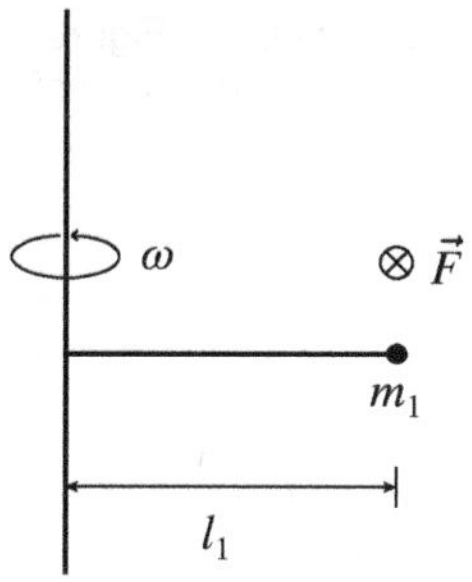

Figure 2.40 A point mass rotating around an axis. The force vector is into the page.

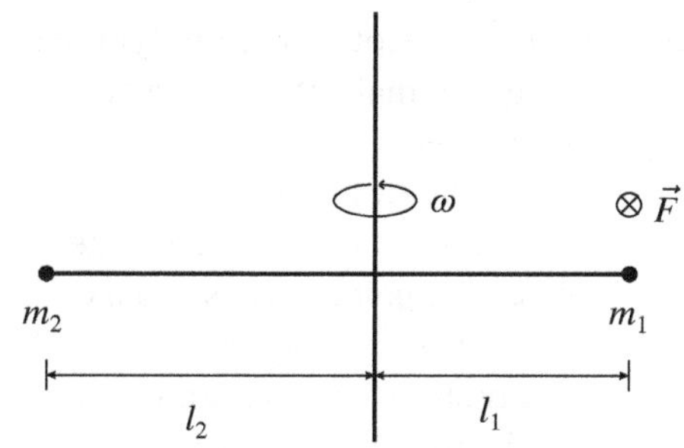

Figure 2.41 A two-mass structure rotating around an axis. The force vector is again into the page.

The next step is to consider a two-mass structure shown in Figure 2.41. If we apply a force in the same way that we did for the first example, rotational mechanics tells us that the angular acceleration will be

$$\frac{d\omega}{dt} = \frac{Fl_1}{m_1 l_1^2 + m_2 l_2^2}. \tag{2.142}$$

Now suppose that we wanted to make a *symmetrical* structure with two equal masses $m = \frac{1}{2}(m_1 + m_2)$ as in Figure 2.42, and our requirement is that the angular acceleration in response to a torque Fl_1 be the same.[21] The answer is that we would choose $l/2$ such that the moments of inertia were the same:

$$\begin{aligned} 2 \cdot \frac{(m_1 + m_2)}{2}(l/2)^2 &= m_1 l_1^2 + m_2 l_2^2 \\ (l/2)^2 &= \frac{m_1 l_1^2 + m_2 l_2^2}{m_1 + m_2}. \end{aligned} \tag{2.143}$$

21 Note that we are keeping the *torque* the same, not the *force*. The force F_{sym} we would need to apply to one of the $\frac{1}{2}(m_1 + m_2)$ masses (perpendicular to the (massless) support beam as in the figure) in order for the torque to be the same is $F_{\text{sym}} = \frac{Fl_1}{l/2}$.

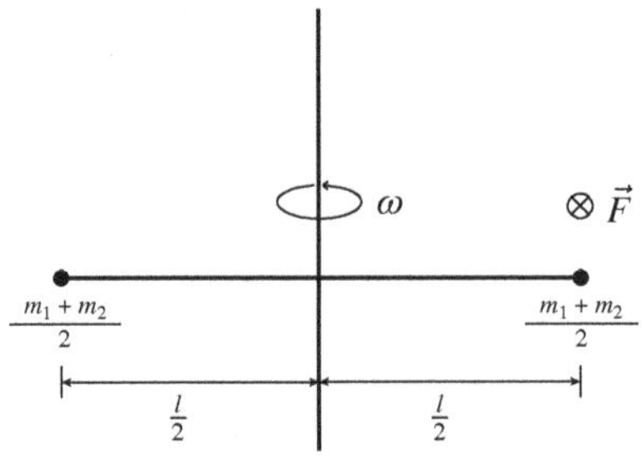

Figure 2.42 A symmetrical version of Figure 2.41 with the same total mass and moment of inertia.

Very interesting: we created a new structure with the same total mass but symmetrical mass distribution, and had to decide how "wide" to make this new structure in order to make it respond identically to applied torques.

Let's now extend this method in a way that makes the ties to our overall problem (remember: determining the "width" of a distribution in either the time or frequency domains) really clear. Suppose that instead of point masses, we have a thin bar whose linear mass density along the long axis is given by $m(x)$. We seek the equivalence shown in Figure 2.43. If we are really sharp with our understanding of integrals, we can put everything together and write down our answer as:

$$\left(\frac{w}{2}\right)^2 = \frac{\int_{-l_2}^{l_1} x^2 m(x)dx}{\int_{-l_2}^{l_1} m(x)dx}$$

$$w^2 = 4 \cdot \frac{\int_{-l_2}^{l_1} x^2 m(x)dx}{\int_{-l_2}^{l_1} m(x)dx}. \tag{2.144}$$

We arrive at last. For our time-domain waveform $x(t)$, we decided that we were interested in the width of $x^2(t)$, which is now analogous to the mass density $m(t)$ in the foregoing. We write down our width in time as

$$(\Delta T)^2 = 4 \cdot \frac{\int_{-\infty}^{\infty} t^2 x^2(t)dt}{\int_{-\infty}^{\infty} x^2(t)dt}, \tag{2.145}$$

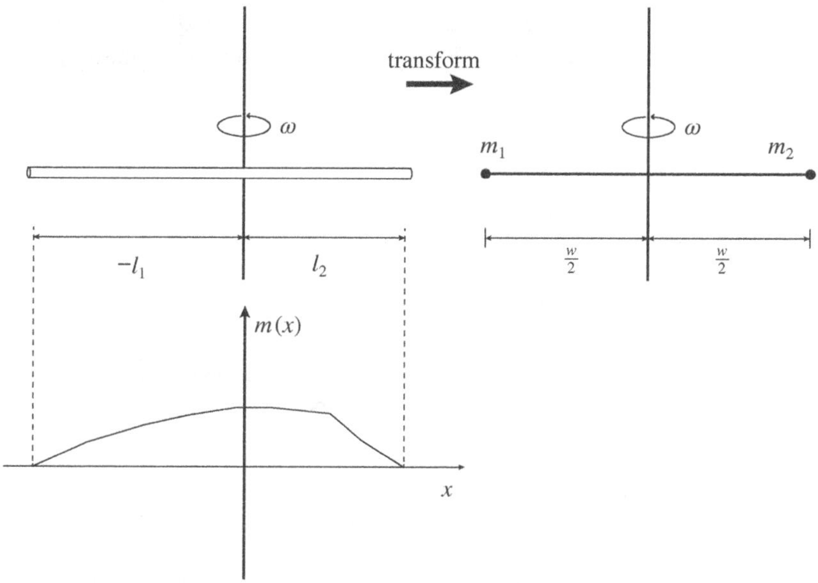

Figure 2.43 Transforming a continuous, asymmetric mass distribution into an equivalent, symmetrical one. In this figure, $m_1 = m_2 = \frac{1}{2}\int_{-l_1}^{l_2} m(x)dx$.

and do similarly for our width or spread in frequency

$$(\Delta\Omega)^2 = 4 \cdot \frac{\int\limits_{-\infty}^{\infty} \omega^2 |X(j\omega)|^2 d\omega}{\int\limits_{-\infty}^{\infty} |X(j\omega)|^2 d\omega}. \tag{2.146}$$

A lot of the heavy lifting is now out of the way. From here, as Siebert writes, "it is an interesting exercise in Fourier manipulation."[22] The first handy fact is that the Fourier transform of $\frac{dx(t)}{dt}$ or $\dot{x}(t)$ is $j\omega X(j\omega)$. It follows from Parseval's theorem that

$$\int\limits_{-\infty}^{\infty} \omega^2 |X(j\omega)|^2 d\omega = 2\pi \int\limits_{-\infty}^{\infty} [\dot{x}(t)]^2 dt. \tag{2.147}$$

We'll use Parseval's theorem in the usual way, too, reminding ourselves that

$$\int\limits_{-\infty}^{\infty} |X(j\omega)|^2 d\omega = 2\pi \int\limits_{-\infty}^{\infty} x^2(t) dt. \tag{2.148}$$

22 Ibid.

Now we set ourselves to computing $(\Delta T)^2(\Delta\Omega)^2$ directly, using Eqs. 2.145 and 2.146, and substituting using Parseval's theorem so that the entire relation is in terms of $x(t)$ and $\dot{x}(t)$:

$$(\Delta T)^2(\Delta\Omega)^2 = 16 \cdot \frac{\int_{-\infty}^{\infty} t^2x^2(t)dt \cdot \int_{-\infty}^{\infty} [\dot{x}(t)]^2 dt}{\left[\int_{-\infty}^{\infty} x^2(t)dt\right]^2}. \tag{2.149}$$

From here we require one more mathematical fun fact, known as the Schwartz Inequality, stated as follows:

$$\left[\int_a^b x(t)y(t)dt\right]^2 \leq \int_a^b x^2(t)dt \int_a^b y^2(t)dt. \tag{2.150}$$

We replace the equality of Eq. 2.149 with an inequality, and apply Schwartz's famous relation to the numerator on the right side:

$$(\Delta T)^2(\Delta\Omega)^2 \geq 16 \cdot \frac{\left[\int_{-\infty}^{\infty} tx(t)\dot{x}(t)dt\right]^2}{\left[\int_{-\infty}^{\infty} x^2(t)dt\right]^2}$$

$$\Delta T \cdot \Delta\Omega \geq 4 \cdot \frac{\left|\int_{-\infty}^{\infty} tx(t)\dot{x}(t)dt\right|}{\int_{-\infty}^{\infty} x^2(t)dt}. \tag{2.151}$$

The final, final, *final* trick is to use integration by parts on the numerator. For integration by parts, we use the fact that for two functions $u(t)$ and $v(t)$, there is the basic differentiation property

$$\frac{d}{dt}[uv] = u\frac{dv}{dt} + v\frac{du}{dt}. \tag{2.152}$$

which helps us solve definite integrals

$$[uv]_a^b = \int_a^b u\frac{dv}{dt}dt + \int_a^b v\frac{du}{dt}dt. \tag{2.153}$$

With regards to our problem, we'll identify $u(t) = tx(t)$ and $v(t) = x(t)$. We then have

$$\int_{-\infty}^{\infty} tx(t)\dot{x}(t)dt = [tx^2(t)]_{-\infty}^{\infty} - \left[\int_{-\infty}^{\infty} x(t)(t\dot{x}(t) + x(t))\right] dt$$

$$2 \cdot \int_{-\infty}^{\infty} tx(t)\dot{x}(t)dt = [tx^2(t)]_{\infty}^{\infty} - \int_{-\infty}^{\infty} x^2(t)dt. \tag{2.154}$$

We can safely assume that $x(t)$ is zero at $\pm\infty$, and so

$$\int_{-\infty}^{\infty} tx(t)\dot{x}(t)dt = -\frac{1}{2}\int_{-\infty}^{\infty} x^2(t)dt$$

$$\left|\int_{-\infty}^{\infty} tx(t)\dot{x}(t)dt\right| = \frac{1}{2}\int_{-\infty}^{\infty} x^2(t)dt. \tag{2.155}$$

All that is left is to substitute, triumphantly, Eq. 2.155 into Eq. 2.151. We are left with the famous Uncertainty Principle with the frequency variable in radians per second as

$$\Delta T \cdot \Delta\Omega \geq 2. \tag{2.156}$$

With the frequency variable f in Hertz, we have $\Delta\Omega = 2\pi\Delta f$, and write

$$\Delta T \cdot \Delta f \geq \frac{1}{\pi}. \tag{2.157}$$

So why is this called the "Uncertainty Principle"? Well, it inherits the name from a very similar relation governing the momentum and position of a particle in quantum mechanics. It turns out that the *wave functions* governing the position x and momentum p of a quantum particle are Fourier transform pairs; it follows that the spread of these two quantities are governed by a relation like Eq. 2.156. In quantum mechanics, it leads down the philosophical rabbit hole of wondering why we cannot know a particle's position *and* momentum simultaneously. In signal processing, the interpretation is much more straightforward. We have shown that fast (short) pulses in time imply high bandwidths in the frequency domain. In other words, bandwidth equals speed.

2.5.2 Implications of the Uncertainty Principle for Communications

With the Uncertainty Principle under our belts, we turn our attention to what it means for communications systems. It happens that a reasonable way to think about modern electronic communications systems is to start with a basic pulse

Figure 2.44 An example pulse shape.

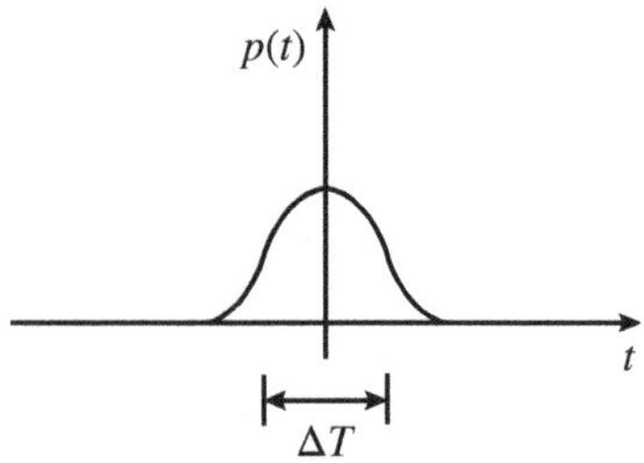

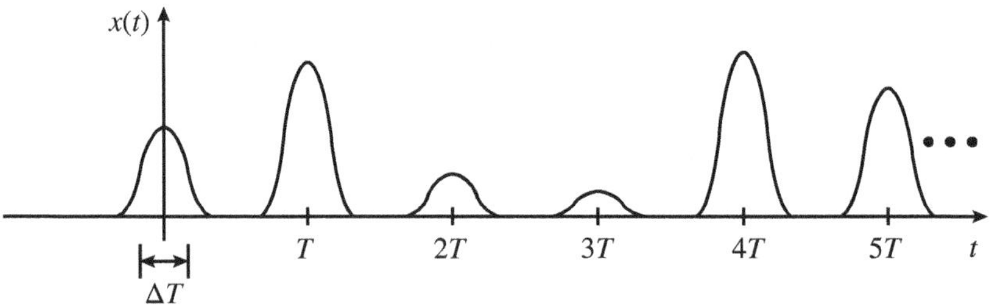

Figure 2.45 A pulse train.

shape in time, $p(t)$, as shown in Figure 2.44. Our transmitted signal is just a series of amplitude-scaled pulses, placed T seconds apart. So, if our data is represented by the sequence $x[n]$, we might have the time waveform shown in Figure 2.45. Mathematically, we express this series of pulses as the sum

$$x(t) = \sum_{n} x[n]p(t - nT). \tag{2.158}$$

For wireless systems, where spectrum is a shared resource, we are particularly interested in how much spectral real estate our pulse train requires. We take the Fourier transform of Eq. 2.158:

$$\begin{aligned}
\int_{-\infty}^{\infty} x(t)e^{-j\omega t}dt &= \int_{-\infty}^{\infty} \sum_{n} x[n]p(t - nT)e^{-j\omega t}dt \\
&= \sum_{n} x[n] \int_{-\infty}^{\infty} p(t - nT)e^{-j\omega t}dt \\
&= \sum_{n} x[n]P(j\omega)e^{-j\omega Tn} \\
X(j\omega) &= P(j\omega) \sum_{n} x[n]e^{-j\omega Tn}.
\end{aligned} \tag{2.159}$$

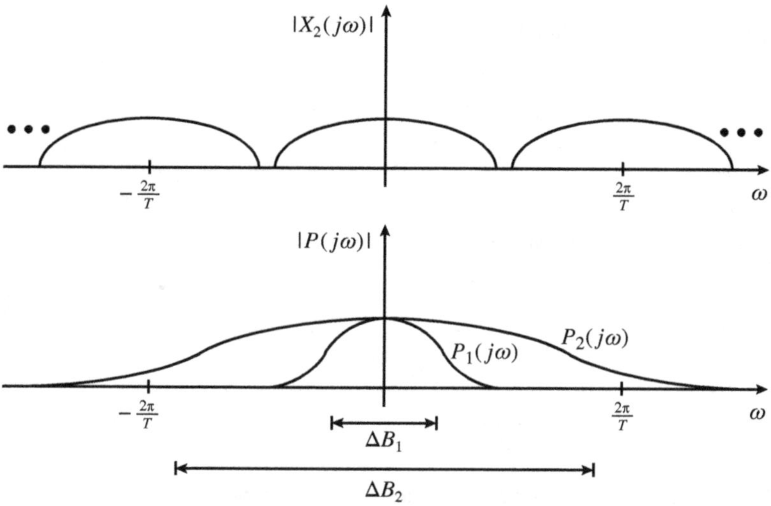

Figure 2.46 In the frequency domain, separate multiplicative components of a pulse train.

It helps to look at the pieces separately, the pieces being $P(j\omega)$ and $X_2(j\omega) = \sum_n x[n]e^{-j\omega Tn}$. Looking at $X_2(j\omega)$, the detailed form is not so important. However, we can pick out one important feature: it is periodic in ω, with period $\frac{2\pi}{T}$. We can thus in general draw $X_2(j\omega)$ as something like the top part of Figure 2.46. Now we can add in the effects of the pulse shape $P(j\omega)$. We see that the Fourier transform of the pulse $P(j\omega)$ will multiply by $X_2(j\omega)$. The bottom part of Figure 2.46 shows two choices of $P(j\omega)$. The Fourier transform of our pulse train would therefore look like Figure 2.47 for $P_1(j\omega)$. For $P_2(j\omega)$, it would look like Figure 2.48. Finally, we can see the implications of the Uncertainty Principle for communications systems. We start with the idea of our data as a succession of scaled but otherwise identical pulses. The narrower our pulses are in time, the more pulses we can fit into a given time window and, therefore, the more bits per second we can send. But since

$$\Delta T \cdot \Delta B_{1,2} \geq \frac{1}{\pi}, \tag{2.160}$$

or,

$$\Delta B_{1,2} \geq \frac{1}{\pi}\frac{1}{\Delta T}, \tag{2.161}$$

a smaller ΔT implies a larger occupied bandwidth. So all other things being equal, *the bit rate we can achieve is proportional to the bandwidth we are allowed to occupy!*

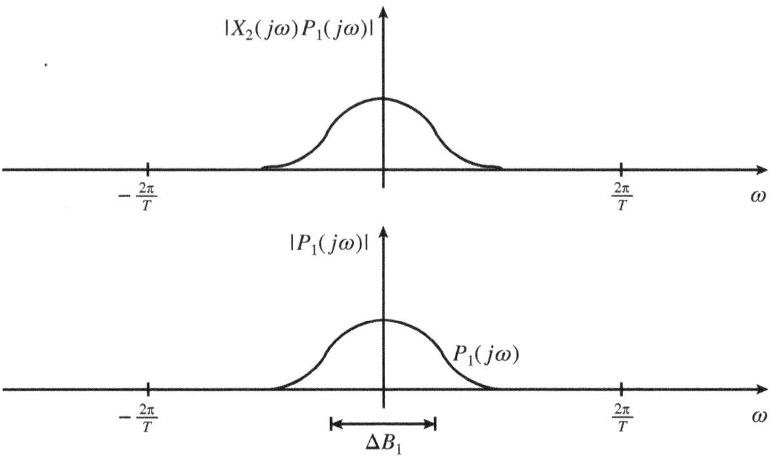

Figure 2.47 If we use P_1 our pulse train occupies a bandwidth of B_1.

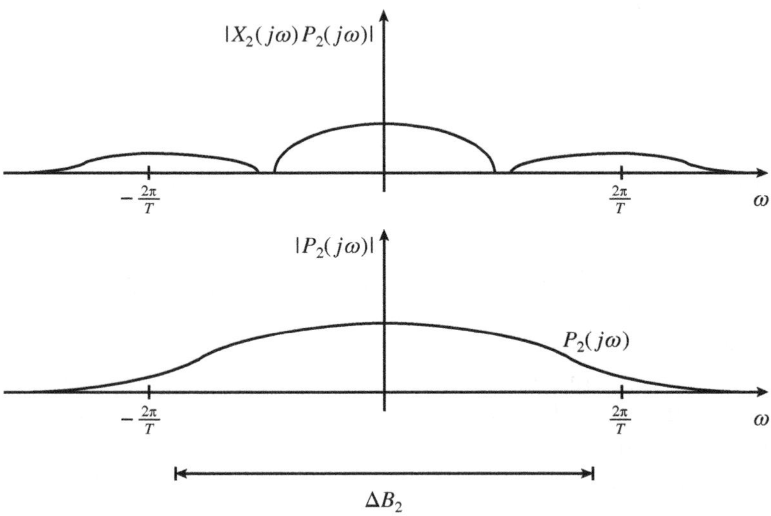

Figure 2.48 If we use P_2 our pulse train occupies a bandwidth of B_2.

We will have *much* more to say about this in Chapter 5. We are coming to a valid conclusion, it turns out, but it is incomplete because we are not accounting for noise in the communications channel. It nevertheless remains that if you have come this far with the Uncertainty Principle, you have made great strides forward in terms of understanding the fundamentals of communications.

3

The Least You Need to Know About Analog Electronics

Electronic circuit design is a huge and fascinating topic. If you come from a circuits background, the notion of devoting a mere chapter may well offend you.

If this is you, then two things. First, and most importantly, there is absolutely no intention to assert that electronic circuit design can be covered comprehensively in a single book, much less a single chapter. As with so many topics in this slender volume, the central thesis is that there are a few principles that are incredibly valuable to both the general practitioner and the specialist alike, and we attempt to treat those here. If you are a systems architect, knowledge of these principles will help you to collaborate meaningfully with the design specialists who realize individual circuit blocks. If you are a design specialist, deep awareness of these principles will arm you to sort out what details are useful to share with the architect and what details are...not.

Second, consider the following seemingly outrageous statement: Most of the time in electronic circuits, the Task is to set a voltage on a capacitor as quickly and as accurately as possible. See Figure 3.1. Our information is often encoded as a voltage, and the node whose voltage needs to be set has some capacitance to an incremental ground. That capacitance may be deliberately placed, or may simply be a parasitic element that cannot be avoided. Either way, if you give it some thought, you will find this simple-minded, generic Task to be shockingly general.

The treatment of analog circuits in this book will likely differ from what you have seen before, as we focus intensely on *energy*. For an electronic machine, our interest is in the energy that Nature requires to accomplish the Task, compared to the energy our machine draws from its power supply ($\int I_{DD}(t) \cdot V_{DD}(t)dt$). At the time of this writing, it is almost always true that the energy our machines draw from the supply *far* exceeds the energy that Nature requires to accomplish the Task. This is inefficiency. We should study it clearly and carefully and understand it thoroughly. We should then reduce that inefficiency to as close to zero as possible.

Energy Fundamentals of Radio, First Edition. Joel L. Dawson.

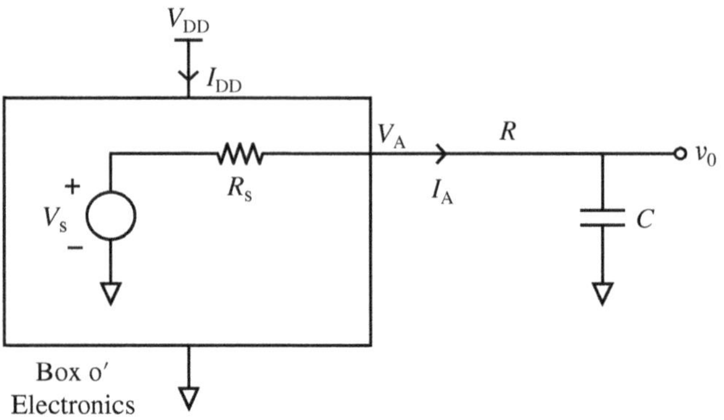

Figure 3.1 A model of the Task for much of analog electronics.

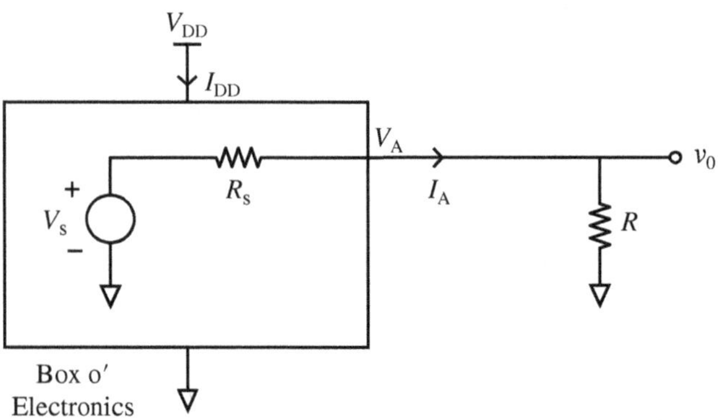

Figure 3.2 A model of the Task when we have to deliver energy.

An important exception to the general scenario described above is when the purpose of an electronic machine is to actually *deliver energy*, rather than merely *represent information*. Examples admittedly abound: a driver circuit that controls an electric motor; circuits that deliver audio signals to a speaker or earphones (acoustic waves are a type of energy propagation); circuits that drive light bulbs (heat and light); and yes, power amplifiers that drive antennas in wireless communications systems (as we shall see, electromagnetic waves are also a form of energy propagation). For this class of circuits, a better rough-cut model is shown in Figure 3.2. The instantaneous power delivered to the load is exactly $V_A \cdot I_A$ joules per second at the load terminals.

We will represent a third broad class of circuits, whose aim is to either set a capacitor voltage or deliver energy, but that *only needs to do so over a narrow range*

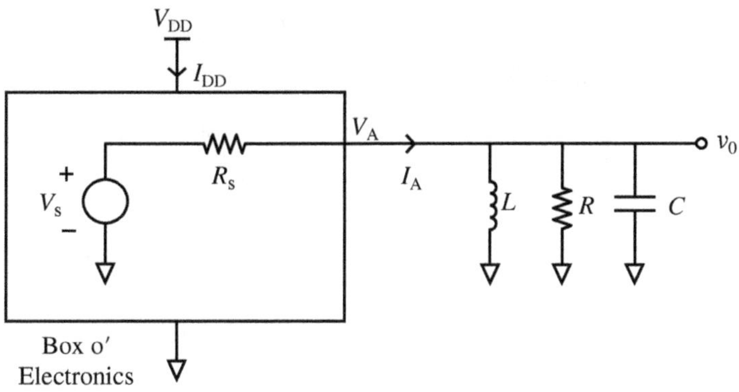

Figure 3.3 A model of the Task when it must be accomplished over a narrow band.

of frequencies. In this case, we will find that it is advantageous to employ resonance. This can be modeled as a parallel *RLC* circuit as shown in Figure 3.3.

In this chapter, we will discuss each of these three types of loads in turn. We will remind you of the standard textbook treatment for each one. Then we'll go off the beaten path and take a hard look at the energy considerations. Going forward in your study of electronics, keep a sharp eye out for where each type of load occurs, and what that means for the energy efficiency of that particular machine.

3.1 The *RC* Circuit

A resistor-capacitor (*RC*) circuit is illustrated in Figure 3.4. Helpful things to understand about this circuit are its response to a step voltage input and to a sinusoidal drive. While we will repeat the development here, you can find these responses derived in practically every book on circuits ever written, as well as in many basic physics textbooks. What you do not often see is careful treatment of energy considerations for this humble circuit. So, review Sections 3.1.1 and 3.1.2 and then really study Section 3.1.3.

3.1.1 Time Domain Behavior

The clearest way to analyze this circuit is to utilize Kirchoff's current law, which states that the total current entering a given node must equal the current leaving that node. In our example, we focus on the output node v_0. We say

$$\frac{v_A - v_0}{R} = C\frac{dv_0}{dt}. \tag{3.1}$$

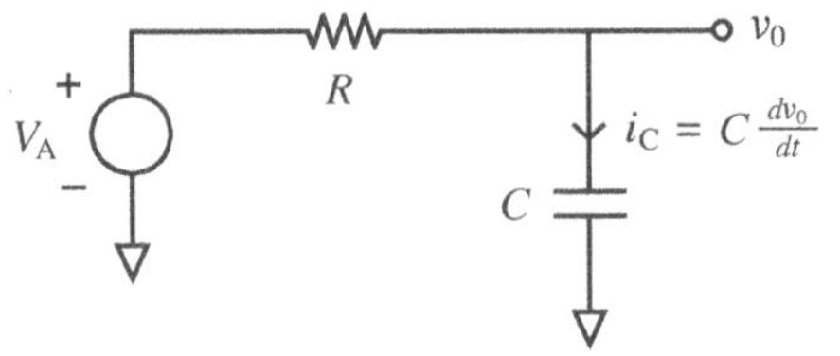

Figure 3.4 The *RC* circuit for analysis.

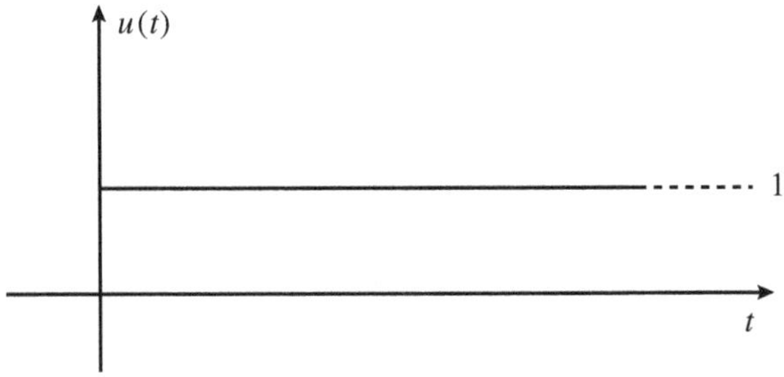

Figure 3.5 A unit step input.

Rearranging into the classic form a differential equation, we arrive at

$$\frac{dv_0}{dt} + \frac{1}{RC}v_0 = \frac{1}{RC}v_A. \tag{3.2}$$

A "step" input for v_A is designated $u(t)$, and has the form shown in Figure 3.5. An awkward thing about differential equations, you may recall, is that we cannot usually solve them in the straight-ahead way that we solve, say, algebraic equations. Instead, we make an (educated) guess as to the form of the solution $v_0(t)$, and then plug it back into the equation to verify that it is correct.

In this case, any one of a number of textbook approaches yields the important result of the step response of this simple circuit. If $v_A(t) = u(t)$, what we find is that

$$v_0(t) = \left(1 - e^{-t/RC}\right) u(t). \tag{3.3}$$

This classic response is shown graphically in Figure 3.6. The product *RC* is often called the "*RC* time constant," and it characterizes the exponential rise of the response. The smaller the product *RC*, the faster output v_0 settles to its final value.

The solution to a first-order *RC* circuit driven by a unit step is one of very, very few solutions that are worth committing to memory. First-order differential equations show up everywhere in science and engineering, either as exact models of the systems being studied or as extremely good approximations. You can take a big step up in fluency with linear systems if you can tuck solutions to first-order equations away in your memory.

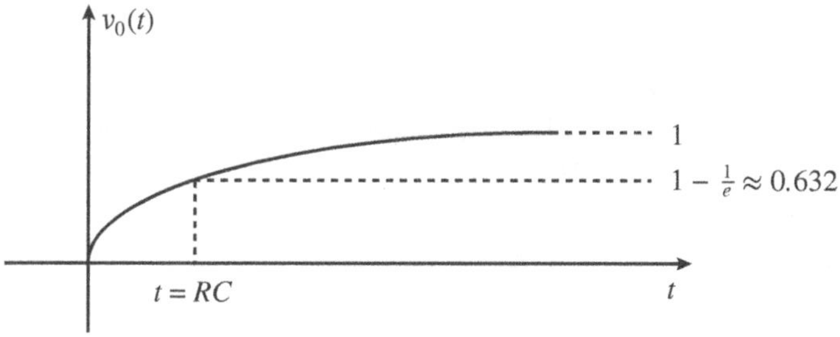

Figure 3.6 Response of an *RC* circuit to a step input.

3.1.2 Frequency Domain Behavior

We ask: for a sinusoidal input drive $v_A = e^{j\omega t}$, what is the output?

From our development in Section 2.3.1, we know that complex exponentials are eigenfunctions of all systems governed by linear differential equations with constant coefficients. So, if the input is $v_A = e^{j\omega t}$, the output *must* be $v_0 = Ae^{j\omega t}$, where A is a complex constant. Our task now is to determine A.

We can do this in a compact sequence of steps:

$$\begin{aligned}
\frac{d}{dt}\left(Ae^{j\omega t}\right) + \frac{1}{RC} \cdot Ae^{j\omega t} &= \frac{1}{RC}e^{j\omega t} \\
j\omega Ae^{j\omega t} + \frac{1}{RC}Ae^{j\omega t} &= \frac{1}{RC}e^{j\omega t} \\
A\left(j\omega + \frac{1}{RC}\right)e^{j\omega t} &= \frac{1}{RC}e^{j\omega t} \\
A\left(j\omega + \frac{1}{RC}\right) &= \frac{1}{RC} \\
A &= \frac{1}{j\omega RC + 1}.
\end{aligned} \tag{3.4}$$

So, if we drive this circuit with a complex sinusoid $e^{j\omega t}$, what we get out is another sinusoid $\left(\frac{1}{j\omega RC+1}\right)e^{j\omega t}$.

It is customary to plot this scaling amplitude A as a function of frequency ω. It is also customary, and useful, to do so on a log–log chart as shown in Figure 3.7. This kind of frequency response leads the *RC* circuit to be often called as a “low pass filter,” because it allows sinusoids with frequencies below the corner frequency $\omega = \frac{1}{RC}$ to pass with an A of close to unity magnitude, while frequencies above the corner frequency are increasingly attenuated.

It is common to analyze systems like this is by exploiting the eigenfunction nature of sinusoids. We know that for an input $x(t)$, we can express it as a sum of sinusoids.

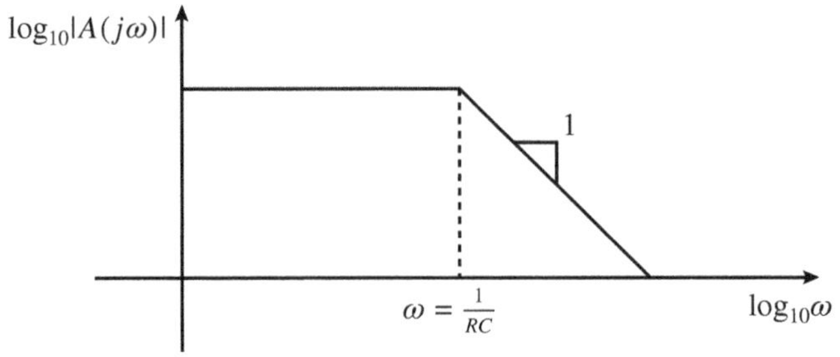

Figure 3.7 Frequency response of the *RC* circuit.

$$x(t) = \lim_{\Delta\omega \to 0} \frac{1}{2\pi} \sum_k X(jk \cdot \Delta\omega) e^{jk \cdot \Delta\omega t} \Delta\omega = \frac{1}{2\pi} \int_{-\infty}^{\infty} X(j\omega) e^{j\omega t}\, d\omega. \tag{3.5}$$

$X(j\omega)$, then, tells us the weight of each of the complex exponentials in the sum. In the instance of the *RC* circuit, any input sinusoid will be scaled by what we'll now call the transfer function $H(j\omega)$:

$$H(j\omega) = \frac{1}{j\omega RC + 1}. \tag{3.6}$$

If you think about it, what we have is the ability to write down the Fourier transform of the output $V_0(j\omega)$ by inspection. That is,

$$V_0(j\omega) = H(j\omega)X(j\omega). \tag{3.7}$$

And therefore,

$$v_0(t) = \frac{1}{2\pi} \int_{-\infty}^{\infty} H(j\omega)X(j\omega)d\omega. \tag{3.8}$$

3.1.3 Energy Considerations for *RC* Circuits

Many readers with traditional engineering backgrounds will have been tempted to skip, or have actually skipped, Sections 3.1.1 and 3.1.2. These readers are urged *not* to skip this section. In traditional programs, energy considerations are treated as a curiosity, an afterthought of interest only to those who are more physics-minded. And it is easy to see why this is the case. The modern electronics of the 2020s came of age in an era when energy was, as a practical matter, free of cost.

Now, this does not mean that energy concerns never impacted the design of our machines. In many electronic systems, all or most of the energy that comes out of the supply leaves the system as heat into the surrounding environment. The short story is that electronic components dissipate heat energy at a rate proportional to the temperature difference between the component and the surrounding environment. They therefore get as hot as necessary to keep heat energy from building up continually, actually melting down or even exploding if that's what it takes(!). Big, power-hungry electronic systems have always required complex mechanical heat sinks to dissipate heat quickly.

On top of heating concerns, the reader is encouraged to remember that the energy that we use to run our electronic circuits *comes from somewhere.* If we plug our systems into a wall socket, for example, there is a through line connecting the energy dissipated in our resistors all the way back to the coal, natural gas, petroleum, solar cells, or wind turbines that are the basis of the local energy infrastructure.

Very good. Consider, then, the *RC* circuit of Figure 3.8, where we have put in a switch that allows us to freeze the action when we want to. We often account for energy in electromagnetic systems by attributing a spatial energy density to the electric and magnetic fields. By this accounting, it turns out that the number of joules of energy stored in the electric field between the capacitor plates is given by

$$E = \frac{1}{2}CV_c^2. \tag{3.9}$$

But now let's ask a different question: how much energy did the voltage source have to provide to store this much energy in the capacitor? If your first answer is $\frac{1}{2}CV_c^2$, it is not out-of-bounds to congratulate yourself on solid instincts around the concept of conservation of energy. However, that answer is incorrect. As current flowed through the resistor to charge the capacitor, energy was being dissipated as heat at an instantaneous rate of $I^2(t)R$ joules per second. It must therefore be true that the energy taken out of the supply is greater than $\frac{1}{2}CV_c^2$. How do we figure this out?

The most straightforward way is to recognize that during the time we were charging the capacitor, the instantaneous power drawn from the supply was always

$$P(t) = V(t) \cdot I(t). \tag{3.10}$$

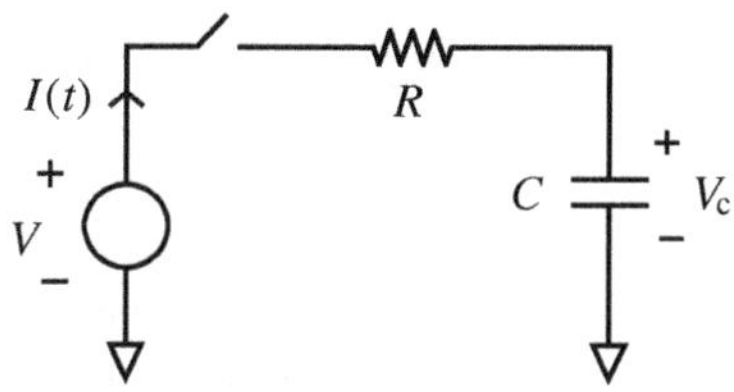

Figure 3.8 Analyzing the energy required to charge the capacitor once the switch closes. We assume in the analysis that the capacitor starts with zero charge.

The total energy taken from the source is, therefore, the sum

$$\begin{aligned} E &= \lim_{\Delta t \to 0} \sum_{n=0}^{\infty} V(n \cdot \Delta t) \cdot I(n \cdot \Delta t) \cdot \Delta t \\ &= \int_0^{\infty} V(t) \cdot I(t)\, dt. \end{aligned} \tag{3.11}$$

Now what is very special in this situation is the constitutive relation between voltage and current for the capacitor. We have

$$I(t) = C\frac{dV_c}{dt}. \tag{3.12}$$

What this special relationship makes happen is that the value of the resistor disappears from our energy calculation.[1] Lowering our heads to do the math:

$$\begin{aligned} E &= \int_0^{\infty} V(t) \cdot C\frac{dV_c}{dt} dt \\ &= VC \int_0^{\infty} \frac{dV_c}{dt} dt \\ &= VC[V_c(\infty) - V_c(0)]. \end{aligned} \tag{3.13}$$

We assume that the capacitor was uncharged before we threw the switch, $V_c = 0$, and that at the end of the charging process the voltage across the capacitor will be V. At the end of it all, then, the energy drawn from the supply is

$$E = CV^2. \tag{3.14}$$

Yikes! We put in CV^2 joules worth of work, and only *half* of that wound up in the electric field between the plates of the capacitor!

What can be done? Well, consider an alternative setup in which we have two voltage sources that we can switch in, one at the full voltage V and the other at $V/2$, as shown in Figure 3.9. We have now broken up our charging strategy: the first step is to connect $V/2$ to the resistor, and wait for it to settle; the second step is to first disconnect $V/2$ and then connect V to complete the charging. The question is now, how much energy was taken total out of our composite power supply?

---

1 If we double the value of the resistor, the current ($C\frac{dV_c}{dt}$) gets cut in half; therefore, the instantaneous power $V(t) \cdot I(t)$ is also halved. But since the total charging time doubles, the integral of the instantaneous power over the interval from 0 to ∞ stays the same. Play this little logic game and you'll convince yourself that our results here are independent of the resistor value. Stop yourself, though, before you get to a resistor value of zero...you're on much safer logical ground if you imagine approaching zero smoothly as a limit rather than just instantiating a zero-value resistor.

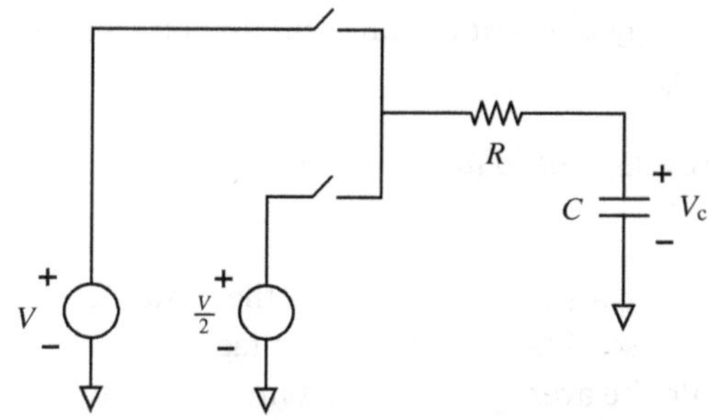

Figure 3.9 Breaking the charging of Figure 3.8 into two steps. First, we connect the *V*/2 voltage source and wait until the capacitor has settled. Then, we open that switch and complete the charging by connecting the *V* voltage source.

If your first reaction is that it should not make any difference, you are (1) totally wrong, but (2) in very good company.[2] It turns out that the math holds surprises. The energy expended in executing step 1 is given by:

$$\begin{aligned} E_1 &= \frac{V}{2} \cdot C\left[\frac{V}{2} - 0\right] \\ &= \frac{1}{4}CV^2. \end{aligned} \tag{3.15}$$

For the second step, we have

$$\begin{aligned} E_2 &= VC\left[V - \frac{V}{2}\right] \\ &= \frac{1}{2}CV^2. \end{aligned} \tag{3.16}$$

The total energy is

$$E_{\text{tot}} = E_1 + E_2 = \frac{3}{4}CV^2. \tag{3.17}$$

So, doing everything in one charging step requires CV^2 joules, while breaking things up into two steps costs us $\frac{3}{4}CV^2$. What?

This may feel suspiciously like some theoretical mischief, but it is not. The basic physical reasoning is that by breaking the charging time into two equal-sized voltage steps, we have doubled our charging time while cutting our charging currents in half. Ohmic losses in the resistor go as $I_R^2 R$, so over the course of our twice-as-long charging cycle, our rate of heat dissipation goes down by a factor of *four*. It works out.

The basic physics here created a momentary craze, particularly in academic circles in the late 1990s and early 2000s, around what we called "adiabatic logic"

2 Or at least, in the company of the author.

or "reversible logic" as a way to do energy-efficient digital computation.[3] For us here, though, the lesson is the following:

> *We pay an energy penalty for insisting that voltages on capacitors move quickly.*

We're going to cement this concept using a sinusoidal drive of the basic *RC* circuit, remembering always that we can assemble any arbitrary voltage drive using a collection of sinusoids. Let's calculate the average power that must be delivered to sustain a sinusoid of amplitude A on the capacitor. That is, suppose the output voltage is given by $V_0 = A\cos\omega_0 t = \frac{A}{2}\left(e^{j\omega_0 t} + e^{-j\omega_0 t}\right)$, as shown in Figure 3.10. From Section 3.1.2, we know that for sinusoidal input $V_\text{I}e^{j\omega t}$,

$$V_0 = \frac{V_\text{I}}{j\omega_0 RC + 1}e^{j\omega_0 t}. \tag{3.18}$$

It follows that if we know the output to be $V_0 e^{j\omega_0 t}$, the input must be

$$V_\text{I} = (j\omega_0 RC + 1)V_0 e^{j\omega_0 t}. \tag{3.19}$$

It follows further that if we know the output to be $V_0 = A\cos\omega_0 t = \frac{A}{2}(e^{j\omega_0 t} + e^{-j\omega_0 t})$, then the input must be given by

$$V_\text{I} = \frac{A}{2}(j\omega_0 RC + 1)e^{j\omega_0 t} + \frac{A}{2}(-j\omega_0 RC + 1)e^{-j\omega_0 t}. \tag{3.20}$$

By Ohm's Law, we have that the current I_I can be written

$$\begin{aligned} I_\text{I} &= \frac{V_\text{I} - V_0}{R} \\ &= \frac{1}{R}\left[\frac{A}{2}(j\omega_0 RC + 1)e^{j\omega_0 t} + \frac{A}{2}(-j\omega_0 RC + 1)e^{-j\omega_0 t} - \frac{A}{2}\left(e^{j\omega_0 t} + e^{-j\omega_0 t}\right)\right] \\ &= \frac{1}{R}\left[\frac{A}{2}j\omega_0 RCe^{j\omega_0 t} - \frac{A}{2}j\omega_0 RCe^{-j\omega_0 t}\right]. \end{aligned} \tag{3.21}$$

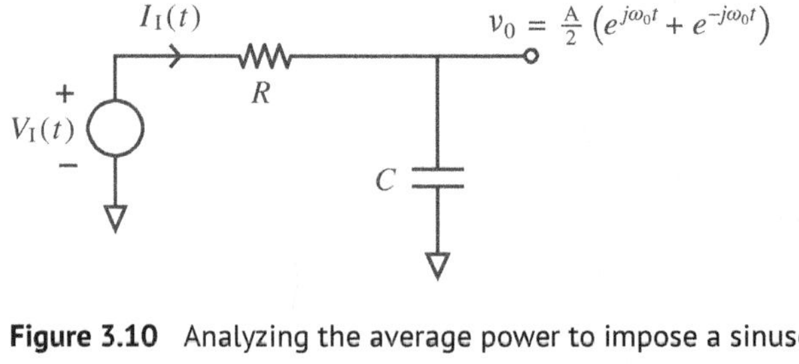

Figure 3.10 Analyzing the average power to impose a sinusoid at the output.

---

3 Ultimately, such systems could not compete on size and speed with more traditional methods.

Now we get the instantaneous power by multiplying Eq. 3.20 by Eq. 3.21, because

$$P(t) = V_I(t) \cdot I_I(t). \tag{3.22}$$

But we're going to look ahead a little bit to simplify the algebra. At the end of the day, what is really interesting are power terms that *do not* average to zero. If you look at the product of Eqs. 3.20 and 3.21, you can see that it should result in terms that are constant and terms in $e^{\pm j\omega_0 t}$. The latter terms, the ones in $e^{\pm j\omega_0 t}$, average to zero and do not interest us: they represent energy that flows toward the load in one-half cycle and returns to the source in the next half cycle. If we are careful, what we find is that the average power, P_{avg} is given by

$$\begin{aligned} P_{\text{avg}} &= -\frac{A^2}{4R} j\omega_0 RC(j\omega_0 RC + 1) + \frac{A^2}{4R} j\omega_0 RC(-j\omega_0 RC + 1) \\ &= \frac{A^2}{2R}\omega_0^2 R^2 C^2 \\ &= \frac{1}{2}A^2\omega_0^2 RC^2. \end{aligned} \tag{3.23}$$

Now for a minor change, which may seem like a flourish but is actually quite important when it comes to writing equations that communicate understanding. Let's recall that $\frac{1}{RC} = \omega_{-3\text{ dB}}$. That is, $\frac{1}{RC}$ is the angular frequency at which the RC circuit causes a sinusoidal drive to lose $\frac{1}{\sqrt{2}}$ of its amplitude. This is also known as the corner frequency or cutoff frequency of this simple filter. Rewriting our average power, we have

$$P_{\text{avg}} = \frac{1}{2}\frac{A^2}{R} \cdot \left(\frac{\omega_0}{\omega_{-3\text{dB}}}\right)^2. \tag{3.24}$$

Time to sit back and reflect a little. If we graph average power versus ω_0, we get Figure 3.11. The humble RC circuit is more interesting than it looks, no? In terms of energy and power, we pay *quadratically* for linear increases in amplitude and frequency to make the capacitor voltage do what we want. This is...extreme. If we want to double the frequency, we have to quadruple the power, which is very harsh.

But wait: it gets worse.

Suppose that we impose a signal on the capacitor with the energy spectral density shown in Figure 3.12. The first step in understanding this situation is to work out for yourself that if we imposed *two* sinusoids on the capacitor, $A_1 \cos \omega_1 t + A_2 \cos \omega_2 t$, then the average power demanded of the source would be

$$P_{\text{avg}} = \frac{1}{2}\frac{A_1^2}{R} \cdot \left(\frac{\omega_1}{\omega_{-3\text{dB}}}\right)^2 + \frac{1}{2}\frac{A_2^2}{R} \cdot \left(\frac{\omega_2}{\omega_{-3\text{dB}}}\right)^2. \tag{3.25}$$

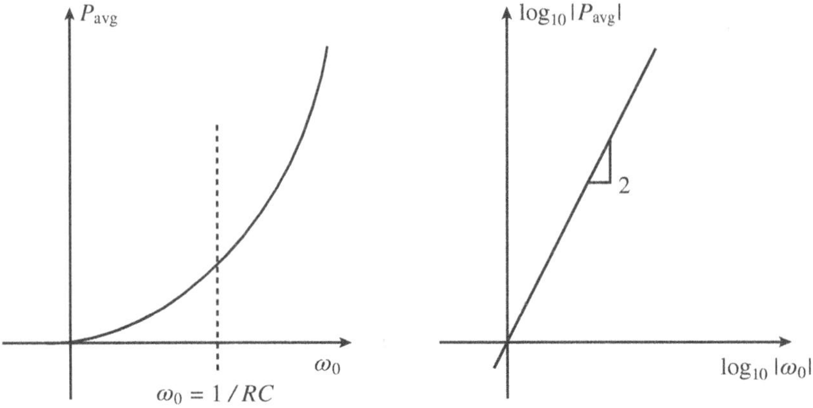

Figure 3.11 The average power required of the source to impose a sinusoid at the output as a function of ω_0. Note that the power is *also* quadratic in amplitude A.

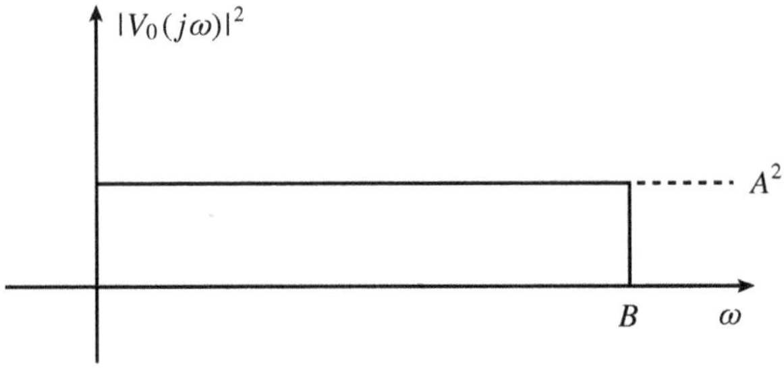

Figure 3.12 Many times, the spectrum that we want to impose on an RC is flat from somewhere near DC to an upper limit B.

If you have really internalized the meaning of integrals, you will recognize that the total average power for an energy spectral density as shown in Figure 3.12 is none other than

$$\begin{aligned} P_{\text{avg}} &= \int_0^B \frac{1}{2}\frac{A^2}{R} \cdot \left(\frac{\omega}{\omega_{-3\text{dB}}}\right)^2 d\omega \\ &= \frac{1}{6}\left(\frac{1}{\omega_{-3\text{dB}}}\right)^2 \frac{A^2}{R}\left[\omega^3\right]_0^B \\ &= \frac{1}{6}\left(\frac{1}{\omega_{-3\text{dB}}}\right)^2 \frac{A^2}{R}B^3. \end{aligned} \tag{3.26}$$

So the power required to enforce a signal of bandwidth B on a capacitor through a resistor actually obeys a cubic power law!

This is a critically important aspect of real-world practical circuit designs. Sometimes, in response to a communications channel that exhibits a low-pass characteristics, signal-processing-minded engineers will use "equalization," or "pre-emphasis," as a way of fighting the attenuation of high-frequency sinusoids that we saw in Eq. 3.6. This section should be eye-opening caution that, especially after we cross over the natural bandwidth of the filter function, we pay dearly in watts to get what we want.

3.2 The Resistive Load: When You Have to *Deliver* Energy

At first, the scenario of driving a resistive load seems almost silly to devote a section to (see Figure 3.13). But what is new since the first time you contemplated this simplest of circuits is that we are now focused on energy and power. So let's consider a charge of q that starts its journey at the top of the resistor of Figure 3.13 and travels all the way to the grounded terminal at the bottom. At the top of the resistor, it had potential energy qV, while at the bottom of the resistor, it had no potential energy at all. From the standpoint of the circuit, qV joules of energy is just *gone.*

Gone, cannot be recovered.

Moreover, the rate at which energy is leaving our circuit, $\frac{dU}{dt}$, is given by the number of coulombs of charge per second completing this calamitous fall, times the height of the fall. That is,

$$\frac{dU}{dt} = P_{\text{diss}} = IV. \tag{3.27}$$

Thanks to Ohm's law, we can write this in two other ways:

$$P_{\text{diss}} = I^2R = \frac{V^2}{R}. \tag{3.28}$$

And where does the energy go? In the case of a resistor, this energy is released as heat at a rate of $IV/I^2R/\frac{V^2}{R}$ joules per second. What is critical to understand is

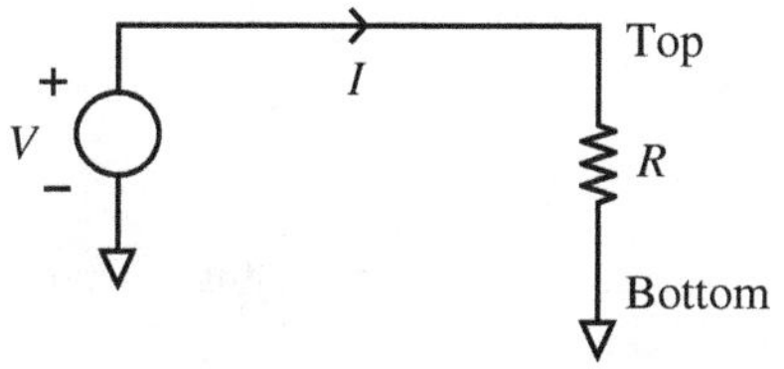

Figure 3.13 Driving a resistor: more interesting than it looks.

that the humble resistor is a useful stand-in for any transducer whose activation results in energy being permanently lost to the circuit. Common examples include audio speakers, light bulbs, electric motors, and, of course, radiating antennas for wireless communications (see Figure 3.14). For any transducer, we can use the principle of the conservation of energy to say that any radiated power from the transducer, P_{rad}, is exactly equal to the product of voltage and current at the transducer terminals:

$$P_{rad} = I(t) \cdot V(t). \tag{3.29}$$

Moreover, if we know the radiated power for a given, say, current, then the value of the resistor that we place as a surrogate for the transducer in our schematic diagram is given by

$$R = \frac{P_{rad}}{I^2}. \tag{3.30}$$

This is why you will see antennas, audio speakers, and the like, characterized by a real impedance like 8 or 50 Ω. For a 50 Ω antenna, if you put in a mean squared current of 1 A^2, you will radiate 50 W of electromagnetic power.

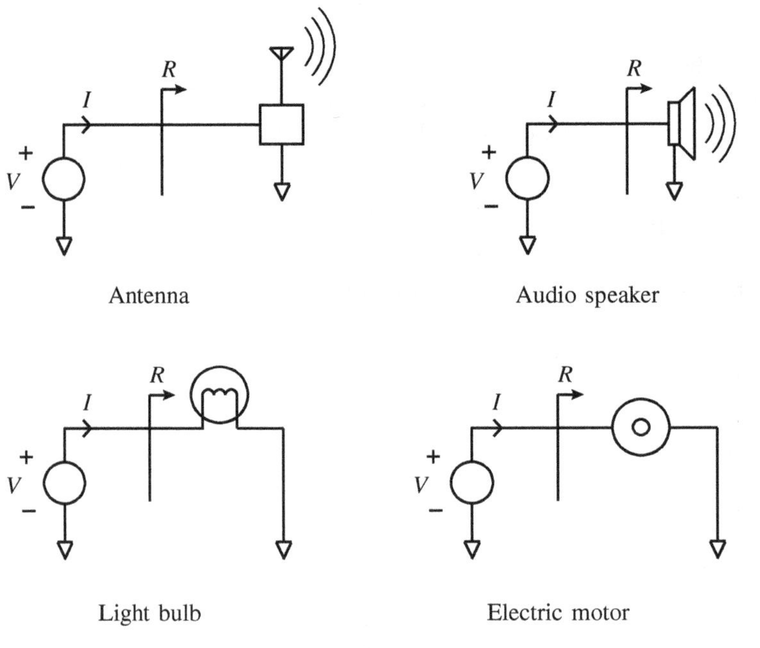

Figure 3.14 Four example transducers. To the source, it appears as though it is driving a resistive load.

Conservation of energy is a great mystery of the natural world. A great commentary on this can be found once again in Richard Feynman's famous Lectures.[4] We quote a portion of it here:

> There is a fact, or if you wish, a *law*, governing all natural phenomena that are known to date. There is no known exception to this law—it is exact so far as we know. The law is called the *conservation of energy*. It states that there is a certain quantity, which we call energy, that does not change in the manifold changes which nature undergoes. That is a most abstract idea, because it is a mathematical principle; it says that there is a numerical quantity which does not change when something happens. It is not a description of a mechanism, or anything concrete; it is just a strange fact that we can calculate some number and when we finish watching nature go through her tricks and calculate the number again, it is the same. (Something like the bishop on a red square, and after a number of moves—details unknown—it is still on some red square. It is a law of this nature.) Since it is an abstract idea, we shall illustrate the meaning of it by an analogy.

The analogy that follows from this point in the quotation is fascinating, and well worth a read.

Finally, it should be said that the equivalent circuit model of any transducer is usually more complicated than a simple resistor. A comprehensive model will usually be a complex little network of inductors, capacitors, and even voltage sources in the case of electric motors. But wherever a transducer takes electrical energy and whisks it away in some other form, that radiation mechanism will show up as a resistor at the heart of the equivalent circuit model.

3.2.1 Maximum Power Transfer

Fine. We now know that delivering power into a "resistor" is serious business. A classic textbook problem from here is to start with a voltage source and a source resistance, as shown in Figure 3.15. In the usual setup, the source resistance R_S is fixed for some reason. We attach a resistive load to the circuit, as in Figure 3.16, and then ask what resistance value R_L we should choose to dissipate maximum power in the load. Remember that what makes this problem interesting is that the load resistance might be a transducer of some sort. We are effectively asking how we can shine the most light, or deliver the loudest sound, or drive the highest axle speed, etc.

4 Feynman, R. Leighton, and M. Sands. *The Feynman Lectures on Physics*. Reading, MA: Addison-Wesley Publishing, 1963. This quote is taken from Volume 1, pages 4-1.

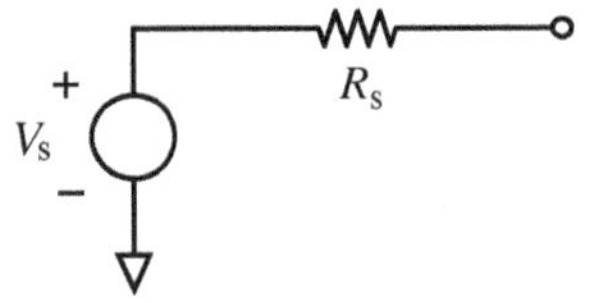

Figure 3.15 A voltage source, with source resistance R_s.

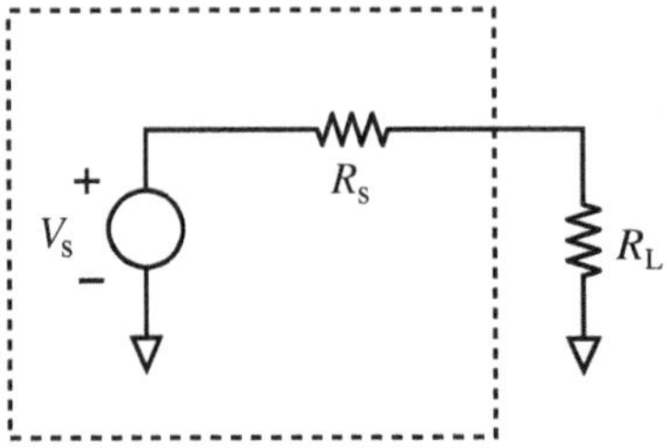

Figure 3.16 Connecting a load resistor. How much power can we transfer to the load?.

The answer to this problem happens to be that we get maximum power transfer if $R_s = R_L$. Accordingly, the maximum power transfer is

$$P_{\max}(t) = \frac{\left[\frac{1}{2}V_s(t)\right]^2}{R_s} = \frac{1}{4R_s}V_s^2(t). \tag{3.31}$$

We've presented this result without proof. If you want to show it yourself, you'll need the basic circuit analysis result that the load voltage is given by:

$$V_L = \frac{R_L}{R_L + R_s}V_s. \tag{3.32}$$

The power delivered to the load eventually simplifies to

$$P_{\text{load}} = \frac{V_L^2}{R_L} = \frac{R_L}{(R_L + R_s)^2}V_s^2. \tag{3.33}$$

From here, the path is a classic extremum problem from calculus, with the derivative of interest being $\frac{dP_{\text{load}}}{dR_L}$. Before you do that, consider two extreme values of R_L in Eq. 3.33. If you make R_L infinite, you can see that the power delivered goes to zero (even though we maximize the voltage). Similarly, if we let R_L go to zero, the power delivered goes to zero (even though we maximize the current). You'd have to be pretty sharp to see $R_L = R_s$ as the "midpoint" between zero and infinity, but there it is.

3.2.2 That Said...The Best Way to Transfer Power Is Not to Have a Source Impedance at All

If you look back at Eq. 3.31, you can see that for a given voltage drive, there is no limit to the power we can supply to the load if R_s is allowed to be vanishingly small. Take the limit of $R_s \rightarrow 0$, and you see it right away.

It is important to point this out, and important to keep in mind, especially in the context of circuits designed for wireless systems. At the time of this writing, it has been the convention for decades to design radio circuit blocks so that their input impedance and output impedance are 50 Ω. There are very good and practical reasons for this, most of which boil down to maintaining modularity in a system where coaxial cables with a characteristic impedance of 50 Ω are standard. Getting deep into the details here is mostly beyond the scope of this book. But...but...when it comes to delivering power to the antenna in a wireless transmitter, the principle of maximum power is frequently misapplied. As in: "Our antenna has an input impedance of 50 Ω; therefore, we must make the output impedance of our power amplifier 50 Ω to maximize power transfer." If we did this, we would deliver as much power to our output-matching network as we did to the antenna, maximizing our efficiency at 50%!

In state-of-the-art wireless transmitters with modern power amplifier structures, we do not do this. Interestingly, though, many engineers absorb the impression that we *do* do this, and it leads to confusion. When you seek to deliver power with as little wasted heat as possible, the maximum power transfer theorem does not really help you.

3.2.3 The Power to Drive a Resistor Is Linear in *B*

Recall that in Section 3.1.3, we looked at the energy cost to impose a flat spectral density on the output node of an *RC* circuit. Our basic conclusion was that it was a disaster, cubic in *B*.

The case of driving a resistor is much more benign. The math:

$$\begin{aligned} P_{\text{avg}} &= \int_0^\infty \frac{1}{2}\frac{|V(j\omega)|^2}{R} d\omega \\ &= \frac{1}{2}\frac{A^2}{R} \cdot B. \end{aligned} \tag{3.34}$$

Linear in *B* is much better than cubic!

3.3 The Parallel *RLC* Circuit

In Section 3.1, we looked at the simple *RC* circuit, and determined that it takes quite a lot of energy to drive as our frequency components approach and/or surpass the $1/RC$ cutoff. Now, it is true that we will sometimes explicitly design *RC* circuits, such as for passive filters. But it is more often true that *RC* filters show up whether we want them to or not. Take the example of the two-stage amplifier shown in Figure 3.17. A perfect voltage amplifier has zero output impedance, but all practical amplifiers have nonzero output impedance R_S. As important, it is true that in any physical realization of a circuit, there is "stray" or "parasitic" capacitance from every node to ground (and to other places as well). Whenever you see a series of system blocks whose key input and output variables are voltages, you would do well to overlay on top of the ideal picture a representation shown in Figure 3.18.

Now it is true that in many communications systems, the signals that we deal with are "narrowband." That is to say, in the frequency domain, they are characterized as a narrow group of sinusoids centered about a "carrier frequency" ω_0 as shown in Figure 3.19. We know from Section 3.1 that the energy penalty for transmitting a $\Delta\omega$-wide swath of frequencies as in Figure 3.19 would be much worse than if we chose a lower center frequency as in Figure 3.20.

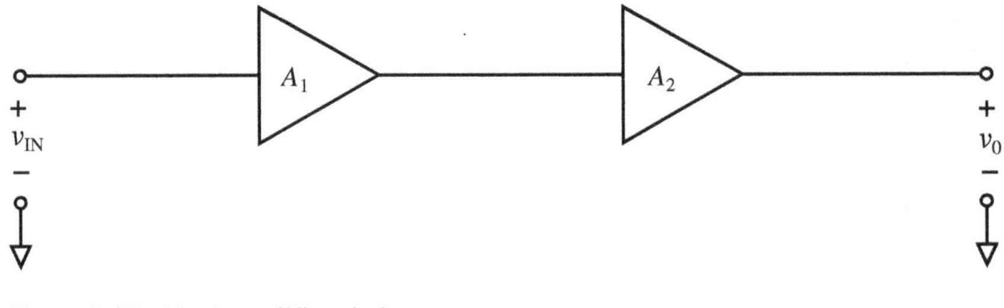

Figure 3.17 Ideal amplifier chain.

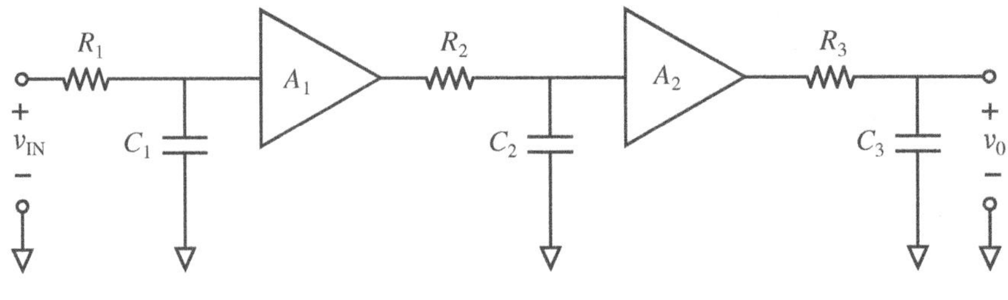

Figure 3.18 Amplifier chain showing finite output impedances and parasitic capacitances.

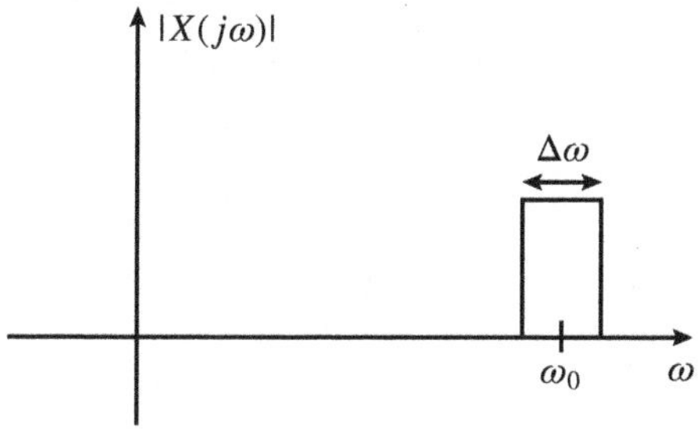

Figure 3.19 An example of a narrowband signal.

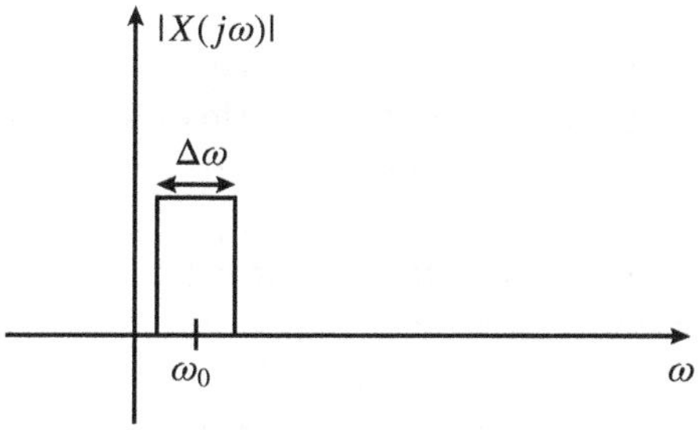

Figure 3.20 An example of a narrowband signal at a lower carrier frequency.

This seems...unfortunate, right? After all, when we developed the uncertainty principle in Section 2.5, we came away with the notion that speed in the time domain was tied to bandwidth and bandwidth alone. It would seem wrong if the center frequency was important.

This line of reasoning leads analog circuit designers to the concept of narrowband resonance. That is, referring back to Figure 3.18, whenever there is an undesired (usually parasitic) capacitance, we *add* an inductance in parallel. The idea is that at the desired center frequency ω_0, $\frac{1}{\omega_0 L} = \omega_0 C$. The admittances cancel, and at that *one* magical frequency we make the troublesome capacitance disappear! We need one more refinement, which is that our inductances and capacitances are never perfectly lossless. We represent these losses with a resistor in parallel with the capacitor and inductor. The result is that, in narrowband design, the capacitors of Figure 3.18 are replaced by parallel *RLC* networks as shown in Figure 3.21. Clever, and used everywhere in circuits for wireless systems. Let's review the time domain and frequency domain characteristics of this super useful design idiom, and then dig deeply into the energy considerations.

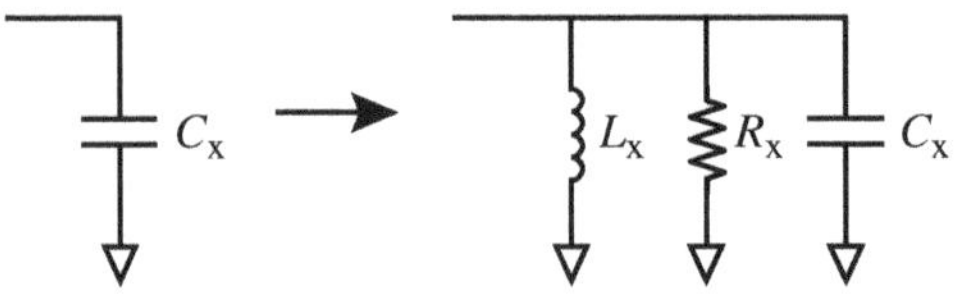

Figure 3.21 Adding an inductor to "resonate out" an inconvenient capacitance. The resistor R_X is not usually explicitly placed as a component, but rather captures loss mechanisms in the imperfect inductor and/or capacitor.

3.3.1 Time Domain Behavior

The simplest setup for driving a parallel resistor–inductor–capacitor (*RLC*) circuit is to use an ideal current source, as shown in Figure 3.22. For any reader concerned about the sudden switch to current source drive, which is all about analytical convenience, it will help to remember that we can always choose between Thevenin equivalents, as shown in Figure 3.23.[5] Note that the shunt resistance in the current source version can be absorbed as a parallel combination with the R of the parallel *RLC*, making the ideal current source setup completely accurate and realistic. Either Thevenin equivalent will give us the same answers, of course, but it turns out that that math is easier with a current source drive.

Fine. We consider the response $v_0(t)$ if the input $I(t)$ is an *impulse* of the form $Q_0\delta(t)$. This is, a pulse that rises from zero and falls back to zero extremely quickly when compared to the dynamics of the *RLC* circuit itself. The picture to have in your mind is that of a mallet hitting a tuning fork, as in Figure 3.24. Under these conditions, only the area under the pulse is important (as opposed to the details of the pulse shape). A defining feature of an impulse is that it has unit area. The effect of driving this circuit with a pulse $I = Q_0\delta(t)$ is that it deposits charge of Q_0 on the capacitor at time $t = 0+$. Depending on any pre-existing charge on the capacitor and/or current in the inductor, the node voltage $v_0(t)$ will evolve according to

$$v_0(t) = e^{-t/2\mathrm{RC}}\,(A_1 \cos \omega_\mathrm{d} t + A_2 \sin \omega_\mathrm{d} t), \tag{3.35}$$

in the so-called "underdamped" case. In this expression, ω_d is given by

$$\omega_\mathrm{d} = \sqrt{\omega_0^2 - \left(\frac{1}{2RC}\right)^2}, \tag{3.36}$$

with A_1 and A_2 set by Q_0 and initial conditions. The frequency ω_0 is special: it is the resonance frequency where the inductor admittance and capacitor admittance cancel. That is:

5 From the output terminals alone, the two circuits are indistinguishable. However, if unloaded, the voltage source version will dissipate no heat, whereas the current source version will dissipate $I_s R_s$ watts. This fun fact shows up in Ph.D. qualifying exams from time to time.

$$
\begin{aligned}
\frac{1}{j\omega_0 L} + j\omega_0 C = 0 \\
\omega_0^2 = \frac{1}{LC} \\
\omega_0^2 = \frac{1}{LC}.
\end{aligned}
\tag{3.37}
$$

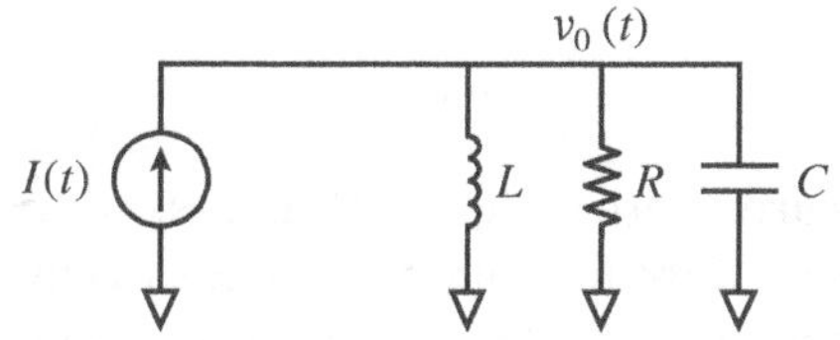

Figure 3.22 Driving a parallel *RLC* circuit with a current source.

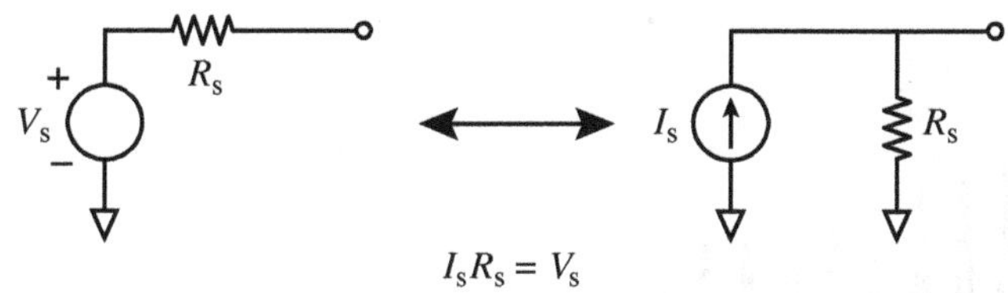

Figure 3.23 Thevenin equivalents: from the output terminals, they are indistinguishable.

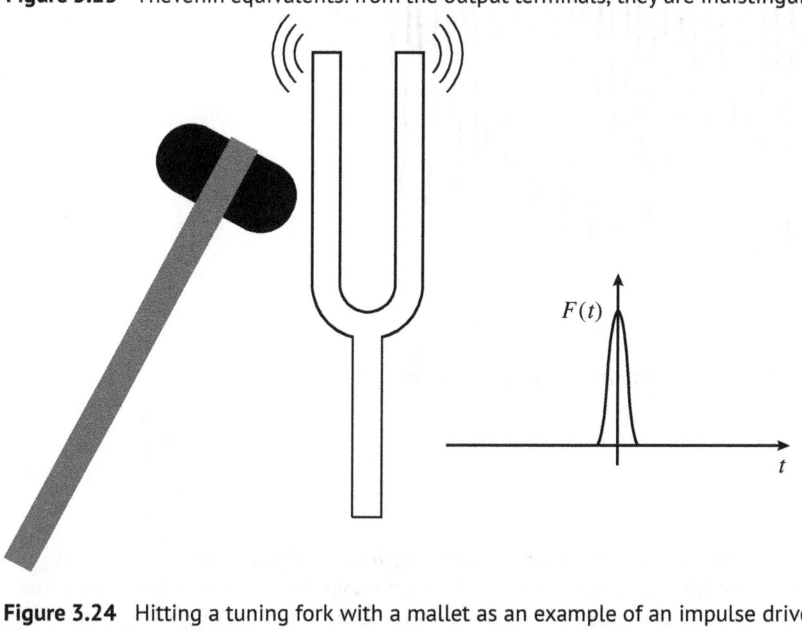

Figure 3.24 Hitting a tuning fork with a mallet as an example of an impulse drive. The idea is that the width of the impulse in time should be small compared to the vibrational period of the tuning fork.

At this one special frequency, from the perspective of the current source, it is as though it is driving only a resistor of value R. This is the magic of resonance. At a single frequency exactly, and over a narrowband approximately, the reactances disappear.[6]

We now move on to driving this parallel *RLC* with sinusoidal currents. But before we do, a visual on the impulse response to go with the equations (see Figure 3.25).

3.3.2 Frequency Domain Behavior

Consider the *RLC* circuit under sinusoidal current drive, $I(t) = e^{j\omega t}$, and ask what the resulting voltage $v_0(t)$ is. It is easily given if we multiply the current $I(t)$ by the impedance $Z(j\omega)$ of the *RLC* combination. Analytically, we help ourselves by

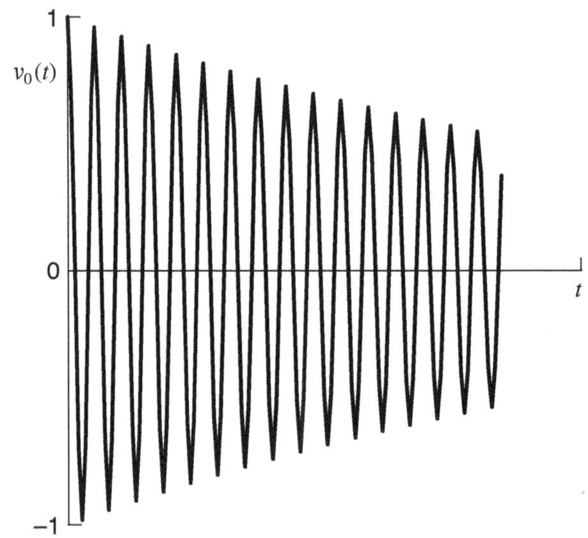

Figure 3.25 Impulse response of a lightly damped resonance.

6 There is more to the story. The branch currents in parallel resonances can get shockingly large. For an excellent discussion on the secret lives of resonance inductors and capacitors, see the third chapter of Lee, Thomas. *The Design of CMOS Radio-Frequency Integrated Circuits*. Cambridge, UK: Cambridge University Press, 1998.

starting with the admittance of the RLC, $Y(j\omega)$, and then inverting using $Z(j\omega) = 1/Y(j\omega)$ after we have performed all our manipulations. Let's start with

$$\begin{aligned} Y(j\omega) &= \frac{1}{R} + j\omega C - \frac{j}{\omega L} \\ &= \frac{1}{R} + \frac{j\omega^2 LC - j}{\omega L} \\ &= \frac{1}{R} + j\frac{\omega^2 LC - 1}{\omega L}. \end{aligned} \tag{3.38}$$

Recall that the resonance frequency, ω_0, is given by $\omega_0 = \frac{1}{\sqrt{LC}}$. This allows us to write

$$Y(j\omega) = \frac{1}{R} + j\frac{\left(\omega^2/\omega_0^2 - 1\right)}{\omega L}. \tag{3.39}$$

Now we are going to do something very clever and practical. In real-world applications, the whole reason we build resonant structures like RLC circuits is because we plan on driving them in a narrow range of frequencies centered around ω_0. We can reflect this in our mathematics by insisting that $\omega = \omega_0 + \delta\omega$, where $|\delta\omega| \ll |\omega_0|$. Follow along:

$$\begin{aligned} Y(j(\omega_0 + \delta\omega)) &= \frac{1}{R} + j\frac{\left((\omega_0 + \delta\omega)^2/\omega_0^2 - 1\right)}{(\omega + \delta\omega)L} \\ &= \frac{1}{R} + j\frac{\left(\frac{\omega_0^2 + 2\delta\omega\omega_0 + (\delta\omega)^2}{\omega_0^2} - 1\right)}{(\omega + \delta\omega)L} \\ &= \frac{1}{R} + \frac{j}{(\omega_0 + \delta\omega)L}\left(\frac{2\delta\omega}{\omega_0} + \left(\frac{\delta\omega}{\omega_0}\right)^2\right) \\ &= \frac{1}{R} + \frac{j}{\omega_0\left(1 + \frac{\delta\omega}{\omega_0}\right)L}\left(\frac{2\delta\omega}{\omega_0} + \left(\frac{\delta\omega}{\omega_0}\right)^2\right) \\ &= \frac{1}{R} + \frac{j}{\omega_0\left(1 + \frac{\delta\omega}{\omega_0}\right)L} \cdot \frac{2\delta\omega}{\omega_0}\left(1 + \frac{1}{2}\left(\frac{\delta\omega}{\omega_0}\right)\right). \end{aligned} \tag{3.40}$$

We have $|\delta\omega| \ll |\omega_0|$, which means $\left|\frac{\delta\omega}{\omega_0}\right| \ll 1$. So we're going to approximate where we can, noting in particular that $1 + \frac{\delta\omega}{\omega_0} \approx 1$, and that $1 + \frac{1}{2}\frac{\delta\omega}{\omega_0}$ is even more well approximated by 1. That simplifies things to

$$Y(j(\omega_0 + \delta\omega)) \approx \frac{1}{R} + \frac{j}{\omega_0 L} \cdot \frac{2\delta\omega}{\omega_0}. \tag{3.41}$$

We have $\omega_0^2 = \frac{1}{LC}$, giving us

$$Y(j(\omega_0 + \delta\omega)) \approx \frac{1}{R} + j2\delta\omega C. \tag{3.42}$$

This means that the impedance, $Z(j(\omega_0 + \delta\omega))$, is given by

$$Z(j(\omega_0 + \delta\omega)) \approx \frac{1}{\frac{1}{R} + j2\delta\omega C} = \frac{R}{1 + j2RC \cdot \delta\omega}. \tag{3.43}$$

All of this is to say that if we want to know $v_0(t)$ for a sinusoidal current drive of $I = e^{j(\omega_0+\delta\omega)t}$, we have

$$v_0(t) = Z(j(\omega_0 + \delta\omega))e^{j\omega t} = \frac{R}{1 + j2RC \cdot \delta\omega}e^{j\omega t}. \tag{3.44}$$

It is useful to plot $\log|Z(j\omega)|$, at least over the limited range where $\left|\frac{\delta\omega}{\omega_0}\right| \ll 1$ (see Figure 3.26). Notice that the total width of the resonance is $\Delta\omega = 1/RC$.

3.3.3 Energy Considerations and *Q*

The parallel *RLC* circuit is an example of resonance: it likes to be driven at a particular frequency ω_0. Driving it at that frequency is almost effortless, in the sense that it requires the least amount of joules per second from a power source to maintain a sinusoidal voltage at that special frequency. It is worth noting that we all have plenty of practical experience with resonance in everyday life. Anyone who has pushed a child on a swing, for example, quickly and naturally defaults to the resonance frequency of that lightly damped pendulum. Not a bad mental picture to carry around with you when you think about resonances of any sort.

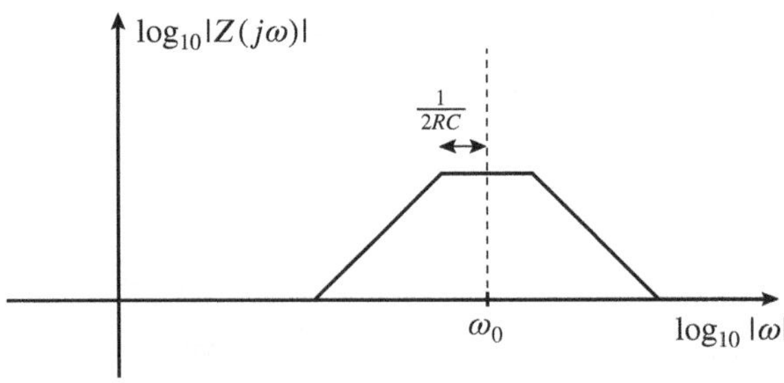

Figure 3.26 Frequency response near resonance for an *RLC* circuit.

When it comes to energy considerations for an *RLC* circuit, it is extremely useful to understand the concept of Q for a physical system. The common definition of Q is stated simply, but a little mystifyingly, as follows:

$$Q = \omega \cdot \frac{\text{Peak Energy Stored}}{\text{Average Power Dissipated}}. \tag{3.45}$$

Many readers will have seen this definition before. But it is difficult to remember because it is a bit of an odd duck. It feels ratio-like, but you have an energy divided by a power so not *really*...but then you have that ω out front (radians per second) that fixes things so that Q itself is unitless (radians evidently don't count as units)...but, yeah.

Another path forward is to recall that

$$\omega = 2\pi f = \frac{2\pi}{T}. \tag{3.46}$$

Therefore,

$$Q = 2\pi \cdot \frac{\text{Peak Energy Stored}}{\text{Average Power Dissipated} \cdot T}. \tag{3.47}$$

One more step:

$$Q = 2\pi \cdot \frac{\text{Peak Energy Stored}}{\text{Average Energy Dissipated Per Cycle}}. \tag{3.48}$$

Now that is something you can really hold on to. We say in words that Q quantifies how close a resonator gets to the ideal of a perpetual motion machine. If it lost none of the energy sloshing back and forth between its energy storage elements,[7] the sloshing would go on forever and we would say that the Q of that resonator is infinite. The further we depart from that ideal, the larger the denominator of Eq. 3.48 grows, and the smaller Q gets.

It turns out that the Q of an *RLC* resonator is given (in three useful forms) by

$$Q = \frac{R}{\omega_0 L} = \frac{R}{\sqrt{L/C}} = \omega_0 RC. \tag{3.49}$$

This actually leads us to an extremely interesting and important fact about resonators. It will look superficially like it only applies to parallel *RLC*s (because we are going to invoke the individual component values), but the result applies to resonators of all sorts. Recall that we discovered that in response to a sinusoidal drive

7 Or between kinetic and potential energy in the case of the child's swing.

(in Section 3.3.2), the circuit had a bandwidth $\Delta\omega = 1/RC$. Looking back at Eq. 3.49, we see that this means

$$Q = \frac{\omega_0}{\Delta\omega}. \tag{3.50}$$

It is worth committing this extremely important result to memory, and internalizing its meaning. High-Q resonators, which dissipate relatively little energy as heat and come close to being perpetual motion machines, *are invariably narrow band!* That is, the bandwidth of the resonance is small compared to the resonant frequency ω_0.

Now as we did with the RC circuit, we will posit a sinusoidal voltage across the resonance $v_0 = \frac{A}{2}e^{j(\omega_0+\delta\omega)t} + \frac{A^*}{2}e^{-j(\omega_0+\delta\omega)t}$ and ask ourselves what power must be delivered by the source in order for this to be the case. Rather than compute this from the instantaneous voltage and current, though, we will use our newly derived definition of Q. If we posit that the voltage waveform across the RLC remains at a constant amplitude, conservation of energy demands that the source *replace exactly* the power that is dissipated.

In this case, the amplitude of the voltage sinusoid is A. The peak energy stored is

$$E_{\text{pk}} = \frac{1}{2}CA^2. \tag{3.51}$$

Using our definition of Q, we know that the energy dissipated over one cycle is

$$\begin{aligned} E_{\text{diss}} &= 2\pi \cdot \frac{E_{\text{pk}}}{Q} \\ &= 2\pi \cdot \frac{\frac{1}{2}CA^2}{Q}. \end{aligned} \tag{3.52}$$

The power dissipation is the energy dissipated per cycle multiplied by the number of cycles per second. That is

$$\begin{aligned} P_{\text{diss}} &= 2\pi \cdot \frac{\frac{1}{2}CA^2}{Q} \cdot f \\ &= \frac{\frac{1}{2}CA^2}{Q} \cdot \omega. \end{aligned} \tag{3.53}$$

That's the answer for one sinusoid of amplitude A.

When we use the parallel RLC in the context of circuits for wireless communications, it is usually true that (1) you design your resonances to be as wide as the bandwidth of your channel B; and (2) when you are sending the maximum number of bits per second, the collection of sinusoids across the resonator look like

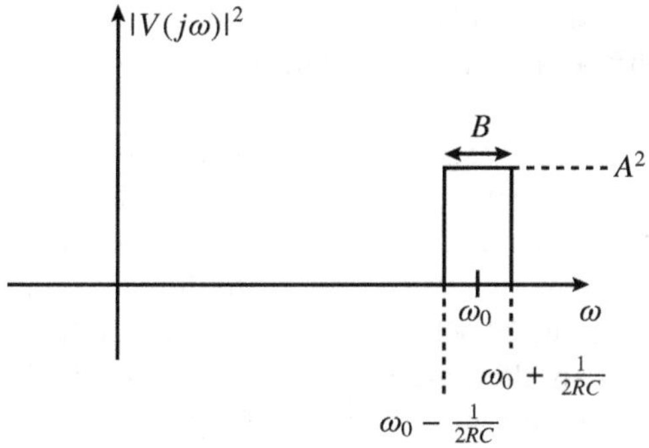

Figure 3.27 The voltage spectral density we seek to impose on the parallel *RLC* circuit.

a flat distribution across the whole band.[8] So a key problem is for us to determine the power cost of maintaining the voltage spectral density shown in Figure 3.27. Proper use of the result in Eq. 3.53 is as follows:

$$
\begin{aligned}
P_{\text{diss-B}} &= \int_{\omega_0-\frac{1}{2RC}}^{\omega_0+\frac{1}{2RC}} \frac{\frac{1}{2}CA^2}{Q}\omega\, d\omega \\
&= \frac{1}{4}\frac{CA^2}{Q}\left[\omega^2\right]_{\omega_0-\frac{1}{2RC}}^{\omega_0+\frac{1}{2RC}} \\
&= \frac{1}{4}\frac{CA^2}{Q}\left[\omega_0^2+\frac{\omega_0}{RC}+\left(\frac{1}{2RC}\right)^2-\omega_0^2+\frac{\omega_0}{RC}-\left(\frac{1}{2RC}\right)^2\right] \\
&= \frac{1}{2}\frac{CA^2}{Q}\cdot\frac{\omega_0}{RC}.
\end{aligned}
\tag{3.54}
$$

Recalling that $B = \frac{1}{RC}$, and that $Q = \frac{\omega_0}{\Delta\omega} = \frac{\omega_0}{B}$, we are left with

$$
P_{\text{diss-B}} = \frac{1}{2}CA^2B^2. \tag{3.55}
$$

Always good to pause at the end of a derivation and make sure that the result makes sense. The units for A^2 must be Volts^2 per radians per second. This means $\frac{1}{2}CA^2B$ has units of energy, and therefore $\frac{1}{2}CA^2B^2$ has units of power. So far, so good.

But otherwise, what does this simple result tell us? What is the physical significance of the B^2 dependence? Well, it tells you that you pay a penalty for bandwidth twice. When you are near the resonance of this structure, you pay in power in

8 This first consideration, that we have optimized the hardware *ahead of time* for the maximum data throughput, is absolutely key but easy to lose track of. If you find yourself surprised by a B^2 dependence in the final result, this is the piece you have forgotten.

proportion to B just as you do for a resistive load as we saw in Section 3.2.3. But here, you get an additional penalty because you must lower the Q enough to accommodate the bandwidth B.

Now in practical circuit design, especially as we get to higher carrier frequencies ω_0, it is not always easy to exercise such exquisite control over the bandwidth and Q of our resonances. You can roughly say that achievable Qs get lower as you go from 1 GHz to 10 GHz to 60 GHz, for example; also, in that realm it may be too much to ask to hardware customize for, say, 20 MHz bandwidth versus 30 MHz bandwidth. We should, therefore, keep an alternate form of Eq. 3.55 handy, in which we leave the Q parameter where it is and show the carrier frequency dependence explicitly:

$$P_{\text{diss-B}} = \frac{1}{2}CA^2B^2 = \frac{1}{2}CA^2\frac{\omega_0 B}{Q}. \tag{3.56}$$

What we see is something that RF circuit designers experience quite strongly in their work. When designing for higher and higher carrier frequencies, design gets "more difficult". What they mean by this is that they require a bigger power consumption budget to drive internal nodes, even when the capacitances at those nodes have been "resonated out" using parallel inductors. We can see in this sense that working at 2 GHz is twice as "difficult" as working at 1 GHz, and also that higher bandwidths require more power than lower bandwidths.

On a final note, some readers may wonder about the propriety of insisting that we impose a flat spectral density in voltage on the output node of the resonance structure. If it sounds complicated, it's because it is, a little bit. Modern communications systems spend significant signal processing resources to "equalize the channel." What this means is that the electronics machinery acts in a closed-loop manner to make sure that what is delivered to the load really is spectrally flat. This is a deep and interesting subject, and beyond the scope of this book. But from the standpoint of understanding global energy trade-offs in the system, Eq. 3.56 is the conceptual gem.

3.4 Energy Considerations Summary

These three circuit structures are surprisingly decent surrogates for much of analog electronics when it comes to thinking about energy. When we choose to represent our information as a voltage, as we normally do, more often than not we will construct signal chains whose intermediate nodes are represented by an RC circuit (extremely common), a pure resistor, or a parallel RLC. What we have determined here is the energy required by Nature to drive these loads over a specified bandwidth (the Task). We found that the RC circuit is the stuff of nightmares, a cubic dependence on bandwidth; the pure resistor is maybe how you thought it always

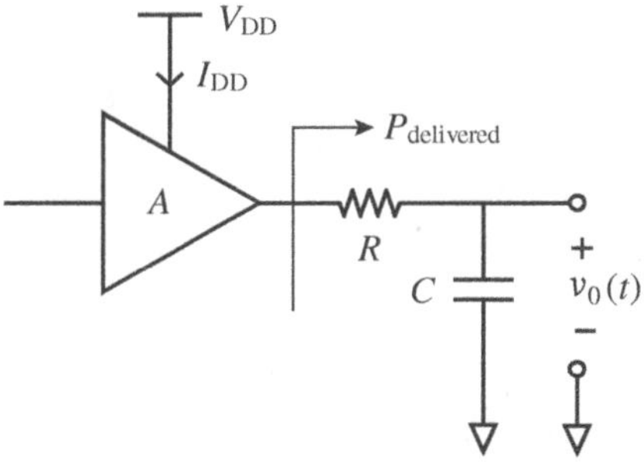

Figure 3.28 The power that the amplifier delivers to the load is often *much* smaller than the power drawn from the power supply.

worked; and the parallel *RLC*, so common in signal chains for wireless systems, has the more sensible but still punishing B^2 dependence.

A *major* omission from the discussion up to this point is any consideration of energy inefficiency when it comes to carrying out the Task. That is, if we have an amplifier driving an *RC* load as shown in Figure 3.28, we have derived how much power the amplifier must deliver, but have politely looked away from the power the amplifier draws from the supply $V_{DD} \cdot I_{DD}$. And we look away because that power is typically quite a lot higher than the energy required by the Task. At the time of this writing, drawing twice as much from the supply as you deliver to the load is a state-of-the-art arrived at only by painful progress over years and only achieved when it is the focus. In other circumstances, the power from the supply is 10×, 100×, or more.

Still, in a book that is not focused on the specialty of detailed circuit design, it is fitting to zoom in on the energy that Nature requires and defer on aspects of implementation (we will explore in a little more depth in Chapter 11). We will see in Chapter 5 that, in broad strokes, if we allow ourselves greater bandwidth B, we can achieve greater data throughput in bits per second. The takeaway from this chapter is that in our electronic machinery, power consumption requirements grow quickly with bandwidth.

4

Noise in Communications Systems

Stepping out of the analytical weeds for a moment, it is fair to ask: *Why should sending information require any energy at all?*

It is a deep and interesting question that is hard to tackle in a completely general way. We can help ourselves out by being specific, and asking why communications using electronic equipment requires energy.[1] It turns out that our electronic communications engines, which internally represent information as voltages, require energy to run because (1) there is capacitance from every circuit node to ground, and (2) because noise is everywhere.

We looked deeply at the energy costs associated with driving capacitors in Chapter 3. To see how noise concerns might have an impact on our energy requirements, it is helpful to return to the resistor–capacitor (*RC*) circuit of Figure 3.4. If there were no noise in the world, we could imagine using very accurate measurements of $v_0(t)$ to make this capacitor a memory cell of very high resolution. For example, if we wanted to encode one million possible values on the capacitor, we would be free to do so even if the source voltage was limited to a range of 0–1 V. In this case, we achieve our aim with discretization steps of 1 μV, and the maximum energy we would ever need to store on the capacitor would be $\frac{1}{2}C(1\text{ V})^2$.

However, suppose that there is noise in the system. And by noise, what we mean is the following: If we take a large number of measurements of v_0 for a fixed V_A, the average of those measurements would, happily, be V_A. But for this particular noise mechanism, suppose that we could only be certain that the measurement would

1 Sometimes, drilling down to the fundamentals of a specific implementation gives us insight that we are then able to generalize. Once you've digested the development here, then, think about communication using other modes such as light or sound waves. Let your imagination run wild. What you find is that the ubiquity of noise (thermal or otherwise), coupled with the fact that any sort of wave propagation is a form of energy propagation...leads to the conclusion that sending bits does require energy(!).

Energy Fundamentals of Radio, First Edition. Joel L. Dawson.

fall in a range $V_A - \frac{\Delta}{2}$ and $V_A + \frac{\Delta}{2}$. It bears repetition: for any true noise process, we are *certain* of both the average *and* the "spread" of the measurements. We just don't know exactly what any particular measurement will *be* ahead of time.

This is the essence of noise, and you can see how it complicates our attempt to use this analog memory cell to encode a million possible states. If $\frac{\Delta}{2}$ is 500 μV, for example, then we cannot, with a *single* measurement, hope to use our memory cell the way we planned. If our goal is to encode a million different levels, we must increase the range of V_A to go between 0 and 1000 V. Our noise process has forced us to have a peak energy storage one million times larger in order to preserve the integrity of our memory cell function.

Or...we can lean on the fact that if we take enough measurements of v_0, the *average* of those measurements will be exactly V_A. The greater the noise spread Δ, of course, the more measurements we will have to average to take advantage of this trick.[2] We get accuracy, apparently, by actively sacrificing speed. If we want speed and accuracy at the same time, we have to pay an energy penalty.

This is a theme that recurs again and again in the world of machines, and we will certainly return to it in this book. For now, it is critical to get a good grounding in the mathematics of noise. Let's get into it.

4.1 The Importance of Thermal Noise

There are a lot of different noise mechanisms in the world. In this book, we focus on thermal noise because it is everywhere, and is the most important and most common noise mechanism you will see in communications systems. Deep understanding of the mathematics of thermal noise is transferable to other noise mechanisms, too, so it's just all around a great place to start.

It is a fact of Nature that the molecules that make up all matter are always buzzing and shaking, and that the average kinetic energy associated with this buzzing is proportional to absolute temperature T. Feynman helps us out again with a beautiful description in Section 39-4, vol.1 of his lectures:[3]

> The mean molecular kinetic energy is a property only of the "temperature." Being a property of the "temperature," and not of the gas, we can use it as a definition of the temperature. The mean kinetic energy of a molecule is thus some function of the temperature. But who is to tell us what scale to use for the temperature? We may arbitrarily define the scale of temperature so that the mean energy is linearly proportional to the temperature. The best way to

2 And it will turn out that you cannot take these measurements arbitrarily close to each other in time. They need to be spaced out in time such that the measurements are uncorrelated. Stay tuned.

3 Ibid.

do it would be to call the mean energy itself "the temperature." That would be the simplest possible function. Unfortunately, the scale of temperature has been chosen differently, so instead of calling it temperature directly, we use a constant conversion factor between the energy of a molecule and a degree of absolute temperature called a degree Kelvin. The constant of proportionality is $k = 1.38 \times 10^{-23}$ Joule for every degree Kelvin. So if T is absolute temperature, our definition says that the mean molecular kinetic energy is $\frac{3}{2}kT$ (The $\frac{3}{2}$ is put in as a matter of convenience, so as to get rid of it somewhere else.)

We point out that the kinetic energy associated with the component of motion in any particular direction is only $\frac{1}{2}kT$. The three independent directions that are involved make it $\frac{3}{2}kT$.

Tying this now to communications systems: all the molecules in our systems, and all the charge carriers that are responsible for our currents, are buzzing according to the above. And so it wouldn't be surprising if, when we measure signals, there is a low-amplitude, additive buzzing that attends the "normal" signals that we expect to see.

Those of you who have performed measurements on circuits in a laboratory know that this is exactly what you see. You are accustomed to calling this phenomenon noise, and so (1) you measure voltages on an oscilloscope by estimating the center of a fuzzy oscilloscope trace; and (2) your pristine peak that a sinusoid puts on a spectrum analyzer looks fine at the top, but near the bottom plunges into the vaguely threatening chaos of The Noise Floor (see Figure 4.1). Many, many electronics engineers have forsworn analog electronics for digital electronics because of this messiness.[4] This is understandable, of course. But underneath the fuzziness of our signals is deep insight into the natural world. In this chapter, we'll explore two aspects of thermal noise that are important for wireless communications. The first is how thermal noise manifests in electronic circuits. The second is the fact that all objects warmer than 0 K to *radiate* in the electromagnetic spectrum.

4.1.1 The Equipartition Theorem for Circuits

Consider the network of Figure 4.2, with various lossy elements (resistors) and lossless elements (inductors and capacitors) connected at random. When we say that this system is "in thermal equilibrium," we mean that the resistors R_1 and R_2 are at the same temperature as each other, and at the same temperature as the surrounding environment.

---

4 Although if you push performance hard enough, you find that noise limits you in digital circuits, too.

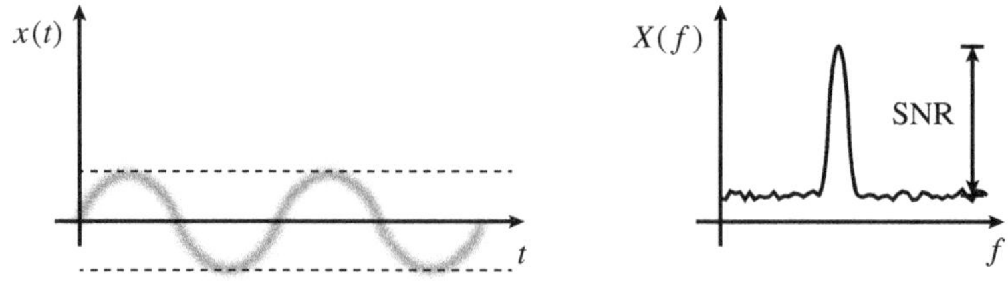

Figure 4.1 A noisy sinusoid on both an oscilloscope (left) and a spectrum analyzer (right).

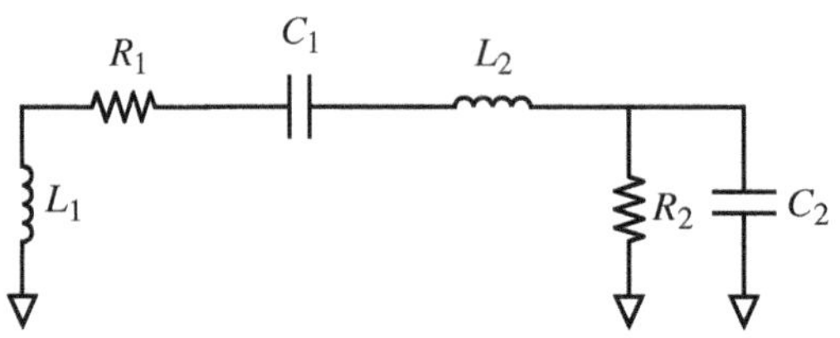

Figure 4.2 A random collection of electronic components, assembled to illustrate the equipartition theorem.

The astounding result of the equipartition theorem is as follows: If we are at thermal equilibrium at temperature T, then each capacitor and inductor will have an average energy stored of $\frac{1}{2}kT$. Simple as that![5]

This means that we can calculate the mean squared voltage across any capacitor or inductor. Since we know the average energy stored on the inductors and the capacitors, we have that

$$\frac{1}{2}C_1\langle V_{C1}\rangle^2 = \frac{1}{2}C_2\langle V_{C2}\rangle^2 = \frac{1}{2}kT, \tag{4.1}$$

and similarly

$$\frac{1}{2}L_1\langle I_{L1}\rangle^2 = \frac{1}{2}L_2\langle I_{L2}\rangle^2 = \frac{1}{2}kT. \tag{4.2}$$

Without doing any circuit analysis whatsoever, we can say that the mean squared voltage across any given capacitor is

$$\langle V_{Ci}\rangle^2 = \frac{kT}{C_i} \tag{4.3}$$

5 If you are confused, or surprised, it is because you are looking for some complicated dependence on the component values and/or details of the network. Learn to cultivate these violent clashes between your intuition and empirical reality. Study, think, and come to peace with how things are. Over time, this greatly strengthens your intuition.

and the mean squared current in any given inductor is

$$\langle I_{Li} \rangle^2 = \frac{kT}{L_i}. \tag{4.4}$$

As a final note, it is important to emphasize that the *mean* voltages and currents will all be zero. That is,

$$\langle V_{Ci} \rangle = 0 \tag{4.5}$$

and

$$\langle I_{Li} \rangle = 0. \tag{4.6}$$

If you're really thinking ahead, you may worry that nonzero-mean squared voltages and currents in the energy storage elements mean that there are nonzero instantaneous currents in resistors R_1 and R_2. And nonzero instantaneous currents in resistors mean that energy is dissipated as heat. Shouldn't we see a steady cooling of the system, getting us ultimately to 0 K?

This is where our insistence that we are in thermal equilibrium with the surrounding environment becomes important. Thermal equilibrium in this system means that every joule of heat energy dissipated from our resistors is *exactly* balanced by a joule of heat absorbed from the surrounding environment. The right picture to have in your mind of thermal equilibrium is of a *continuous, precisely balanced exchange of energy* that leaves all objects at the same, unchanging temperature.

4.1.2 All Objects Radiate...

All objects whose temperature is above 0 K radiate electromagnetic energy. Consider what this means for a radio receiver out in the world, as shown in Figure 4.3. All the objects in the picture radiate in the electromagnetic spectrum according to Planck's famous law:

$$B_\omega(\omega, T) = \frac{\hbar\omega^3}{4\pi^3 c^2} \cdot \frac{1}{e^{\hbar\omega/kT} - 1}, \tag{4.7}$$

or in regular frequency units

$$B_f(f, T) = \frac{2hf^3}{c^2} \cdot \frac{1}{e^{hf/kT} - 1}. \tag{4.8}$$

The units of B are Watts per square meter per steradian per hertz. Wikipedia puts it well: "An infinitesimal amount of power $B_f(f, T) \cos\theta dA d\Omega df$ is radiated in the direction described by the angle θ from the surface normal from infinitesimal surface area dA into infinitesimal solid angle $d\Omega$ in an infinitesimal frequency band of width df centered on frequency f."[6] The other thing to know about this law is

6 https://en.wikipedia.org/wiki/Planck%27s_law#Black-body_radiation.

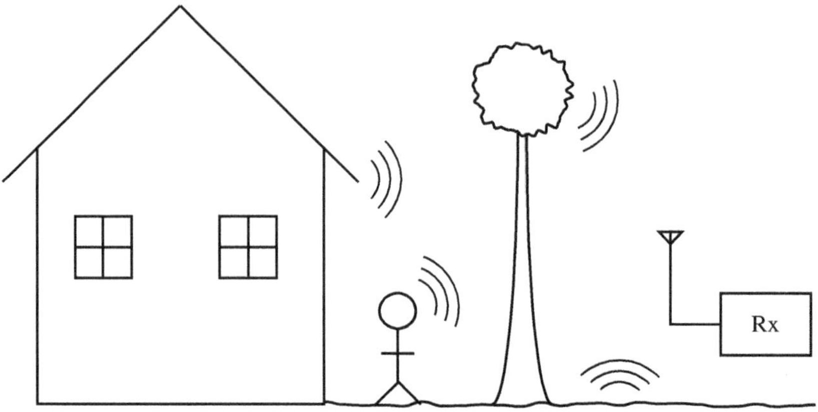

Figure 4.3 All objects radiate in the electromagnetic spectrum.

that it is for an idealized blackbody radiator, and it captures the theoretical maximum radiation that an object will radiate at a given temperature. Real objects will fall short of that maximum by a factor of ϵ, called the emissivity, which is a unitless constant between 0 and 1.

Now based on Section 4.1.1, Eqs. 4.7 and 4.8 look perhaps more complicated than they should. It will help to come clean and admit that the equations in our discussion of the equipartition theorem were technically approximations. The full expression of Eq. 4.1 actually looks like

$$\frac{1}{2}C_1\langle V_{C1}\rangle^2 = \frac{1}{2}C_2\langle V_{C2}\rangle^2 = \frac{1}{2}\frac{hf}{e^{hf/kT} - 1}. \tag{4.9}$$

But we will see that Nyquist had a point in his original paper when he went with Eq. 4.1 as the useful form. Start with the Taylor expansion of $e^{hf/kT}$ to first order, for which we have

$$e^{hf/kT} \approx 1 + hf/kT \tag{4.10}$$

if $hf/kT \ll 1$. If we assume a temperature T of 300 K, and plug in values for Planck's constant h and Boltzmann's constant k,

$$\begin{aligned} h &= 6.626 \times 10^{-34}\text{J/Hz} \\ k &= 1.38 \times 10^{-23}\text{J/K}, \end{aligned} \tag{4.11}$$

we can write Eq. 4.10 as

$$e^{hf/kT} \approx 1 + 1.60 \times 10^{-13}\text{Hz}^{-1} \cdot f. \tag{4.12}$$

We see that in order for hf/kT to equal unity, f must equal 6.25 THz. This is more than one hundred times the frequency that is useful for cellular communications.

We therefore conclude that the approximation in Eq. 4.10 is justified for communications applications, and rewrite Eq. 4.9 as:

$$\begin{aligned}\frac{1}{2}C_i\langle V_{Ci}\rangle^2 &= \frac{1}{2}\frac{hf}{e^{hf/kT}-1}\\ &\approx \frac{1}{2}\frac{hf}{1+hf/kT-1}\\ &= \frac{1}{2}kT,\end{aligned} \tag{4.13}$$

as before.

Applying the same approximation to Eqs. 4.7 and 4.8, we arrive at

$$B_\omega(\omega, T) = \frac{\omega^2}{4\pi^3c^2}\cdot kT \tag{4.14}$$

and

$$B_f(f, T) = \frac{2f^2}{c^2}\cdot kT. \tag{4.15}$$

4.1.3 ...But You'll Only Notice This Using Radio Telescopes

Upon reflection, though, Eqs. 4.14 and 4.15 fly in the face of your experience if you have done careful measurements with receivers in the laboratory. In particular, you have probably never associated an f^2 dependence with thermal noise.

This is because it takes extraordinary care, and often cryogenic cooling, to build receivers whose own circuit noise does *not* swamp out the radiation of objects within view of the receiving antenna. We typically call such sensitive receivers *radio telescopes*,[7] and they are so good when pointed up at the night sky they are sensitive to cosmic background radiation with associated temperatures of below 3 K. Most receivers built for communications in the 1–10 GHz range have noise floors dominated by their own circuit noise of the type that we return to in Section 4.3. Thermal circuit noise is spectrally flat.

4.2 Dealing with Probabilistic Signals

For deterministic signals specified, for example, as $x(t)$, what we mean is that for <u>any</u> time instant t_0, we know for certain what the value of the signal is: it is $x(t_0)$.

---

7 Kraus. *Radio Astronomy*, New York, NY: McGraw-Hill Book Company, 1966. A particularly helpful plot of sky noise temperatures can be found on page 237.

With probabilistic signals like noise, this is not our situation. The outcome of measuring a probabilistic signal at a particular time instant t_0 is not known ahead of time. But here's what we *can* know:

- The mean value of the measurements will be constant. We'll call it μ.
- The $\overline{\text{mean}}$ squared deviation from the mean. We'll call this σ^2, as in, $\sigma^2 = \overline{(x - \mu)^2}$.
- What knowledge of a measurement at time $t_1, x(t_1)$, tells us about a measurement at some time later t_2. For most common stochastic processes, this will be purely a function of the time separation $|t_1 - t_2|$. Also, for most common stochastic processes, the larger the time separation $|t_1 - t_2|$, the less "related" $x(t_1)$ is to $x(t_2)$. In the vernacular, we say that two measurements are "uncorrelated" if knowledge of measurement $x(t_1)$ does not alter in any way the probabilistic distribution of $x(t_2)$.

More to come on all of this.

4.2.1 "Normal" or Gaussian Distributions

The usual approach to discussing normal distributions is through an abstract, mathematical treatment, followed by an application (if time). If you have been reading this book in its presented order, however, it will not surprise you to see a return to our trusty *RC* circuit. We will gain insight into normal distributions through this familiar example.

We examine the circuit in Figure 4.4, which is the familiar *RC* circuit but with a switch between the resistor and the capacitor. When this switch is closed as shown, the capacitor voltage is moving all the time because of thermal agitation of charge carriers in the resistor. We don't know in advance what the exact voltage on the capacitor will be at any given time, but thanks to the amazing equipartition theorem we know that the mean squared deviation from the mean (in this

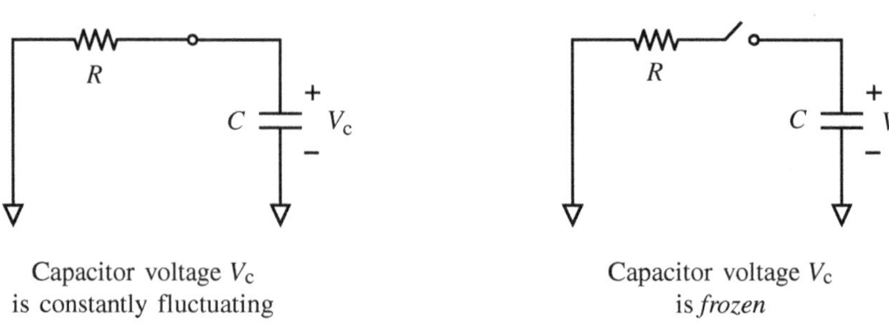

Figure 4.4 Examining thermal noise in a simple *RC* circuit.

case, 0 V) is $\frac{kT}{C}$. When the switch is open, however, the voltage across the capacitor *freezes*. It stays at whatever voltage it was in the instant before the switch was opened. By opening the switch, then, we interrupt the time evolution of the capacitor voltage to take "samples," or snapshots of its stochastic evolution. We will call each closing of the switch, and subsequent opening of the switch and reading off of the frozen voltage, a "measurement." And we will assume that we wait long enough between measurements that the result of one measurement in no way affects the probability distribution of the next measurement.

Now we run an experiment as follows:

1. Close the switch and allow the capacitor voltage to settle.
2. Open the switch and measure the voltage V_c with a perfect voltmeter. Write down the result.
3. Repeat steps (1) and (2), until we have one million measurements.

What would we expect this group of measurements to look like? The answer is that we expect our measurements to have a "normal" or "Gaussian," distribution, with a mean or average value of zero, a "standard deviation" of $\sqrt{kT/C}$, and a "variance" or "mean squared" value of kT/C.

If you are a practicing engineer, you have heard and used the vocabulary of Gaussian distributions for your entire professional life. Indeed, the language of Gaussian distributions is the default for almost any sort of statistical analysis of experimental data, regardless of the experiment, and regardless of whether the underlying stochastic mechanism (if it is even stochastic!) has any relation at all to a Gaussian distribution. This is actually okay. The language of Gaussian distributions is *useful*, and thus there is much to be gained by getting your understanding solid through application to a familiar example.

Fine. When we say that a random variable (in our case V_c) "has a Gaussian distribution," what we mean is that the value of any given measurement is governed by a *probability density function* of the form:

$$f(x) = \frac{1}{\sigma\sqrt{2\pi}} e^{-\frac{1}{2}\left(\frac{x-\mu}{\sigma}\right)^2}. \tag{4.16}$$

In this expression, σ is the standard deviation, σ^2 is the variance or mean squared deviation from the mean, and μ is the mean. So for our V_c, the governing probability density function is

$$f(V_c) = \frac{1}{\sqrt{kT/C}\sqrt{2\pi}} e^{-\frac{1}{2}\frac{V_c^2}{kT/C}}. \tag{4.17}$$

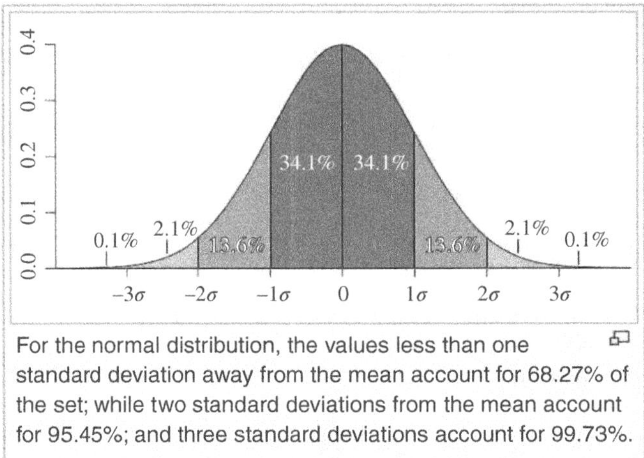

Figure 4.5 Graph showing probability distribution for a Gaussian random variable. *Source*: Ainali/Wikipedia Commons/CC BY 3.0/https://en.wikipedia.org/wiki/Normal_distribution.

And what exactly does it mean for our measurements to be "governed" by a probability density function? It means that *the probability of an individual measurement falling between V_1 and V_2* is given by:

$$P = \int_{V_1}^{V_2} \frac{1}{\sqrt{kT/C}\sqrt{2\pi}} e^{-\frac{1}{2}\frac{V_c^2}{kT/C}} dV_c. \tag{4.18}$$

If we stretch our integration limits to $V_1 = -\infty$ and $V_2 = +\infty$, the normalization constant $\frac{1}{\sqrt{kT/C}\sqrt{2\pi}}$ ensures that the integration totals to unity, as it should.

A useful working definition of a Gaussian distribution to carry around in your head is as follows. Approximately 68% of your measurements will fall in the range $\pm\sigma$ on either side of the mean (so in the range $\pm\sqrt{kT/C}$ V in the current example); 95.5% of your measurements will fall within $\pm 2\sigma$ of the mean; 99.7% of them will fall within $\pm 3\sigma$ of the mean.

An excellent graphical illustration of a Gaussian distribution can be found on Wikipedia[8]; see Figure 4.5. Even better is the following table from the same source; see Figure 4.6. All of this is to say that when we talk about noise, it needn't remain some abstract, theoretical concept in your mind. In the real world, noise introduces an element of probability in our otherwise deterministic measurements. To really make it concrete, let's go back to fill in some numbers for our little sample-and-hold circuit of Figure 4.4, but now with $R = 1000\ \Omega$ and $C = 1$ pF. Using our trusty

8 https://en.wikipedia.org/wiki/Normal_distribution.

n	$p = F(\mu + n\sigma) - F(\mu - n\sigma)$	i.e. $1 - p$	or 1 in p
1	0.682 689 492 137	0.317 310 507 863	3.151 487 187 53
2	0.954 499 736 104	0.045 500 263 896	21.977 894 5080
3	0.997 300 203 937	0.002 699 796 063	370.398 347 345
4	0.999 936 657 516	0.000 063 342 484	15 787.192 7673
5	0.999 999 426 697	0.000 000 573 303	1 744 277.893 62
6	0.999 999 998 027	0.000 000 001 973	506 797 345.897

Figure 4.6 Table taken from Wikipedia showing probability distribution for a Gaussian random variable.

expression for the variance, $kT/C = \sigma^2$, and filling in for Boltzmann's constant $k = 1.38 \times 10^{-23}$ J/K and $T = 300$ K, we have that

$$\begin{aligned}\sigma^2 &= 4.14 \times 10^{-9}\ \text{V}^2 \\ \sigma &= 64.3\ \mu\text{V}.\end{aligned} \tag{4.19}$$

If we are careful with our one million measurements with our perfect voltmeter, we would expect to find:

1. The average of our voltage measurements would be zero.
2. 683 000 of those measurements would fall in the range of ± 64.3 μV.
3. About 2700 of those measurements would fall either higher than 193 μV or lower than -193 μV.
4. We would be unlikely to see measurements above 322 μV or below -322 μV. (The chances of falling outside of the 5σ range is about 1 in 1.7 million.)

Finally, understand the implications of modifying this circuit to include a voltage source, as shown in Figure 4.7. The *only* effect is to shift the center of the probability distribution function. The mean, or average, of our one million measurements is now V_{DC}, and 683 000 of our measurements now fall in the range $V_{DC} \pm 64.3$ μV.

4.2.2 The Meaning of Correlation

Correlation is a very important concept when it comes to dealing with probabilistic variables. The basic idea behind correlation is that we want to *quantify* how well one random variable tracks another random variable. When one goes up, does the other go up more often than not? What about when one of them goes down? We say that two variables are uncorrelated when observation of one tells

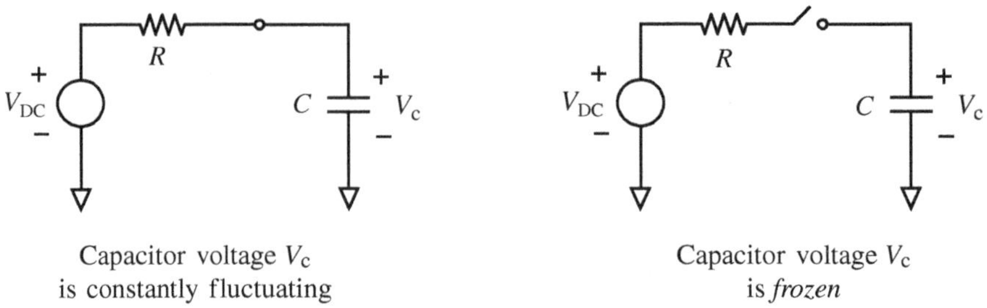

Figure 4.7 Examining thermal noise in a simple *RC* circuit.

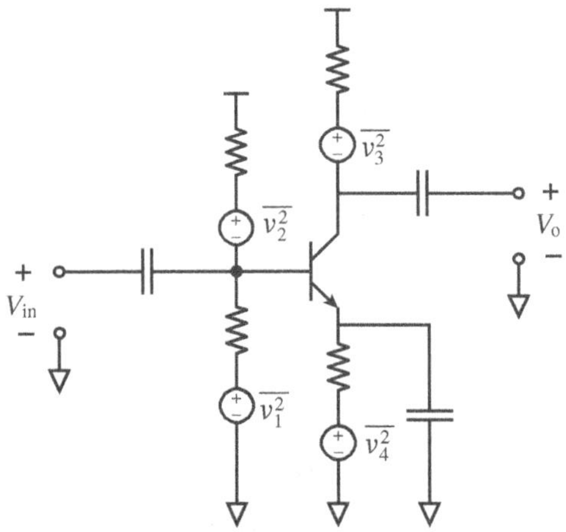

Figure 4.8 An example single-stage transistor amplifier. We model thermal noise in resistors as a series voltage source with noise variance $\overline{v_n^2} = 4kTR\Delta f$. See Section 4.9.

you absolutely nothing about the other: when one goes up, the other goes up or down with equal likelihood.

Why do we care? Well, suppose we had a simple one-stage electronic amplifier as shown in Figure 4.8. Suppose further that we wanted to calculate the noise at the output due to thermal fluctuations of charge carriers in all four resistors, plus the noise mechanisms associated with the bipolar transistor itself. How would we handle such a problem?

It turns out to be a fairly complicated calculation, and it is not necessary to go into great detail here. But there is one conceptual aspect that is easy and important to grasp. Our problem is to combine the effects of the noise mechanisms associated with the four resistors and the transistor. To do so, it helps to imagine a spectrum of assumptions about how these effects would combine at the output:

1. We could assume that the noise fluctuations due to each noise source have a *positive* correlation. What we mean is that noise fluctuations in one resistor occur in a way that *reinforces* the noise fluctuations from every other resistor when the effects of these fluctuations are computed at the output. This corresponds to the mean squared noise voltage at the output being as high as it could possibly be. This would be a diabolical worst case when it comes to noise.
2. On the other hand, we could assume an extremely fortuitous and generalized *negative* correlation, in which the various noise sources maximally canceled each other when their effects were summed at the output. The mean squared noise at the output is as small as it could possibly be, and we have a overwhelming evidence that this is our lucky day.
3. We could assume that the various noise sources combine constructively as often as they combine destructively and are therefore what we call completely uncorrelated. The mean squared noise would be smaller than in scenario no. 1, and larger than scenario no. 2. This is, in fact, what we go with.

The foregoing is a solid conceptual basis for thinking about correlation. When it comes to quantitative analysis, we use a parameter called a correlation coefficient to capture the range of possibilities between extreme coincidence, zero coincidence, and total opposition. It may not surprise you that our convention is for the correlation coefficient to be +1 for extreme correlation (case no. 1 above), −1 for anti-correlation (case no. 2), and 0 for no correlation at all.

Great. We're going to approach the problem of constructing a numerical correlation constant "from below," using simple example signals first. Our simple example signals will be discrete time, can assume only a value of +1 or −1, and at any given time index will assume one of those values with equal probability. Consider, then, the pairs of signals in Figure 4.9. Note that these examples have been carefully constructed. In pair no. 1, the signals are identical, and so our correlation coefficient ought to give its highest +1 in this case. In pair no. 2, the signals agree half of the time and disagree half the time, so our correlation coefficient should be zero. In pair no. 3, the signals are inversions of one another. We expect our correlation measure in this case to give a satisfying −1 result.

Our simplifying decision to limit signal values to ±1 is very suggestive of a way forward. For a given index m, if we take the product $x_a[m] \cdot x_b[m]$, a +1 result tells us that the signals went the same way, whereas a −1 result tells us that they went in opposite directions. We could...then...take the average over all m for $x_a[m] \cdot x_b[m]$, and see if the shoe fits:

$$\mathrm{corr}(x_a, x_b) = \frac{1}{N} \sum_{m=0}^{N-1} x_a[m] \cdot x_b[m]. \tag{4.20}$$

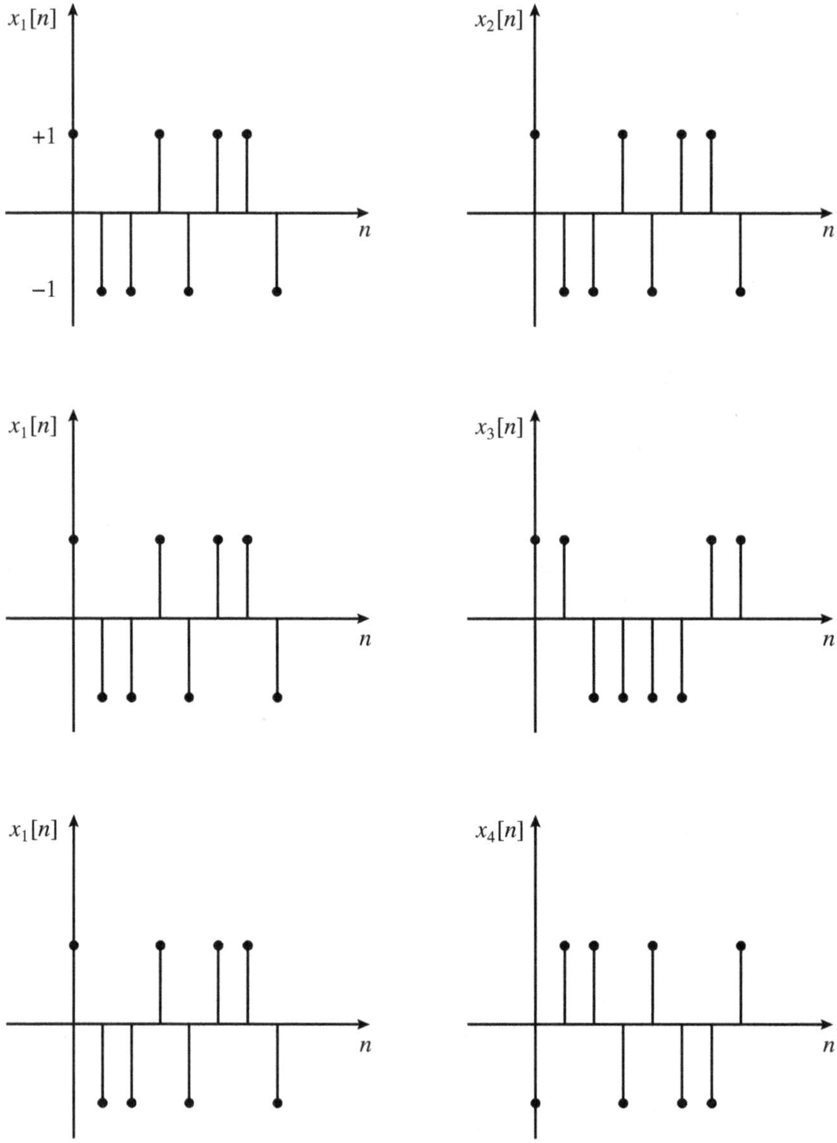

Figure 4.9 Example signals for thinking about correlation.

How well does this work? Looking through our made-up examples:

$$\begin{aligned}\operatorname{corr}(x_1, x_2) &= \frac{1}{8} \cdot (1 + 1 + 1 + 1 + 1 + 1 + 1 + 1) = 1 \\ \operatorname{corr}(x_1, x_3) &= \frac{1}{8} \cdot (1 - 1 + 1 - 1 + 1 - 1 + 1 - 1) = 0 \\ \operatorname{corr}(x_1, x_4) &= \frac{1}{8} \cdot (-1 - 1 - 1 - 1 - 1 - 1 - 1 - 1) = -1. \end{aligned} \tag{4.21}$$

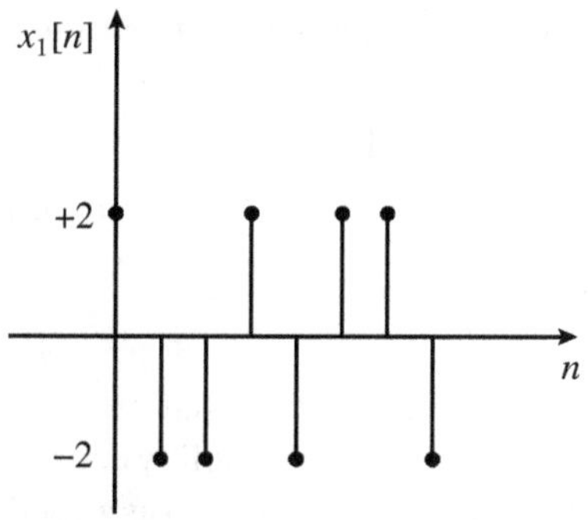

Figure 4.10 Scaling our original $x_1[n]$ by a factor of 2.

Progress! But we are not quite done. Suppose that we left x_2, x_3, and x_4 the same, but for x_1, for each sample we amplitude scale by a factor of 2. A *good* correlation coefficient would not be fooled by a mere scaling of one (or both) of the signals and would instead give the same results as above. We must find a way to *normalize*, in other words.

The root mean squared value of the signals turns out to be the normalization factor that we need:

$$\text{corr}(x_a, x_b) = \frac{1}{N}\sum_{m=0}^{N-1} \frac{x_a[m]}{\sqrt{\frac{1}{N}\sum_{k=0}^{N-1} x_a^2[k]}} \cdot \frac{x_b[m]}{\sqrt{\frac{1}{N}\sum_{l=0}^{N-1} x_b^2[l]}}. \tag{4.22}$$

Let's check this with $\text{corr}(x_1, x_2)$, but where we have scaled x_1 by a factor of two according to Figure 4.10.

$$\begin{aligned}\text{corr}(x_1, x_2) &= \frac{1}{8} \cdot \sum_{m=0}^{7} \frac{x_1[m]}{\sqrt{\frac{1}{8} \cdot 32}} \cdot \frac{x_2[m]}{\sqrt{\frac{1}{8} \cdot 8}} \\ &= \frac{1}{16} \cdot \sum_{m=0}^{7} x_1[m] \cdot x_2[m] \\ &= \frac{1}{16} \cdot (2 + 2 + 2 + 2 + 2 + 2 + 2 + 2) \\ &= 1. \end{aligned} \tag{4.23}$$

It passes the test! If you check the other two cases, you'll find further reassurance that our normalization choice is the correct one.

Up until this point our example signals have been binary: for any given sample, the signal could only have been +1 or −1. We did this because it made the logic of calculating a correlation coefficient easy to follow. But it turns out that going forward we need not be so restrictive. For N measurements of a pair of random variables, we may estimate the correlation coefficient using Eq. 4.22, but rewritten here with the factors of $\frac{1}{N}$ properly canceled out:

$$\mathrm{corr}(x_a, x_b) = \sum_{m=0}^{N-1} \frac{x_a[m]}{\sqrt{\sum_{k=0}^{N-1} x_a^2[k]}} \cdot \frac{x_b[m]}{\sqrt{\sum_{l=0}^{N-1} x_b^2[l]}}. \tag{4.24}$$

This formula assumes that we are dealing with random variables that are zero mean, but otherwise there are no restrictions on the form of the probability density functions for the $x_i[m]$s. Going forward, then, you can internalize Eq. 4.24 as a good mathematical definition of "correlation," sometimes called "cross correlation" to differentiate from "autocorrelation," which we will describe in Section 4.2.3. Most of the time when we fret about correlation, it will be to show that the variables in question are uncorrelated. In this happy circumstance, the correlation coefficient of Eq. 4.24 is zero.

4.2.3 The Meaning of Autocorrelation

Autocorrelation is one of those topics where it may help to just start with the math. For a discrete-time signal $x[n]$, the autocorrelation function $R_{xx}[m]$ is given by

$$R_{xx}[m] = \sum_n x[n] \cdot x[n-m]. \tag{4.25}$$

Remember that $x[n-m]$ is just $x[n]$ delayed by m samples. In words, then, the autocorrelation function tells us how correlated a given signal is with a delayed copy of itself.[9]

What is the point of the autocorrelation function? What question are we trying to answer? We are accustomed at this point with probabilistic signals to dealing with statistical moments like the average or the variance (mean and mean squared value). But what these statistical moments do not tell you is ... how rapidly in time the signals are moving around! It is worth pausing to consider how you would construct a mathematical tool to help you with this problem. After some consideration, you may well conclude that keeping track of the correlation of the signal with a delayed copy of itself, with that delay as a parameter, is a pretty clever solution.

Right? Consider what it would mean for a signal to be well correlated with a delayed version of itself for large values of m. You would say that this random process does not move much in between sample instants, or that the random process moves slowly relative to the sample period. By way of contrast, suppose that for an

9 Do not be disturbed by the absence of the carefully derived normalization convention that we used in Section 4.2.2. We are usually more interested in the *shape* of the autocorrelation function than we are in exact numerical values.

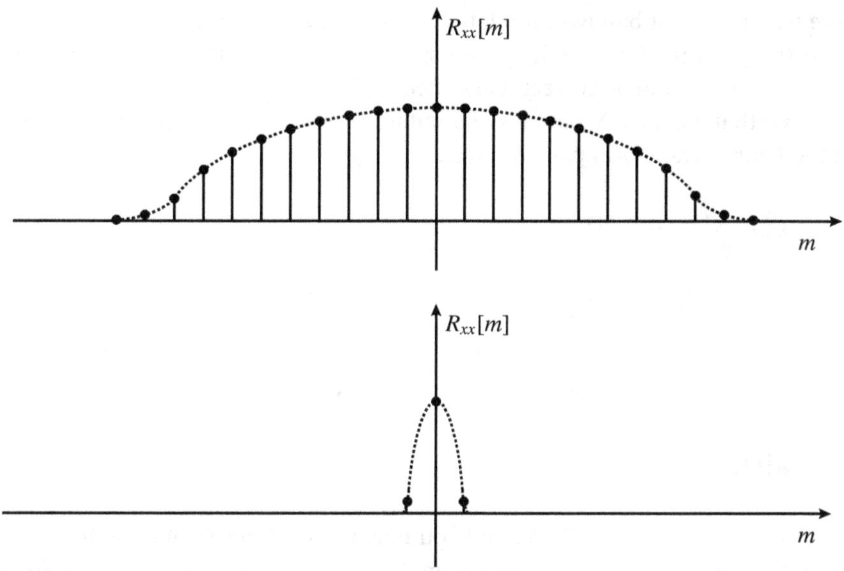

Figure 4.11 Autocorrelation functions for a random process that evolves slowly (top) versus quickly (bottom).

$m = \pm 1$, the autocorrelation function fell to zero and stayed that way for all $|m| > 1$. You would conclude correctly that, in between samples, the underlying random process is evolving so rapidly that adjacent samples are completely uncorrelated with each other.

Graphically, we sketch in Figure 4.11 two extremes of the autocorrelation function. We say that the example on top represents a random process that is evolving slowly relative to the sample period, while the example on the bottom represents a process that evolves rapidly.

4.2.4 Autocorrelation and Power Spectral Density as a Fourier Transform Pairs

The world of probabilistic signals and stochastic processes has its own complete vocabulary and mindset which, given the scope of this book, we will mostly avoid. Our mindset when a signal is the result of a random process is that we will always run a "trial" of sufficient duration such that, say, the sample average is exactly equal to the expected value or mean of the underlying process.

Now we make the connection in Section 4.2.3 that sequences characterized by a "broad" autocorrelation function correspond to signals that "move slowly." Similarly, signals that have a narrow autocorrelation function move quickly. If you put that together with the discussion of the Uncertainty Principle in Section 2.5, where we found that bandwidth relates to speed...well, you might guess that the autocorrelation function and the power spectral density (PSD) are Fourier transform pairs. You would be correct! Let's check this.

Suppose that we have N samples of a random process as the signal $x[n]$. Recall that the Fourier transform pair would be

$$\begin{aligned} x[n] &= \sum_{k=0}^{N-1} C_k e^{j\left(\frac{2\pi}{N}\cdot k\right)n} \\ C_k &= \frac{1}{N}\sum_{n=0}^{N-1} x[n] e^{-j\left(\frac{2\pi}{N}\cdot k\right)n}. \end{aligned} \tag{4.26}$$

Our autocorrelation function is

$$R_{xx}[m] = \sum_{n=0}^{N-1} x[n]\cdot x[n-m]. \tag{4.27}$$

(Bookkeeping is important, here, and you may worry about finite-length signals that cannot completely overlap in the sum. If you are worried about this, the fix is that we don't do things like this unless $N \gg m$.)

We test our hypothesis by simply computing the Fourier transform of $R_{xx}[m]$:

$$\begin{aligned} C_k &= \frac{1}{N}\sum_{m=0}^{N-1} R_{xx}[m] e^{-j\left(\frac{2\pi}{N}\cdot k\right)m} \\ &= \frac{1}{N}\sum_{m=0}^{N-1}\sum_{n=0}^{N-1} x[n]\cdot x[n-m] e^{-j\left(\frac{2\pi}{N}\cdot k\right)m}. \end{aligned} \tag{4.28}$$

Now here is the trick: It turns out that it does not matter whether we run our summations over m first and then over n, or vice versa. Because we know the answer we are trying to get to,[10] which is that autocorrelation function and the PSD are Fourier transform pairs, we are motivated to interchange the order of our summations:

$$C_k = \frac{1}{N}\sum_{n=0}^{N-1}\sum_{m=0}^{N-1} x[n]\cdot x[n-m] e^{-j\left(\frac{2\pi}{N}\cdot k\right)m}. \tag{4.29}$$

There's actually one more trick that makes the accounting easier. Our mindset is that we have N samples total, and that outside of the index $0 \le n \le N-1$, $x[n]$ is zero. So, we don't run the risk of any rogue infinities if we raise our summation limits to $\pm\infty$ like so:

10 Foreknowledge of the final answer *always* helps when you are doing derivations!

$$C_k = \frac{1}{N} \sum_{n=-\infty}^{\infty} \sum_{m=-\infty}^{\infty} x[n] \cdot x[n-m] e^{-j\left(\frac{2\pi}{N} \cdot k\right)m}. \tag{4.30}$$

Now,

$$C_k = \frac{1}{N} \sum_{n=-\infty}^{\infty} x[n] \sum_{m=-\infty}^{\infty} x[n-m] e^{-j\left(\frac{2\pi}{N} \cdot k\right)m}. \tag{4.31}$$

Let's take a look at that second piece first:

$$\sum_{m=-\infty}^{\infty} x[n-m] e^{-j\left(\frac{2\pi}{N} \cdot k\right)m}. \tag{4.32}$$

The thing about sums and integrals is that you can change anything you want as long as you don't change the result of the sum. Check for yourself the validity of the next step, substituting $p = n - m$:

$$= \sum_{p=\infty}^{-\infty} x[p] e^{-j\left(\frac{2\pi}{N} \cdot k\right)n} e^{+j\left(\frac{2\pi}{N} \cdot k\right)p} \tag{4.33}$$

$$= \sum_{p=-\infty}^{\infty} x[p] e^{-j\left(\frac{2\pi}{N} \cdot k\right)n} e^{+j\left(\frac{2\pi}{N} \cdot k\right)p}$$

$$= e^{-j\left(\frac{2\pi}{N} \cdot k\right)n} \sum_{p=-\infty}^{\infty} x[p] e^{+j\left(\frac{2\pi}{N} \cdot k\right)p}$$

$$= e^{-j\left(\frac{2\pi}{N} \cdot k\right)n} \sum_{p=-\infty}^{\infty} x[p] e^{-j\left(\frac{2\pi}{N} \cdot (-k)\right)p}$$

$$= e^{-j\left(\frac{2\pi}{N} \cdot k\right)n} \sum_{p=0}^{N-1} x[p] e^{-j\left(\frac{2\pi}{N} \cdot (-k)\right)p} \tag{4.34}$$

Based on our definition of the DFT, reiterated in Eq. 4.29, we can write this expression in terms of the DFT of $x[p]$, which we'll label C_{kx}:

$$= e^{-j\left(\frac{2\pi}{N} \cdot k\right)n} C_{-kx} \cdot N. \tag{4.35}$$

Remembering that C_{-kx} and C_{kx} are complex conjugates, we can write instead

$$= e^{-j\left(\frac{2\pi}{N} \cdot k\right)n} C_{kx}^{*} \cdot N. \tag{4.36}$$

Returning to our original goal, which was to compute the DFT of the autocorrelation function in Eq. 4.28, we have

$$\begin{aligned} C_k &= \frac{1}{N} \sum_{n=-\infty}^{\infty} x[n] e^{-j\left(\frac{2\pi}{N} \cdot k\right)n} C_{kx}^* \cdot N \\ &= \frac{1}{N} \sum_{n=0}^{N-1} x[n] e^{-j\left(\frac{2\pi}{N} \cdot k\right)n} C_{kx}^* \cdot N \\ &= N \cdot C_{kx} \cdot C_{kx}^* \\ &= N \cdot |C_{kx}|^2. \end{aligned} \tag{4.37}$$

It is customary, and absolutely fine, to drop the constant N and just note that

$$C_k \propto |C_{kx}|^2. \tag{4.38}$$

In other words, the Fourier transform of the autocorrelation function is essentially the PSD of $x[n]$! We thus validate our guess from earlier in the section.

In continuous time, if $x_N(t)$ is a noisy signal, the convention is to label the PSD $S_{xx}(f)$, and define it as

$$S_{xx}(f) = |X_N(f)|^2. \tag{4.39}$$

Then, for the autocorrelation function $R_{xx}(\tau)$ (where the continuous variable τ takes the place of the separation m in discrete time) we have

$$\begin{aligned} R_{xx}(\tau) &= \int_{-\infty}^{\infty} S_{xx}(f) e^{j2\pi f\tau} df \\ S_{xx}(f) &= \int_{-\infty}^{\infty} R_{xx}(\tau) e^{-j2\pi f\tau} d\tau. \end{aligned} \tag{4.40}$$

In closing this section, it is worth recalling our discussion of the equipartition theorem for circuits in Section 4.1.1. We rightly thought it remarkable that we could write down by inspection the mean squared currents and voltages for inductors and capacitors for systems in thermal equilibrium. That *is* remarkable, but it is an incomplete picture. We are also interested in how quickly our random variables are evolving in time. A very convenient characterization of the speed of evolution for a noisy process turns out to be the autocorrelation function. And, again, the autocorrelation function and PSD turn out to be Fourier transform pairs.

4.3 Characteristics of Thermal Noise in Circuits

We care about thermal noise because it is everywhere. Given that, we care about its spectral characteristics for reasons described in Section 2.3.1. If we are lucky, in our communications machinery we can separate the spectral content of the noise from that of our signal. Therefore, we must know what our noise looks like in the frequency domain. In this section, we examine thermal noise in both the frequency and time domains.

4.3.1 In the Frequency Domain

In the frequency domain, the PSD for thermal noise turns out to be flat.[11] If we model a resistor as a noiseless resistance in series with a noise voltage source, as in Figures 4.8 and 4.12, then it turns out that the mean squared noise voltage at the output is given by

$$\overline{v_0^2} = 4kTR\Delta f. \tag{4.41}$$

Again, k is Boltzmann's constant and is about 1.38×10^{-23} J/K, and T is the absolute temperature in degrees Kelvin.

A couple of housekeeping items. First, sometimes people are confused by the Δf in the formula, and their confusion is only heightened when someone points out that this the "observation bandwidth." A useful picture to have in your head is to augment Figure 4.12 with an ideal, noiseless, brick-wall filter as in Figure 4.13. The observation bandwidth is now B.

When you think about it, the fact of an observation bandwidth is not so unreasonable. Any voltage measurement system will have a finite bandwidth, whether it be a voltmeter, an oscilloscope, a spectrum analyzer, or an amplifier stage. What might still trouble you is that a perfectly flat, brick-wall filter may seem unnecessarily simplistic. We'll get to this in Section 4.4. Accept for now that any real-world filter, in all of its varying frequency response glory, can be adequately and accurately represented by an equivalent flat filter of the type shown in Figure 4.13. It will turn out that we can do this because thermal noise itself is spectrally flat.

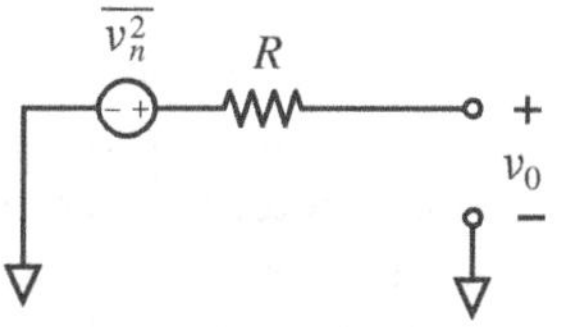

Figure 4.12 Model for a resistor with thermal noise. Note that we first invoked this model in Figure 4.8.

11 At least, for frequencies normally of interest in wireless communications. These will be well below the 6.25 THz discussed in Section 4.1.2.

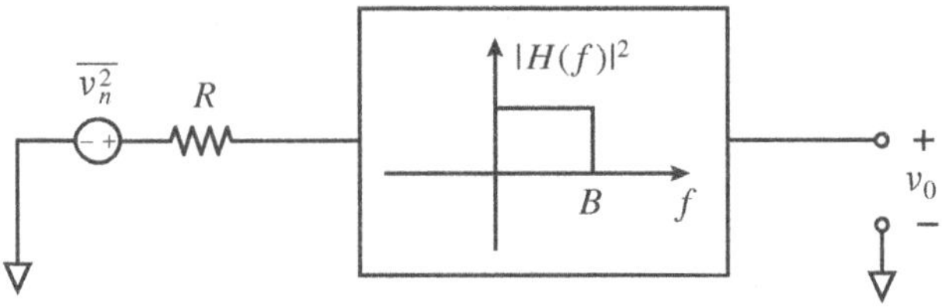

Figure 4.13 Augmenting our noisy resistor model with an observation bandwidth B.

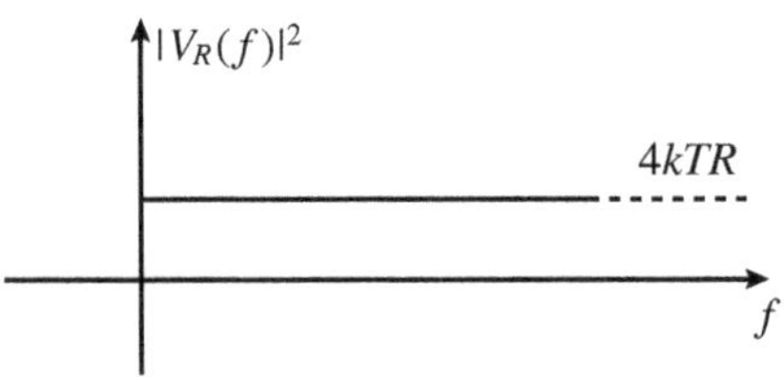

Figure 4.14 The power spectral density for our noisy resistor.

Let's check in once again on the "spectrally flat" piece. It will not surprise you by now to know that the no-holds-barred version of Eq. (4.41) is actually

$$\overline{v_0^2} = \frac{4Rhf}{e^{hf/kT} - 1} \cdot \Delta f. \tag{4.42}$$

But at frequencies normally of interest in wireless communications, Eq. 4.41 is a righteous, rigorous approximation. In our heads, therefore, the PSD is as depicted in Figure 4.14. In closing out this section, it is worth pointing the reader to two original papers on thermal noise, one by Johnson[12] and one by Nyquist,[13] both written in 1928 and indeed practically companion papers in the same issue of the *Physical Review*. What stands out about this writing is the authorial mindset of Nyquist and Johnson, which is that they were elucidating the key concepts of thermal noise as new and unfamiliar. This stands in contrast to writing on the topic today, nearly a century later, when authors lean on a general familiarity with the concepts and take a certain logical framework for granted. This is noted neither to praise Nyquist and Johnson nor to criticize contemporary authors, as it is a most natural evolution. But it can be supremely frustrating for a novice approaching thermal noise for the first time, coming to grips with these concepts here in the early 21st century. *This* author has done his very best to spell things out in a way

---

12 J. B. Johnson, "Thermal agitation of electricity in conductors," Physical Review, vol. 32, pp. 97–109, 1928.
13 H. Nyquist, "Thermal agitation of electric charge in conductors," *Physical Review*, vol. 32, pp. 110–113, 1928.

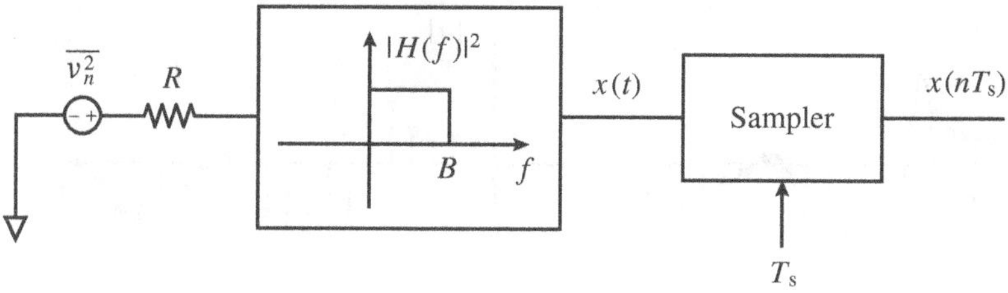

Figure 4.15 Sampling the output from our noisy resistor.

that is legible to a true novice. But if you find that he falls short, take a look at these two pioneering papers.

4.3.2 In the Time Domain

We have spoken of "how quickly a noisy signal moves around," and now is a good time to be concrete about what that means. For a conceptual framework, we'll augment Figure 4.13 from the last section to now include a sampler as well as a filter (see Figure 4.15). (Now is also a good time to remind ourselves that we need not separate in our minds "continuous time" from "discrete time": for sufficiently fast sampling (small T_s), we can make an arbitrarily good approximation to CT. For this particular development, it is easiest and clearest to use the formalism of discrete time for the math.)

Notice that this example is very similar to the setup in Section 4.2.1, only a little more abstract and idealized. In Section 4.2.1, it was the filter that limited the bandwidth of the noise process. (The filtering function is anything but "brickwall," of course, but we'll get that in Section 4.4.) In its place, we have a flat filter with bandwidth B and add a sampler. Our samples are the signal $x[n] = x(nT_s)$, and about this signal we will make two observations and ask one important question. The observations are: (1) $x[n]$ will be zero mean; (2) it's mean squared value will be $\overline{x^2} = 4kTRB$. The question is, how far apart in time do the samples need to be in order to be uncorrelated from each other?

In pictures, what we are asking is if we are taking our samples of the random process as in the two scenarios depicted in Figure 4.16. Really slow down and absorb the contrast. With the lower axes, if we were just looking at the discrete-time samples, we would say qualitatively that knowing one sample tells us relatively little about what will happen in the next sample. In the top axes, however, while we might concede that we are dealing with a probabilistic signal, we would also say that we would expect any given $x[n+1]$ to be "reasonably near" its immediately preceding $x[n]$. We would therefore say that the lower axes illustrate adjacent samples being uncorrelated, and therefore *the autocorrelation function being a narrow*

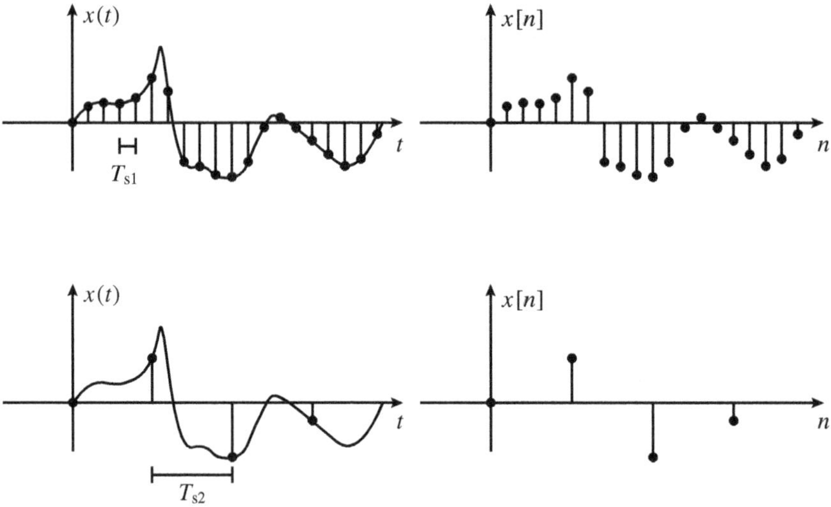

Figure 4.16 Sampling a noisy process quickly (upper traces) vs. slowly (lower axes).

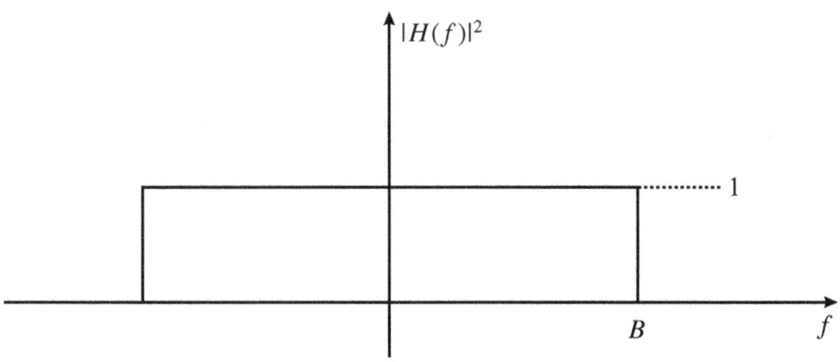

Figure 4.17 Ideal low-pass filter.

peak. The *shorter* the T_s that results in samples like in the lower axes, the *faster* the random process evolves in time.

It turns out that Section 4.2.4 gives us all the tools we need to make analytical sense of the timescale of evolution in random processes. We know that the PSD and the autocorrelation function are Fourier transform pairs. We can start therefore with the PSD of $x(t)$, utilizing negative frequencies, in Figure 4.17. We then use the inverse Fourier transform to get the autocorrelation function $R_{xx}(\tau)$

$$R_{xx}(\tau) = \frac{\sin 2\pi B\tau}{\pi\tau}, \tag{4.43}$$

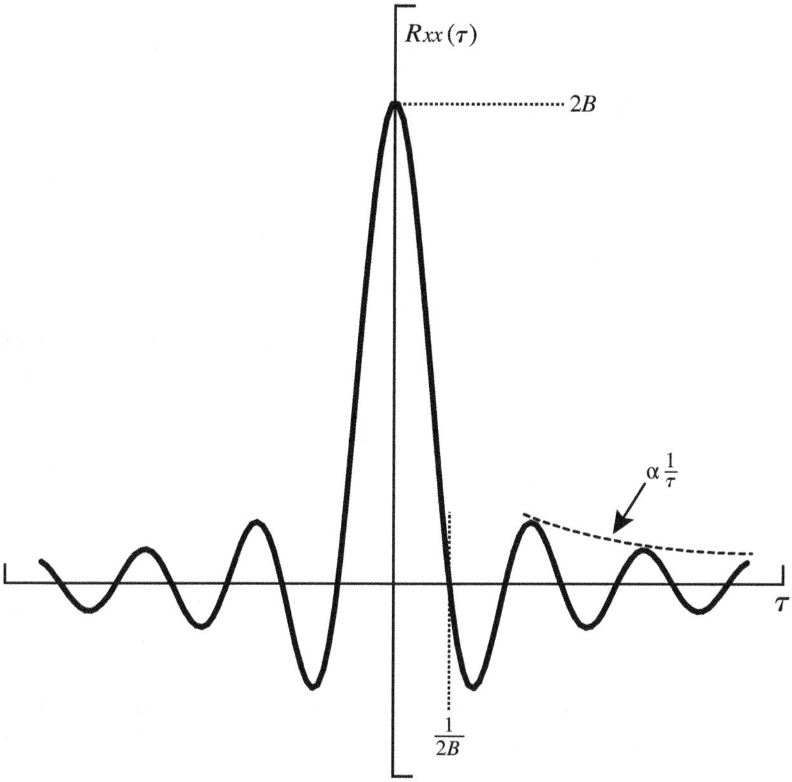

Figure 4.18 The sinc function, which is the inverse Fourier transform of the ideal low-pass filter of Figure 4.17.

which is shown graphically in Figure 4.18. Very roughly speaking, based on the first zero crossings in the autocorrelation function, we might say that two adjacent samples cease to be meaningfully correlated when they are separated by a time difference of $1/2B$ or more. The higher the bandwidth B, the shorter time T_s we need to wait for one sample to be uncorrelated with the next (or previous) sample.

4.4 What Happens When Noise Gets Filtered

Consider now a setup where we have a thermal noise source, such as a resistor, and then filter the output as in Figure 4.19. If we denote the PSD of the resistor as $S_{xx}(f)$, then it turns out the PSD at the filter output is given by

$$S_{yy}(f) = |H)(f)|^2 S_{xx}(f). \tag{4.44}$$

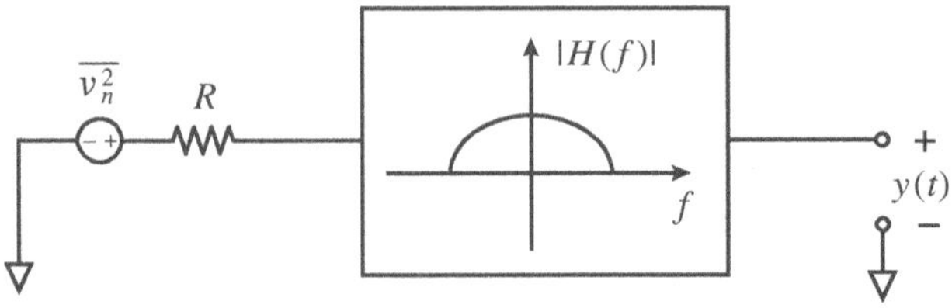

Figure 4.19 A noisy resistor with a filtering function that is not flat.

You may be familiar with the concept of using filtering to "reduce the amount of noise." Let's now take a look at exactly what we mean. For Figure 4.19, suppose that we take a series of measurements of y at intervals far enough apart in time that the measurements are uncorrelated. On this series of measurements, we expect a mean of zero and a mean squared value of

$$\overline{y^2} = \int |H)(f)|^2 S_{xx}(f)df. \tag{4.45}$$

The important thing to note is the effect of the filter function $H(f)$ on the evaluation of the integral in Eq. 4.45. Specifically, a low-pass filter, such as the one shown, will lower the overall result of the integration, and therefore reduce the statistical variance $\overline{y^2}$ of our measurements. And reducing that variance *is exactly what we mean when we talk of "reducing the noise."* If the variance were zero, then every one of our measurements would be zero!

Now, the spectral flatness of thermal noise for most frequencies of interest allows for a useful simplification when dealing with filtering functions. In actual calculations, we love simple, brick-wall filters (filters that are absolutely flat in the passband and zero otherwise) because they make the integral in Eq. 4.45 easy to do. In analysis, therefore, we typically replace our real filters with equivalent brick-wall filters. An equivalent brick-wall filter (an equivalence made possible, remember, by the fact that thermal noise is spectrally flat) has a height that is equal to the peak value of the filter we are replacing. Its width is then chosen such that the brick-wall filter passes the same amount of noise power. A picture is so much better than words, here. See Figure 4.20. To make the substitution work, we define a noise bandwidth B_N according to

$$B_N = \frac{1}{|H_{\text{peak}}|^2} \int_0^{\infty} |H(f)|^2 df. \tag{4.46}$$

Worth remembering.

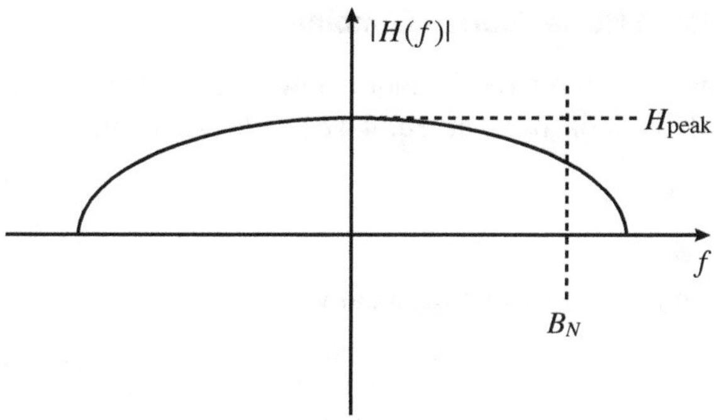

Figure 4.20 Substituting the real filtering function with one that is flat in the pass band.

4.5 What Happens When Noise Gets Averaged

Another concept that gets tossed around in the laboratory is that if you have a noisy measurement, you can "reduce the noise" by taking a bunch of measurements and averaging them. Built into this concept is the assumption, once again, that you are taking your measurements far enough apart in time that they are uncorrelated. Consider the random variable x_{avg}, which is an average of N measurements of the (zero-mean) random variable x:

$$x_{\text{avg}} = \frac{x_1 + x_2 + x_3 + \cdots + x_N}{N}. \tag{4.47}$$

We have an interest in the average squared value[14] $\langle x^2_{\text{avg}} \rangle$. We have

$$\langle x^2_{\text{avg}} \rangle = \langle \frac{x_1 + x_2 + x_3 + \cdots + x_N}{N} \cdot \frac{x_1 + x_2 + x_3 + \cdots + x_N}{N} \rangle. \tag{4.48}$$

Now, taking measurements far enough apart that they are uncorrelated means that the *average* of the cross terms $x_i x_j$, for which $i \neq j$, is zero, which also means that the *sum* of these terms is also zero (if N is large enough). Moreover, we already know $\langle x_i^2 \rangle$: it's just the variance σ_x^2. So we have

$$\langle x^2_{\text{avg}} \rangle = \frac{N \cdot \sigma_x^2}{N^2} = \frac{\sigma_x^2}{N}. \tag{4.49}$$

In words: The random variable x_{avg}, which itself is the average of N random variables x_i, has a variance that is N times smaller than the variance σ_x^2. This is the sense in which averaging reduces the noise.

14 Brackets here mean "average."

4.5.1 Notice How Uncorrelated Noise Sources Combine

Hidden in the foregoing is an important result about how uncorrelated noise sources combine additively. If we look back at Eq. 4.47 but leave off the $1/N$, we have

$$x_{\text{sum}} = x_1 + x_2 + x_3 + \cdots + x_N. \tag{4.50}$$

If we follow similar algebraic steps from this point, what we find is that

$$\langle x_{\text{sum}}^2 \rangle = N \cdot \sigma_x^2. \tag{4.51}$$

Say, for example, you are trying to combine the effect of a number of noise sources at a particular output node, and you know that the noise sources are uncorrelated. The right way to apply superposition is to calculate the noise variance at the output due to each of the noise sources individually, then *add those variances together*. The rms noise voltage will be the square root of that sum.

4.6 Harmonizing Averaging and Filtering

We have now discussed two ways of reducing the impact of thermal noise: we can filter, or we can average uncorrelated samples. It is worth explicitly relating these two techniques to each other within the same analytical framework. Consider the system in Figure 4.21. Now suppose we define the filtering function g[n] according to

$$y[n] = \frac{1}{N} \sum_{k=0}^{N-1} x[n-k]. \tag{4.52}$$

That is to say, the current output $y[n]$ is the numerical average of the current sample $y[n]$ and the previous $N-1$ samples, a so-called "moving average" filter. If we are to examine the mean squared value of the $y[n]$'s, we find that we must separate our analyses into two cases based on the relationship between the sampling period T_s and the signal bandwidth B:

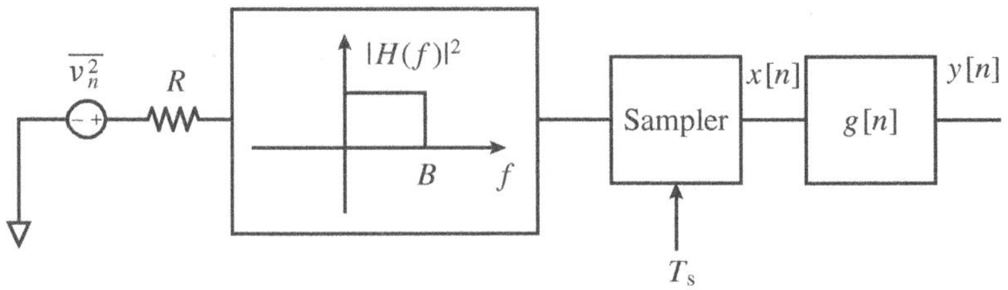

Figure 4.21 Placing a discrete-time averaging filter after the sampler.

1. Sampling slower than the Nyquist rate, so $f_s < 2B$ or $T_s > \frac{1}{2B}$.
2. Sampling at or above the Nyquist rate, so $f_s > 2B$ or $T_s < \frac{1}{2B}$.

For the slower sampling case, everything is messed up by aliasing, and our moving average filter has no practical meaning with regard to frequency selectivity. The right thing to do is to use the autocorrelation function as a guide to figure out how far apart in time you need your samples to be such that they are uncorrelated *enough*.[15] From there, you can proceed in the mean squared calculation according to Section 4.5. The moving average "filter" will even do the right thing.

For the faster sampling case, you are back in the domain of treating the moving average filter as, in fact, a frequency-selective filter. To follow the steps of Section 4.5, though, you must first use Eq. 4.52 to determine the *impulse response* of this averaging filter. Then the Fourier transform of that impulse response gives you the filter function. This filter function finally allows you to come up with a brick-wall equivalent.[16]

4.7 The All-important Signal-to-noise Ratio

This is a big one. The ratio of the signal power to the noise power is absolutely central to the study of communications links of all types. It will not surprise you to learn that when the signal-to-noise ratio (SNR) is *high*, you can transmit and receive large quantities of information in a short amount of time.[17] What catches many people off guard is the fact that even when the noise power is greater than the signal power, all is not lost, the communications link is not ipso facto broken. Under this unhappy circumstance, it may take you a long time to send your message. But if you are patient it is possible.[18]

The mechanics of SNR are straightforward. In practice, we are almost always going to find that a particular node voltage in a circuit will be a superposition of

---

15 And don't get too excited about the zero crossings in the autocorrelation function, by the way. Remember they are only in those exact locations for a perfect brick-wall filter. As a practical matter in the laboratory, they are not useful. If your brick-wall filter is an approximation or equivalent to give you the observation bandwidth, the good news is you can still use Eq. 4.43 as a guide. Just focus on the $1/\tau$ "envelope" when you are figuring out how long to wait between samples. Every doubling of interval time cuts the autocorrelation coefficient in half.

16 For a fantastic, friendly reference on the nuts and bolts of linear system theory, see Z.Z. Karu. *Signals and Systems Made Ridiculously Simple*. Huntsville, AL: ZiZi Press, 1995.

17 Although it may surprise you to learn that achievable data rates grow very slowly with increasing SNR once you get above 0 dB (signal power equal to the noise power). This is an extremely important, and not necessarily intuitive, result to build into your understanding (see Section 5.1).

18 Again, Section 5.1.

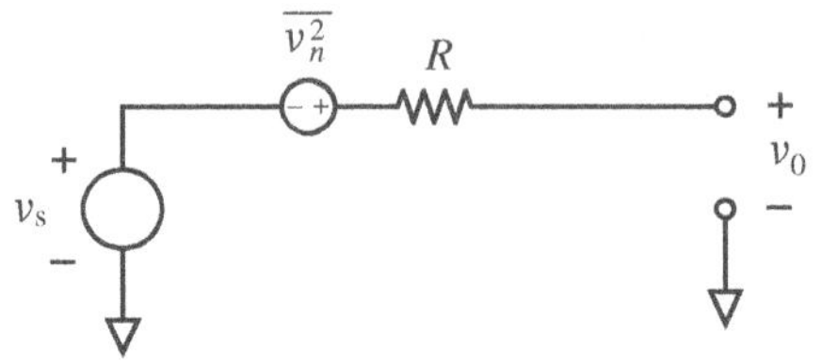

Figure 4.22 The output voltage here is the superposition of signal v_s and noise voltage v_n.

a signal voltage v_s and a noise voltage v_n, illustrated in Figure 4.22. The SNR is simply the ratio of the signal energy (power) to the noise energy (power). That is,

$$\mathrm{SNR_{lin}} = \overline{v_s^2}/\overline{v_n^2}. \tag{4.53}$$

(It may be helpful to refer back to Section 2.4.1.)

In practice, what we find is that the *range* of SNRs we encounter is quite large. At the output of an audio-quality data converter, for example, it is not crazy to encounter SNRs of 10^9 or more. On the low end, a wireless link experiencing channel fading might suffer an SNR that is less than unity, maybe 10^{-1}, 10^{-2}, or worse. The attentive reader will notice that we've already given away the game by speaking in powers of 10. It is quite natural, given the wide range, to characterize SNRs *logarithmically*. We formalize according to

$$\mathrm{SNR_{log}} = \log_{10}\left(\overline{v_s^2}/\overline{v_n^2}\right). \tag{4.54}$$

But wait! We are not done. It is nearly universal in the discussion of SNR to express it in decibels like so:

$$\mathrm{SNR_{dB}} = 10 \cdot \log_{10}\left(\overline{v_s^2}/\overline{v_n^2}\right) = 20 \cdot \log_{10}\left(\frac{\sqrt{\overline{v_s^2}}}{\sqrt{\overline{v_n^2}}}\right). \tag{4.55}$$

You may be asking, "What is that last step?! Why?" Well, when you define things this way, an SNR of 10 dB is 10× higher than an SNR of 0 dB, and an SNR of 20 dB is 10× higher still. That is to say, a change in the SNR of 10 dB corresponds to a change by a multiplicative factor of 10 in the underlying ratio. Very convenient. If this annoys you, sit with it for as long as you need. You will come around.

That said, we pause here for a cautionary note. While logarithmic units like decibels are convenient, it is sometimes difficult to keep track of what it all means when we are talking about power consumption. Suppose we are designing a communications link and setting our transmit power. For a given noise floor, we debate between requiring an SNR of 40 dB versus an SNR of 43 dB. What is important to keep track of is that we are deciding between a certain transmit power (for 40 dB SNR) and *double* that transmit power (43 dB)! It takes an extremely disciplined mind to look at those choices and not see the latter as a mere 7.5% increase.

Many readers will chuckle at that last caution. And you should: it *is* kind of funny. But it is also deadly serious. Hazy, casual bookkeeping with logarithmic units is a major cause of energy waste in communications systems, simply because logarithmic units soften energy costs that would otherwise set off alarm bells. Watch out for this.

4.8 The Effective Number of Bits, or ENOB

As discussed in Section 2.2.4, it is a fact of modern computational engines that we deal almost exclusively in discretized amplitude levels for signals. A reasonable model to keep in your head is that of an ideal quantizer, which takes as an input a "true" signal $v(t)$, and quantizes the signal perfectly on a linear scale of a certain number of levels. We represent this graphically in Figure 4.23. The top of Figure 4.23 shows one way of mapping a continuously variable signal to a discrete, in this case *binary*, encoding. The bottom graph shows the *quantization error* as a function of the input voltage amplitude.

Now from an analytical standpoint, this nonlinear mapping from input voltage to output code presents significant problems. We don't have a nice, neat transfer function that allows us to use the Fourier transform methods in Chapter 2, for example. What to do?

The answer is that we pull out a tried and true trick, useful in all manner of analytical emergencies. We close our eyes, and substitute for the nonlinear system a linear system that seems close enough! Then we go about our lives, checking now and again that the errors introduced by our fraudulent substitution do no practical harm.

This is absolutely serious, and amazing. Since there are no perfectly linear systems in Nature, dear reader, your *entire history applying the theory of LTI systems is a testament to the power of this technique.*

How do we do this force fitting in the case at hand? The first thing is to assert that the obviously deterministic distortion introduced by the quantizer is, instead, random. We'll call it "quantization noise." Note that this assumption falls apart completely in the case of a DC signal, for which the quantization error manifests as a static offset. It falls apart again for periodic signals, for which the same errors will occur at the same point over and over again in the repeating signal waveform, leading to tones or "spurs" in the frequency domain.

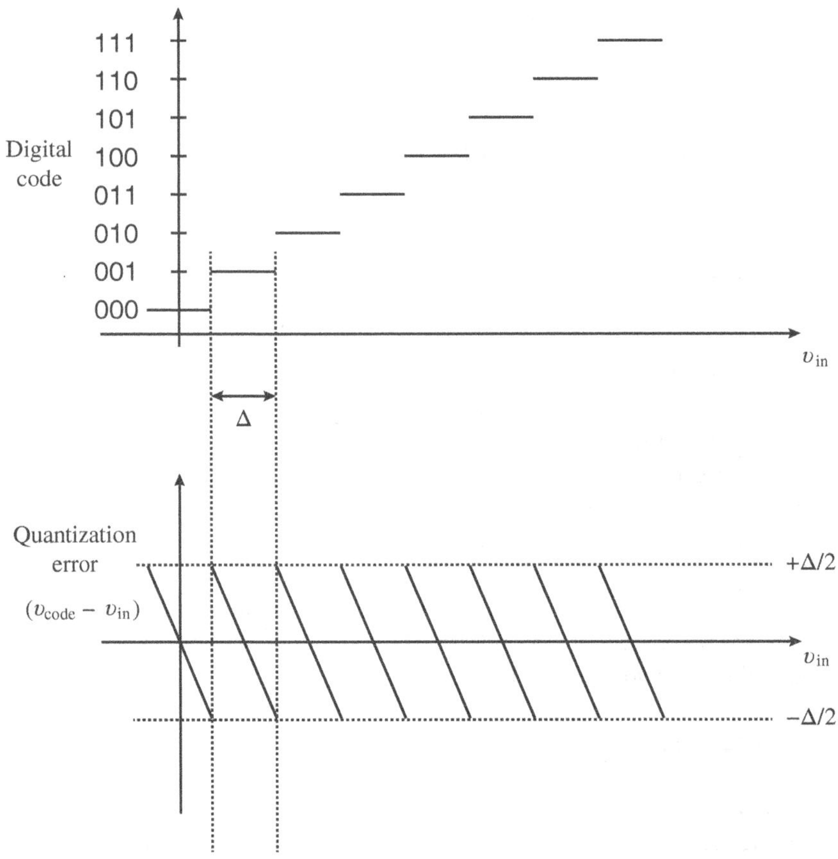

Figure 4.23 Mapping a continuously variable input (v_{in}) to a quantized, binary code.

Nevertheless, for sufficiently busy signals such as those that carry data traffic in a communications system, we do not do too badly if we:

1. Model the quantization effect as noise additive to the output of a perfectly linear gain block, as illustrated in Figure 4.24.
2. Assume that at any given time instant t, the contribution to the quantized output of e_q is a uniformly distributed random variable between $-\Delta/2$ and $+\Delta/2$ (see Figure 4.25).
3. Assume that in the frequency domain, the PSD of the quantization noise is flat, similar to thermal noise. If this is a sampled data system, as it usually is when we quantize, and we quantize a new sample every T_s seconds, then the single-sided PSD is as shown in Figure 4.26.

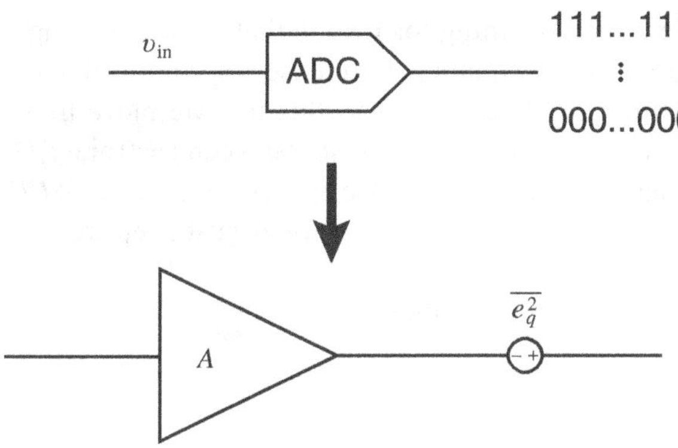

Figure 4.24 We model a quantizer as a perfectly linear gain followed by a thermal-like noise source. The gain A determines how big a voltage change is required at the input in order to change the output code by one bit. (One of the most common quantizers in electronics is an analog-to-digital converter, which takes an analog voltage as an input and provides digital codes as an output.)

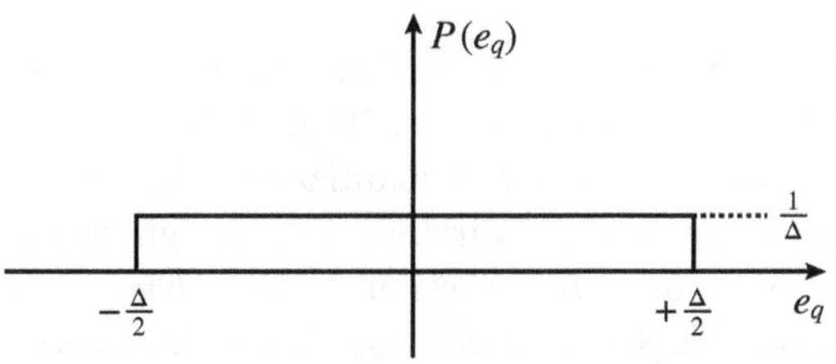

Figure 4.25 The probability density function for the additive quantization error.

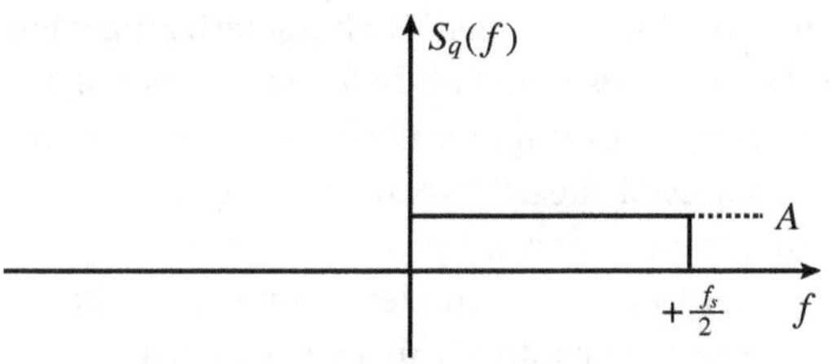

Figure 4.26 How we represent quantization noise in the frequency domain.

For that last item, we can use Parseval's theorem to determine A:

$$\int_0^{f_s/2} S_q(f)\,df = \overline{e^2}$$

$$\int_0^{f_s/2} A\,df = \overline{e^2}$$

$$A = \frac{\overline{e^2}}{f_s/2}. \tag{4.56}$$

We'll compute $\overline{e^2}$ shortly. The important thing for now is that we are making a simplifying assumption, which is that the quantization noise is spectrally flat just like thermal noise. Another thing to notice is that in this text we move freely between the single-sided PSD, where the noise energy lies between $f = 0$ and $f_s/2$, and the double-sided PSD, where we say that the noise energy lies between $+f_s/2$ and $-f_s/2$. When you actually need a numerical value for A, just keep track of which convention you are following.

Now, let's compute that important noise variance, $\overline{e^2}$.

$$\begin{aligned}\overline{e^2} &= \int_{-\Delta/2}^{+\Delta/2} \frac{1}{\Delta} \cdot e^2 de \\ &= \frac{1}{3\Delta}\left[e^3\right]_{-\Delta/2}^{+\Delta/2} \\ &= \frac{1}{3\Delta}\left[\frac{\Delta^3}{4}\right] \\ &= \frac{\Delta^2}{12}. \end{aligned} \tag{4.57}$$

We are going to use this result for quantization noise variance to introduce an important and useful concept called the "effective number of bits," or ENOB (pronounced EE-nob). The idea is that the finer the degree of quantization, the smaller the Δ and therefore the lower the quantization noise variance. The thought experiment is then that when we look at noisy data, *regardless of the underlying noise mechanism*, and determine or measure its SNR, we can use that SNR to determine how many bits an ideal quantizer would need to represent our data with the same SNR.

There are at least two motivations for doing an ENOB characterization. Far and away, the most common is for keeping designers of data converters honest. Sometimes, an analog-to-digital converter, for example, will give as its output, say, 12 digital bits. But if you grab a long series of measurements from the output of the data converter and do a spectral analysis, normally you see a "noise" floor attributable entirely to the data converter that is nevertheless higher than that of an ideal 12-bit converter. In the laboratory we capture this by saying something like, "the data converter has an ENOB of 11.1." In other words, it gives you 12 bits of data, but an SNR consistent with 11.1 bits of quantization.[19]

Another way that ENOB can be useful is in considering how wide your digital data path needs to be for representing signals in your particular system. If your channel has an SNR of 20 dB, for example, we will find that it is probably not necessary to have 20-bit analog-to-digital converters on the receiver front end.

---

19 Don't get thrown by the noninteger number of bits. The idea of ENOB is to give a continuous scale for rating the degree of imperfection of our data converter.

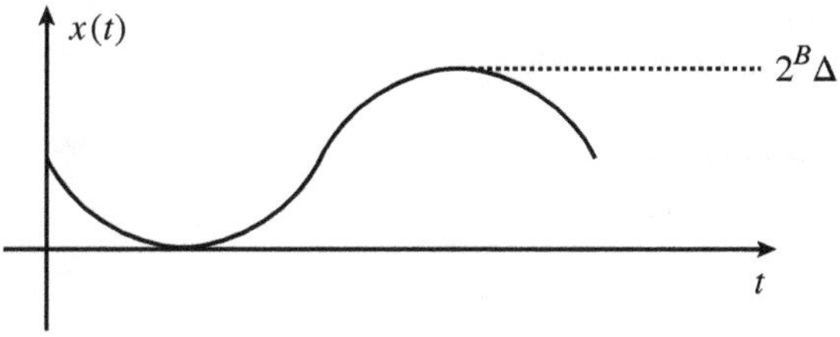

Figure 4.27 A full-scale sinusoid input to a quantizer.

So let's figure out mathematically what ENOB is. It is tradition to determine ENOB by exciting our quantizer with a full-scale sinusoid as shown in Figure 4.27. The amplitude of the sinusoid is $\frac{1}{2} \cdot 2^B\Delta$. The mean squared value of a sinusoid of amplitude A is $\frac{1}{2}A^2$, so in this experiment, the signal power turns out to be

$$P_{\text{sig}} = \frac{1}{2}\left(\frac{1}{2} \cdot 2^B\Delta\right)^2. \tag{4.58}$$

We have already worked out the noise power in Eq. 4.57, so we can use that result to express the SNR:

$$\text{SNR} = \frac{\frac{1}{2}\left(\frac{1}{2} \cdot 2^B\Delta\right)^2}{\frac{\Delta^2}{12}} = 1.5 \times 2^{2B}. \tag{4.59}$$

Taking the common log and multiplying by 10, we put the SNR in dB according to:

$$10 \cdot \log_{10}(\text{SNR}) = 10 \cdot \log_{10}(1.5) + 10 \cdot 2B \log_{10}(2)$$
$$10 \cdot \log_{10}(\text{SNR}) = 1.76 + 6.02 \cdot B. \tag{4.60}$$

This is the famous ENOB formula. The procedure in the field is that we feed as an input to our system a full-scale sinusoid, take the Fourier transform or DFT of the output, and measure the SNR. We then assign the system an ENOB based on Eq. 4.60.

In this text we rely heavily on equations, but ENOB is one concept where a table conveys a better feel (see Table 4.1). The author has found it useful to memorize one row (say, 62 dB corresponds to 10 bits), and then all other rows are accessible by adding/subtracting roughly 6 dB to/from the SNR for every increment/decrement in the number of bits. For example, if 10 bits correspond to 62 dB, then 9 bits must correspond to 56 dB, 11 bits must correspond to 68 dB, etc. Note that this trick even works for noninteger ENOBs, as going up 0.5 bits in ENOB corresponds to going up $0.5 \cdot 6 = 3$ dB in SNR.

Table 4.1 Representative ENOB values. An example with a noninteger ENOB is given in the second row.

SNR	ENOB
50 dB	8 bits
53 dB	**8.5 bits**
56 dB	9 bits
62 dB	10 bits
68 dB	11 bits
74 dB	12 bits
80 dB	13 bits
86 dB	14 bits
92 dB	15 bits
98 dB	16 bits
104 dB	17 bits
110 dB	18 bits
116 dB	19 bits
122 dB	20 bits

The reader is encouraged to use this section to think deeply about the bit as an information-theoretic concept. It has shown up once already (see Section 2.2.4) and it is foundational to the understanding of the Shannon limit in Chapter 5.

4.9 The *RC* Circuit as a Noisy System

Now let's go back. We started this chapter off with an exploration of the equipartition theorem for circuits (see Section 4.1.1). We found that in thermal equilibrium, we can write down the mean squared noise voltage across a capacitor or the mean squared noise current through an inductor with astounding ease. We then reflected that *just* knowing the mean squared voltage or current leaves you wanting, in that you have no sense of how rapidly the noise voltage/current is fluctuating in time. The rapidity of the fluctuation problem required an understanding first of autocorrelation (Section 4.2.3). We discovered that autocorrelation and PSD are Fourier transform pairs (Section 4.2.4), and tied everything together with an examination of what happens when noise gets filtered (Section 4.4).

We'll now come full circle and revisit our stand-in for much of electronics, the *RC* circuit, as a real-world example of a noisy system. Recall that we model the noise of a resistor as a series noise voltage source (see Figure 4.12). At the output terminal, the mean squared noise voltage is given by

$$\overline{v_n^2} = 4kTR\Delta f. \tag{4.61}$$

That is to say, for every 1 Hz of observation bandwidth, we add $4kTR$ of statistical variance to the voltage measurements at the output node. It goes almost without saying that the standard deviation of our measurements is

$$\sqrt{\overline{v_n^2}} = \sqrt{4kTR\Delta f}. \tag{4.62}$$

Furthermore, notice that with thermal noise, we are talking about absolute bandwidth and not fractional bandwidth. That is, there is as much noise power between 1 Hz and 2 Hz as there is between 1 and 1.000001 MHz.

We'll now make use of two prior results for our analysis. The first is that if you pass a noise signal with PSD $S_{xx}(f)$ through a linear filter with transfer function $H(f)$, the PSD at the output is given by $S_{yy}(f) = |H)(f)|^2 S_{xx}(f)$ (Eq. 4.44). The second is that we will use a brick-wall filter substitution for the actual filter function of the *RC* circuit, using the procedure described in Section 4.4.

Our *RC* circuit modified to include a thermal noise source is shown in Figure 4.28. Based on Section 3.1.2, we may write down $|H(f)|^2$ as

$$|H(f)|^2 = \frac{1}{(2\pi RCf)^2 + 1}. \tag{4.63}$$

We plot $|H(f)|^2$ in Figure 4.29. If we now apply the brick-wall filter concept to this filter function, and dust off some old definite integral skills, what we find is that

$$\Delta f = \frac{1}{4RC}. \tag{4.64}$$

The noise variance for the capacitor voltage is therefore none other than

$$\overline{v_0^2} = 4kTR \cdot \Delta f = 4kTR \cdot \frac{1}{4RC} = \frac{kT}{C}. \tag{4.65}$$

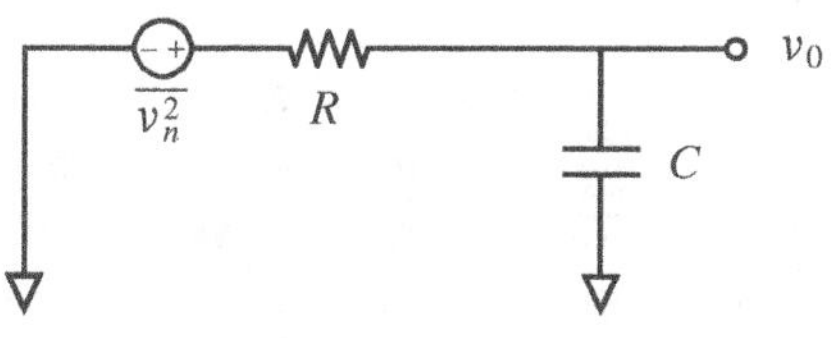

Figure 4.28 An *RC* circuit with a noisy resistor.

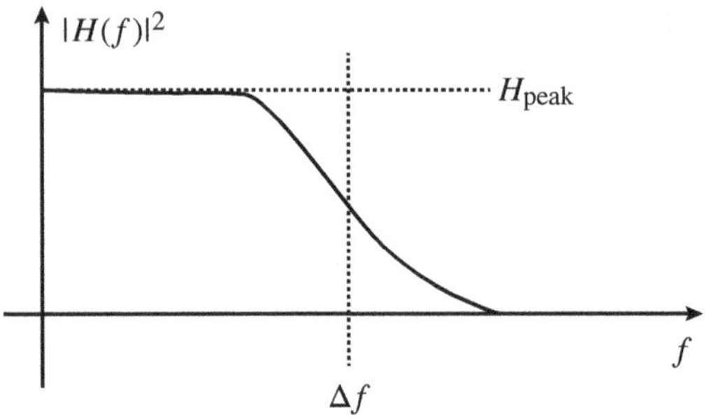

Figure 4.29 Frequency response of an *RC* filter, this time on linear instead of logarithmic axes.

Amazingly, this agrees with the requirements of the equipartition theorem of Section 4.1.1.

Now, we are savvy enough to know that just knowing the noise variance does not tell us the time scale of the noise fluctuations, and also that the autocorrelation function is connected to the PSD through the Fourier transform. The PSD of the capacitor noise voltage is given by

$$S_{yy}(f) = \frac{4kTR}{(2\pi RCf)^2 + 1}. \tag{4.66}$$

For the autocorrelation function, we *could* lower our heads and work out the inverse Fourier transform of S_{yy}. However, a simpler way to get the information we need is to use the Uncertainty Principle as written in Eq. 2.157, which is $\Delta T \cdot \Delta f \geq 1/\pi$. What we seek here is the time scale over which successive measurements of the output voltage will be uncorrelated. With $\Delta f = 1/4RC$, the Uncertainty Principle tells us that $\Delta T \geq 4RC/\pi$. You will not do badly for filter functions involved here if you replace the inequality like so:

$$\Delta T \approx \frac{4RC}{\pi}. \tag{4.67}$$

4.10 The Thermal Noise Floor in Radio Receivers

Let's close out this chapter with a discussion of how you are likely to see thermal noise in practical RF work. The typical setup to have in your mind is the very front end of the receiver, as shown in Figure 4.30. For reasons that we will get to in Chapter 8, the receive antenna "looks" like a resistor and a noisy voltage source, and is shown in Figure 4.31. It is very common for the antenna radiation resistance R_A to be 50 Ω, although other choices are sometimes made. For the noise

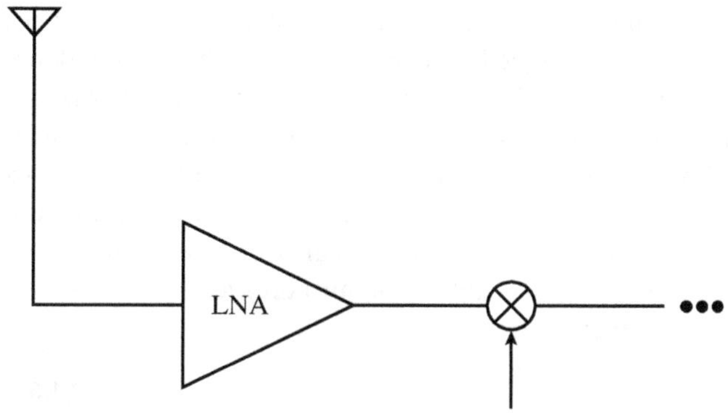

Figure 4.30 A greatly simplified receiver front end.

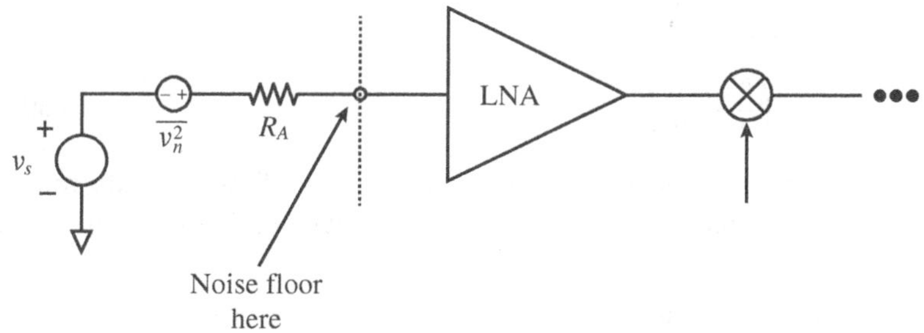

Figure 4.31 A greatly simplified receiver front end, replacing the antenna with its equivalent signal voltage, source resistance, and thermal noise.

voltage $\overline{v_n^2} = 4kTR\Delta f$, the relevant temperature T is that of the objects within the transmit/receive view of the antenna.[20]

Now from here on out, *everything* is about tradition and convenience, which tends to be useful more often than not because the traditions started in a good place. You will hear system designers and link budget mavens talk about a "noise floor," which is the pristine noise floor in the signal chain before we corrupt it with all of our circuit blocks. For the purposes of this noise floor, we imagine the noise temperature to be 300 K. We are careful to say "imagine," here, because you'll

20 It is helpful to reflect on what we mean by thermal equilibrium in this context. Consider the objects whose radiation reaches the receive antenna, and then consider sources of thermal noise within the receiver that couple and transmit via the antenna. Thermal equilibrium means that there is an even, exactly balanced exchange of energy between the thermal radiators in the environment and the thermal noise sources within the receiver. If we stipulate that we are in thermal equilibrium, what we mean is that the relevant objects in the environment and the thermal noise sources within the receiver are at the same temperature.

recall from Section 4.1.3 that the noise coming from any actual environment is not spectrally flat. But most receivers designed for communications have noise floors dominated by their own circuit noise, which is spectrally flat, and moreover would correspond to antenna noise temperatures in the thousands of degrees Kelvin. The usefulness of a baseline characterized by a noise temperature of 300 K derives mainly from its near universality. Everyone speaks the same language, and we are able to use this baseline to quickly compare the performance of different receivers.

Great. The maximum noise power available from our noisy resistor, if we were to match for maximum power transfer, is

$$P = kT \cdot \Delta f. \tag{4.68}$$

In RF work, it is very common to deal with power generally as we dealt with SNR in Section 4.7, which is to use logarithmic units. In this case, we use decibels relative to 1 mW, or dBm. That is

$$P_{\text{dBm}} = 10\ \log_{10}\left(\frac{kT \cdot \Delta f}{1\ \text{mW}}\right). \tag{4.69}$$

We now make some modifications that will eventually allow us to do rapid mental calculations. Instead of expressing Δf in Hertz, we break it up into

Table 4.2 Handy common log numbers.

Bandwidth B_m	**Common log**
1 MHz	6.0
2 MHz	6.3
3 MHz	6.5
5 MHz	6.7
10 MHz	7
20 MHz	7.3
30 MHz	7.5
50 MHz	7.7
100 MHz	8
200 MHz	8.3
300 MHz	8.5
500 MHz	8.7

$\Delta f = 1\ \text{Hz} \cdot B_m$, where B_m is a dimensionless multiplier whose value is the same as the number of Hertz in Δf. Making the substitution, we have:

$$\begin{aligned} P_{\text{dBm}} &= 10\ \log_{10}\left(\frac{kT \cdot 1\ \text{Hz} \cdot B_m}{1\ \text{mW}}\right) \\ &= 10\ \log_{10}\left(\frac{kT \cdot 1\ \text{Hz}}{1\ \text{mW}}\right) + \log_{10}(B_m). \end{aligned} \tag{4.70}$$

For T of 300 K, this equates to

$$P_{\text{dBm}} = -174\ \text{dBm} + 10\ \log_{10}(B_m). \tag{4.71}$$

This is why you will hear practitioners talk about the thermal noise floor as having a "spectral density of −174 dBm per Hertz at room temperature." If you now want to figure out how much baseline noise power is in the bandwidth that you care about, all you need to do is figure out the common logarithm of the bandwidth in question and then do some addition. For example, if your bandwidth is 10 MHz, its common logarithm is 7. So the baseline noise power in the band is

$$P_{\text{dBm}} = -174 + 10 \cdot 7 = -104\ \text{dBm}. \tag{4.72}$$

It helps for mental calculation to just know a few common logarithms. Table 4.2 is an aid to this end.

5

Understanding the Shannon Limit

For a practicing engineer seeking to broaden their understanding of communications, there is perhaps no better place to start than with the Shannon capacity limit. This may be news to engineers who experience communications theory only through harsh, uncompromising performance specifications on their designs. The author was years into his own circuit design training before he even heard of the Shannon limit. Nevertheless, those important specifications that can make circuit design so difficult are often grounded in this esoteric, but simple, information theoretic result.[1]

Completely on the other side of the spectrum, you have communications theorists and system architects who are divorced from physical implementation details. For this crowd, the Shannon limit was practically the first thing they learned. Maximizing capacity in the network is the prime directive, and it drives all sorts of top-level decisions, such as what modulation strategies to support, how to allocate spectral resources and/or time slots to multiple users, etc. What is *usually* missing from discussions among this set is the difficulty of implementing their network architectures in physical hardware. To be more precise: the more ambitious your performance goals in terms of network capacity, the more energy your network will consume. If we are careful, we can augment the theorists' innate understanding of Shannon capacity with an appreciation for the *energy costs that Nature requires*. Mercifully, this can be done without necessarily building expertise in implementation and circuit design.

No matter where you start, you have much to gain by examining the famous Shannon limit with new eyes. Let's get into it.

1 C. E. Shannon, "Communication in the Presence of Noise," *Proc. of the I.R.E*, vol. 37, pp. 10–15, Jan. 1949.

Energy Fundamentals of Radio, First Edition. Joel L. Dawson.

5.1 Putting the Shannon Limit to Work

If you decide to read carefully only *one* section of this book, this is the right one (Section 5.1). Understanding the form of the Shannon limit and its implications will profoundly alter your perspective, whether you know how it is derived or not. If you get curious as to how this remarkable result comes about, Sections 5.3 and 5.4 are for you.

5.1.1 Exploring the Shannon Limit with Numbers

Let's kick things off by playing with numbers. The Shannon limit is simply stated. For a communications link, the absolute highest number of bits we can transfer from transmitter to receiver, with arbitrarily low error, is given by

$$C = B \cdot \log_2 (\mathrm{SNR} + 1), \tag{5.1}$$

where

- C is the "capacity" in bits per second.
- B is the bandwidth of the channel in hertz.
- SNR is the signal-to-noise ratio in linear power units (e.g. watts/watts, not dB).

Now suppose that we have a 20 MHz bandwidth channel, which is kind of middle-of-the-road bandwidth for a wireless link at the time of this writing. It is instructive to fill out a table of channel capacities corresponding to different SNRs as in Table 5.1. Interesting, yes? A couple of things jump out that you should internalize right away:

Table 5.1 Calculating the Shannon limit for a 20 MHz channel.

SNR (W/W)	SNR (dB)	Capacity (bits/sec)
10^{-3}	−30 dB	2.9×10^4
10^{-2}	−20 dB	2.9×10^5
10^{-1}	−10 dB	2.8×10^6
$10^0 (= 1)$	0 dB	2.0×10^7
10^1	10 dB	6.9×10^7
10^2	20 dB	1.3×10^8
10^3	30 dB	2.0×10^8
10^6	60 dB	4.0×10^8

1. Even for negative SNR scenarios, where the noise power is *greater* than the signal power, the Shannon limit says that you can still communicate! Don't allow yourself to just blow by this as a natural consequence of the analytical form. If you didn't know the correct answer, left only to your own intuition, you might conclude that the task of communicating crosses some impossibility threshold when the noise power exceeds the signal power. But that turns out to be far from true.[2] If Table 5.1 is to be believed, then even when the noise power is 1000× as strong as the signal power, we can still communicate at a rate of 29 kilobits per second! Calculate for yourself what SNR would allow a maximum bit rate of one bit per second for a 20 MHz channel.[3]
2. Pause to take in the awesome compressive nature of the logarithm in Eq. 5.1. If you didn't already know the answer, you would probably guess that a 10× increase in transmit power, corresponding to a 10× increase in the SNR, ought to make the link "ten times as good." Which is to say, we can successfully convey 10× the number of bits in a given time interval. But examination of Table 5.1 reveals that this is only true when we are in the region where the noise power greatly exceeds the signal power. Take a look at the step between SNR = −30 dB and SNR = −20 dB. Here we can see that the link capacity goes up almost exactly by the expected order of magnitude. But now look at the bottom two rows in the table. For a given noise floor, a jump in SNR from 30 to 60 dB means we have increased our signal power (and therefore our system power consumption) by 1000×. But you can see that the capacity has barely budged, moving only a factor of two. This, friends, is one of the worst bargains in Nature. Call it what you will, paying exponentially for linear gain, or paying linearly for logarithmic gain, it is just awful. You can see that the trade-off transitions from reasonable to terrible roughly in the SNR = 0 dB to SNR = 10 dB range, where for a 10× increase in power consumption we get a 3.5× increase in capacity.

Very interesting, and for many of you, quite surprising. While you internalize this numerical behavior, another thing to understand about the Shannon limit is that it is not tied to implementation methods. We are focused in this book on electronic radio transceivers, of course. But a dramatic illustration of this remarkable limit comes to us from audio playback systems. Consider Figure 5.1. The question: how much information can we expect to get from the recording system to the human brain using the "channel" consisting of the amplifier, the speaker, gases in the atmosphere, and your ears?

2 We'll examine one practical way to handle this in Section 5.2.

3 Ans: for 20 MHz BW, an SNR of 3.47×10^{-8}, or −74.6 dB, does the trick. For those of you keeping track at home, that is a noise power 28.8 *million* times higher than the signal power.

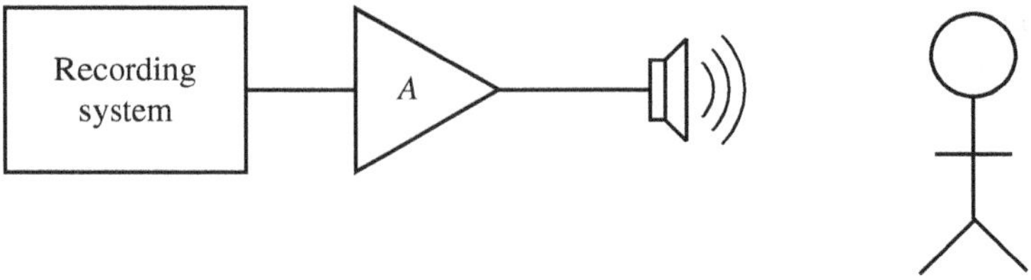

Figure 5.1 Considering an audio playback setup from the standpoint of the Shannon capacity limit.

Yikes. Where to start? Well, since this chapter is on the Shannon limit, we may as well start with $C = B \cdot \log_2(\text{SNR} + 1)$. Then remember the First Rule of Shoehorning Theory Into Practice, which is simply stated: start with the easy part. In a choice between B and SNR, as we have here, that is definitely B. It turns out that the frequency range of human hearing is from about 20 Hz on the low end to about 20 kHz. This gives us a quick victory:

$$B \approx 20\ \text{kHz}. \tag{5.2}$$

Halfway there!

Now for the SNR, let's say we are in an anechoic chamber, the quietest possible environment for human hearing. At its very best, the human ear can process about 120 dB of dynamic range under such circumstances. It turns out that to avoid permanent hearing loss, we need to back off so that the average power is about 90 dB above the threshold of hearing (which we assign to 0 dB). So, for maximum information transfer, we are looking at an SNR of about 90 dB.

Where does this leave us?

$$C = (20\ \text{kHz}) \cdot \log_2 \left(10^9 + 1\right) = 0.6\ \text{MB/ sec}. \tag{5.3}$$

Well! Nice and tidy, but a little uncomfortable without a way to check our modeling assumptions. How do we tell if our model here is reasonable?

A good place to start is with the performance specifications of a real-world system, preferably one where the engineers were highly incentivized to optimize for performance. It turns out that in the early 1980s, engineers at Sony and Philips were leading the way on a new way to store music: the compact disc (CD). What made CDs new was that they were a *digital* storage medium, as opposed to a cassette tape or long play record. Legend has it that the engineers wanted to make their CD such that it could contain all of Beethoven's ninth symphony without having to change discs (depending on the performance, this could be as long as 74 minutes). Because the cost of a CD was tied to its physical size, and because its physical size scaled with its storage capacity, the entire industry was highly motivated to store only the number of bits per second that can make it to the human ear *and no more*. This is not *proof* that they relied on Shannon to figure this out, but if we do the math and the results come out close, it is probably not a coincidence.

So what does the math say? CDs sampled audio at 44.1 ksamples/sec,[4] and the standard bit depth was 16 bits. Looks like the bits per second flowing through the audio channel to the ear was 44.1 ksamples/ sec $\times$ 16 bits/sample = 0.7 MB/ sec! Not an exact match, but close.

There is every reason to believe that the engineers at the time were fully aware of the Shannon limit and Nyquist rate sampling when they drew up the technical specifications. A point worth internalizing, though, is that there are many ways to get to fundamental limits when the pressure to optimize is sufficiently high. Iterative, careful experimentation, for example, will also get you there. Explicit knowledge of fundamental limits just helps you to know where you are going ahead of time.

5.1.2 Using the Shannon Limit

When you want to use the Shannon limit to get analytical insight, it helps to do a careful review of the behavior of the function

$$\log_2(x+1), \tag{5.4}$$

as well as of the behavior of the first derivative, or slope

$$\frac{d}{dx}\log_2(x+1) = \frac{1}{\ln 2} \cdot \frac{1}{x+1} = \frac{1.44}{x+1}. \tag{5.5}$$

Graphically, we have Figure 5.2. Roughly speaking, we may divide the function into two regions. We will call the first Region 1, which is where the SNR is low (below unity; $x = \text{SNR} < 1$, $10 \cdot \log_{10}(\text{SNR}) < 0$ dB). We will call the second Region 2, where the SNR is above 0 dB. Let's look at each of these regions in turn.

Region 1: Signal Power Less Than Noise Power

In this region, the first derivative of the logarithm is pretty constant, at $\frac{d}{dx}\log_2(x+1) \approx \frac{1}{\ln 2} = 1.44$. This means that the channel capacity in this SNR region is well approximated by

$$C \approx 1.44 \cdot B \cdot \text{SNR}. \tag{5.6}$$

In the fairly general case of a thermal noise floor, we may write the SNR as P_{sig}/kTB, so

$$C \approx 1.44 \cdot \frac{P_{\text{sig}}}{kT}. \tag{5.7}$$

4 The range of human hearing goes up to 20 kHz. Looks like both Nyquist and Shannon had influence here!

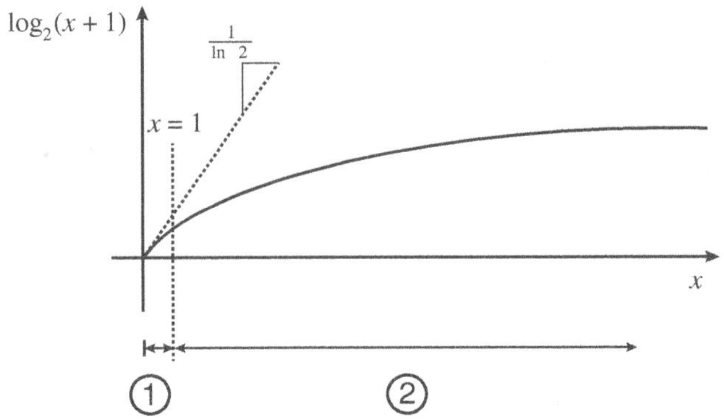

Figure 5.2 Dividing the Shannon capacity into two regions based on SNR.

This means we can express the required signal power in terms of the desired capacity as follows:

$$P_{\text{sig}} \approx kT \cdot \frac{C}{1.44}. \tag{5.8}$$

In this region, then, the math gives at least two inputs to our conceptual understanding:

1. As we increase the signal power, the capacity in bits per second *does* go up on a one-to-one basis. That is, if we want to double the number of bits per second, we must double our signal power. This is probably how you thought communications links worked all the time. But it is not. Only when the signal is buried beneath the noise floor does capacity go 1:1 with increased signal power.
2. Our normal equivalence between bandwidth and capacity does not apply here. The 1:1 equivalence between capacity and bandwidth is probably pretty fixed in your mind, and for good reason (as we shall see when we look at Region 2). But it turns out that a precondition for "bandwidth equals bits/sec" is a healthy SNR. Lacking this, we see from Eq. 5.7 that channel capacity depends only on power and is independent of bandwidth.

Region 2: Signal Power Greater Than Noise Power

In this region, we may approximate the Shannon limit as

$$C \approx B \cdot \log_2\left(\frac{P_{\text{sig}}}{kTB}\right). \tag{5.9}$$

Let's look at how much signal power is required to achieve capacity C_0 given channel bandwidth B. Rearranging the above formula for signal power, we get simply

$$P_{\text{sig}} = kT \cdot B \cdot 2^{C_0/B}. \tag{5.10}$$

So...when the SNR is high, the required signal power goes up exponentially with desired capacity for a fixed bandwidth.

It bears repeating that this is just about the worst energy/performance trade-off in Nature. The only thing worse would be for the capacity to actually be independent of signal power. Awful.

It is also important to pause here and consider the real-world implications for communications infrastructure. At the time of this writing, much of the world is excited about building out fifth generation or "5G" networks. Many promises have been made about the greatly increased capacity of these networks. Perhaps the most famous among these promises was articulated by a San Diego-based chip company as the "1000× Challenge," increasing network capacities by a factor of 1000. Consider:

- Suppose that 1000× the bandwidth was made available. The power to run the networks of the world, already consuming more than 3% of the world's electricity, would grow by 1000×.
- But we know that spectral real estate is finite. Suppose, then, in service of meeting the 1000× Challenge, "only" a 10× increase in bandwidth is possible. This means that C_0/B in Eq. 5.10 goes up by a factor of 100. Play around with that on your calculator of choice.

There is much to say and think about this simple trade-off, which is not widely discussed outside the communications theory community. But everyone in telecommunications feels the effects of Eq. 5.10. 5G networks in particular have been the focus of an almost frantic exploration of "new" techniques, including but not limited to mm-wave communications, extreme beamforming and tracking of mobile users, and super dense symbol constellations to take advantage of high SNRs when transmitting at high powers. The sudden incentive to implement in real-world systems ideas that have kicked around academia for years comes from the reality that if you don't do something different, energy costs skyrocket and eat real-world profits. Whenever you hear about a new technique in wireless communications, a useful habit is to see if you can figure out *if*, and then *how*, it eases the energy penalty in Eq. 5.10. It is a great way to build insight.

5.1.3 Repeat: We *Do* Require Energy to Move Bits Through a Network!

This is a theme we return to, having first contemplated it at the start of Chapter 4. A bit is, after all, an information theoretic concept, and carries no mass. Why should energy be required at all to "move" one? Well, we see from the presence of T in Eq. 5.10 that thermal noise is a major culprit. If the environment could be cooled to 0 K,[5] evidently, our problems would go away! But we can't do that. And so in a world where thermal noise is ubiquitous, and we are stuck at temperatures that support life, it costs energy to transmit bits.

We can actually go further and make a point about energy consumption in electronic machines, which are at the moment the best and fastest machines we have for communications. Information internal to these machines is represented as voltages, and there is always capacitance from each of these information bearing nodes to ground. The lesson of Section 3.1.3 is that moving those voltages results in energy being uselessly dissipated as heat. Nature does not give us a way out. Because of thermal noise, then, and because of the existence of capacitance in the world, we must accept an energy penalty for every bit that we wish to send.

5.2 Detecting Signals Buried in Noise

There is one concept that will bother many readers, which is the idea that it is possible to read a signal that is absolutely buried in noise. If you do not understand the difficulty, consider an oscilloscope trace that captures only bandlimited thermal noise in Figure 5.3. Now consider that same trace, but this time with a sinusoid added whose signal power is 1/1,000,000 the total noise power in Figure 5.3, shown in Figure 5.4. The two figures look pretty similar, yes?[6]

The key insight from this section is that when communicating in the presence of (a lot of) noise, *it is vital to have some idea of what the signal is going to look like*

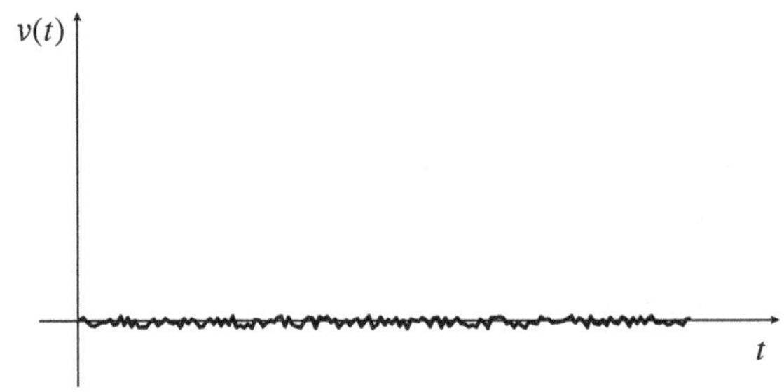

Figure 5.3 Bandlimited noise as seen on an oscilloscope.

5 −273.15° C, or −459.67° F. Water freezes at 0° C, or 32° F. Car batteries famously struggle to start their mostly (at the time of this writing) internal combustion engines when the temperature is a toasty −22° F.

6 We are using as examples extremely low SNRs for clarity (using here an SNR of −60 dB). The truth is that in the laboratory, SNRs of −10, 0, or even +10 dB look distressingly messy. In the present example, to the unaided eye there is no difference between Figures 5.3 and 5.4.

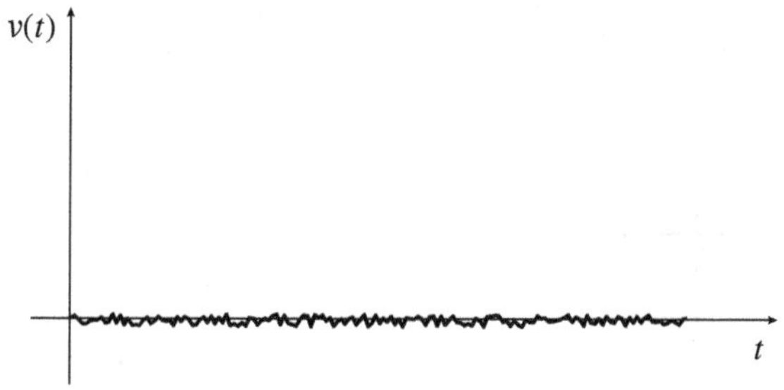

Figure 5.4 Bandlimited noise as seen on an oscilloscope, this time with a sinusoid added whose power is 1/1,000,000 the power in the noise. There is no visible difference between this and Figure 5.3.

ahead of time. Of course, you are not going to know what the overall message is ahead of time: then there would be no need to communicate. But modern digital communications are built on a foundation of knowing that all of our messages are pieced together from a finite library of symbols, and those symbols are known ahead of time and agreed upon by both the transmitting and receiving parties.

Let's get concrete. It would be possible, for example, to agree that there are two "symbols" in our communication system, as shown in Figure 5.5. With a firm agreement in place that we will compose our message using only these two symbols, it turns out we have a powerful signal processing technique for extracting that message from a noisy environment. That technique is our old friend correlation, and we will get into some of the mechanics next.

5.2.1 Using Correlation to Extract Signal from Noise

We begin by revisiting the tricks of using averaging and filtering to determine the DC value of a signal, treated together in Section 4.6. The upshot is that if $x[n]$ is a signal with a DC component, and we desire to know what that DC component is, we can determine it to arbitrary accuracy if we pass $x[n]$ through the averaging filter

$$y[n] = \frac{1}{N}\sum_{k=0}^{N-1} x[n-k]. \tag{5.11}$$

The larger we make N, the better we do.

Now we are going to look at this averaging function slightly differently. First let's rewrite Eq. 5.11 as, instead,

$$y[n] = \frac{1}{N}\sum_{m=n-(N-1)}^{n} 1 \cdot x[m]. \tag{5.12}$$

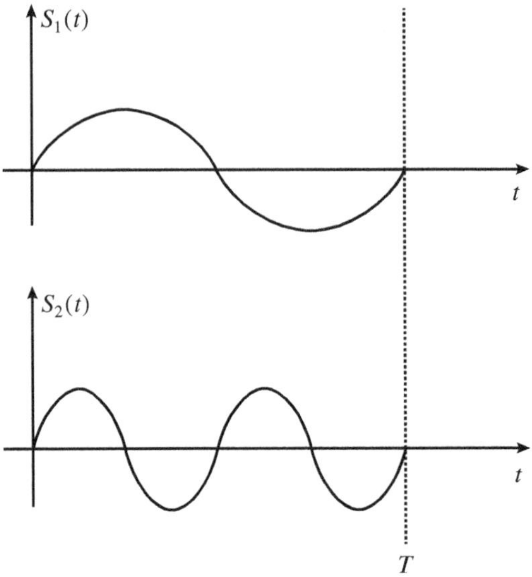

Figure 5.5 A two-symbol library.

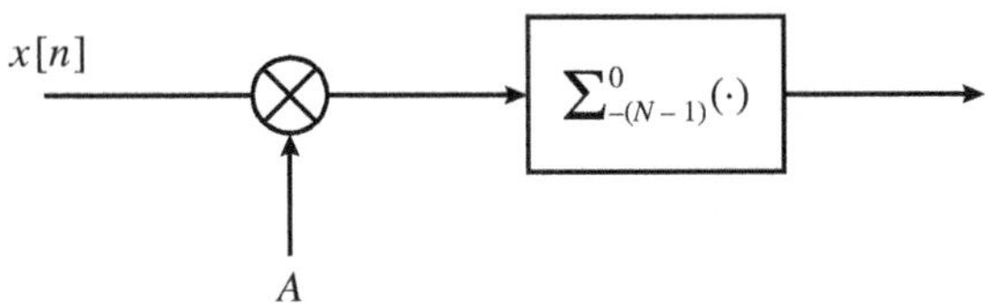

Figure 5.6 Using correlation to detect DC content.

We see that we can take a different view of the averaging function in Eq. 5.11, and instead treat it as a *special case of correlative detection*. That is, when we are looking for the DC component of the signal $x[n]$, we can find it by correlating $x[n]$ with the constant signal $c[n] = 1$. Compare Eq. 5.12 with our definition of correlation in Eq. 4.24. To within a normalization constant, they are the same.

We illustrate this trick in Figure 5.6. Our question was, does the signal $x[n]$ have a DC component? It turns out that we could answer that question by correlating $x[n]$ with a DC signal. We knew what we were looking for, and we used correlation to see if it was buried in there somewhere. No matter what else is going on in $x[n]$, if there is a DC component, and if we choose N to be large enough, we can extract the value of that DC component from the signal. Another name for this trick is *matched filtering*.

5.2.2 Correlation-based Detection

Matched filtering is a way to take advantage of the fact we know ahead of time that we are using a finite library of predetermined symbols to represent our message. The job of the receiver in our system is to determine, at the end of any given time slot of length T, which symbol was transmitted.

For example, consider a system in which we have agreed upon two possible symbols, shown in Figure 5.7. The bedrock correlative sampler is shown in Figure 5.8. The integrate-and-dump block performs the following function:

$$y(t) = \frac{1}{T}\int_{0}^{T} v_{\mathrm{in}}(t)dt. \tag{5.13}$$

The integrator is cleared (set to zero, or dumped) at the end of each integration period, and for the purposes of this little engine, $t = 0$ is the start of a symbol period.

Now we will consider the case for when $x(t)$ is Symbol 1, and calculate the detector output when we correlate first with Symbol 1, and then with Symbol 2. The key trigonometric identity will be

$$\sin \omega_1 t \cdot \sin \omega_2 t = \frac{1}{2}\left[\cos\left(\omega_1 - \omega_2\right) t - \cos\left(\omega_1 + \omega_2\right) t\right]. \tag{5.14}$$

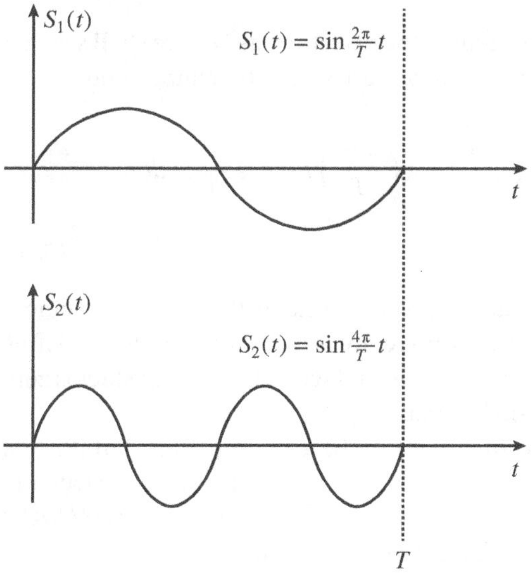

Figure 5.7 A two-symbol library, with a little more detail.

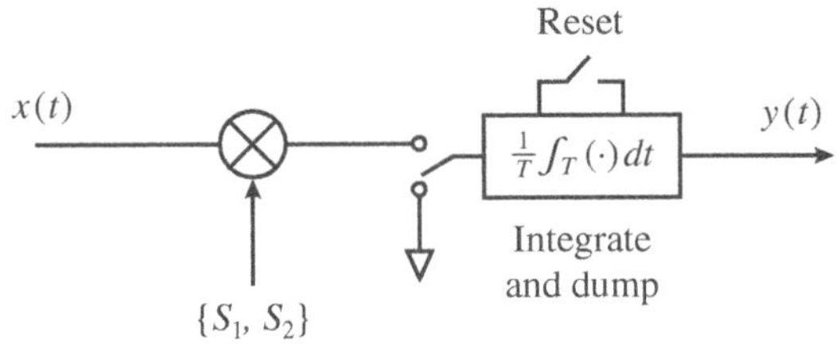

Figure 5.8 A correlative detector.

We run our correlative detector first with Symbol 1:

$$\begin{aligned}\frac{1}{T}\int_0^T \sin\frac{2\pi}{T}t\cdot\sin\frac{2\pi}{T}t\,dt &= \frac{1}{2T}\int_0^T\left(\cos 0\cdot t-\cos\frac{4\pi}{T}t\right)dt\\ &= \frac{1}{2T}\int_0^T\left(1-\cos\frac{4\pi}{T}t\right)dt\\ &= \frac{1}{2T}\cdot T\\ &= \frac{1}{2}.\end{aligned} \tag{5.15}$$

Don't get stuck on the fact that the output of the detector was 1/2; we shall see that the important fact is that it was nonzero. If we correlate with Symbol 2 now:

$$\begin{aligned}\frac{1}{T}\int_0^T \sin\frac{2\pi}{T}t\cdot\sin\frac{4\pi}{T}t\,dt &= \frac{1}{2T}\int_0^T\left(\cos\left(\frac{-2\pi}{T}\right)t-\cos\frac{6\pi}{T}t\right)dt\\ &= 0.\end{aligned} \tag{5.16}$$

Interesting. Looking at Eqs. 5.15 and 5.16, what we say is that our two-element symbol library has the property that its symbols are *orthogonal*. That is, we define our dot product as $x(t)\cdot y(t)=\frac{1}{T}\int_{\langle T\rangle}x(t)\cdot y(t)dt$, and when that dot product is zero we say the signals $x(t)$ and $y(t)$ are orthogonal.

Pretty neat. Now orthogonality is important in the art of signaling, but it is not the *most* important thing. The most important thing is what a detector structure enables in the presence of thermal noise, or additive white Gaussian noise (AWGN for those of you enamored of acronyms). Let's look into this.

We go back and modify Figure 5.8 as shown in Figure 5.9. We have our signal $x(t)$, but now we are accounting for additive noise in the communications channel. For the purposes of analysis, we are going to redraw yet again, being careful

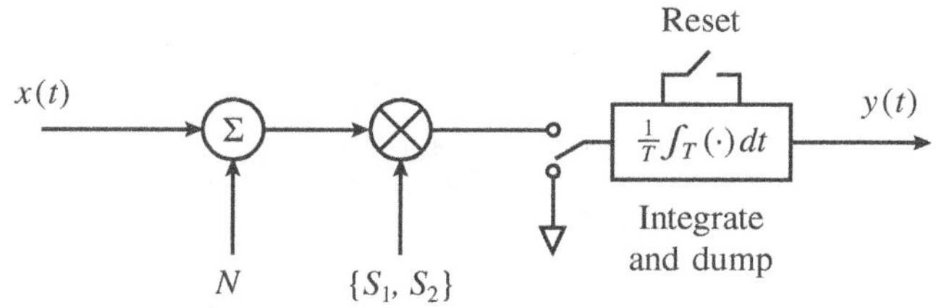

Figure 5.9 A correlative detector with thermal noise in the channel.

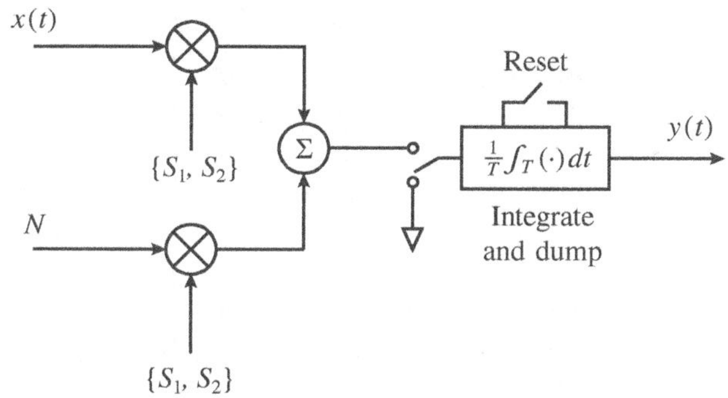

Figure 5.10 First redrawing of Figure 5.9 for analysis.

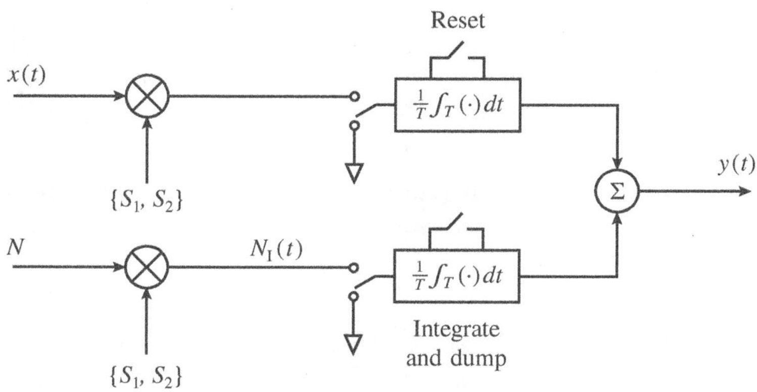

Figure 5.11 Second redrawing of Figure 5.9 for analysis.

with each new manipulation to keep the results of summations and multiplications unchanged. See Figure 5.10, and then move logically to Figure 5.11.

From here, there are quite a lot of analytical false trails to follow. Ultimately, what we want is a comparison between the signal power of $S(t)$ and the noise power of $N_0(t)$. *A good detection machine will reward patience (increasing T) with*

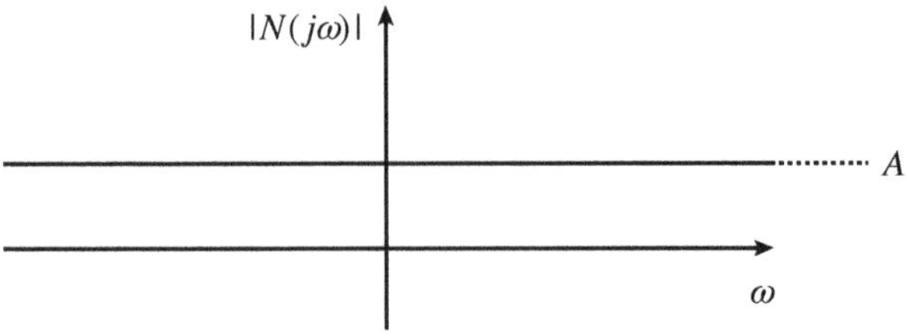

Figure 5.12 Energy spectral density for $N(j\omega)$.

improved SNR. For the author, the way through the wilderness begins with figuring out the PSD of N_I, which is the product of probabilistic noise signal $N(t)$ and a choice of deterministic symbols $S_1(t)$ or $S_2(t)$. A tricky thing about dealing with probabilistic signals in analysis is that we can't get all the way down to specifics, like instantaneous value of a signal contaminated by noise. But if we confine ourselves to averages of the sort that are needed to compute signal power to noise powers, we are fine.

So if we look at the Fourier transform of a particular instance of $N(t)$, $N(j\omega)$, what we expect is that for all reasonable frequencies, its magnitude is approximated as flat, as shown in Figure 5.12. This we know, and throughout our analysis we will confine ourselves to needing only this fact.

Looking at now N_I, we have that it is the *convolution* $S_i(j\omega) * N(j\omega)$:[7]

$$N_I(j\omega) = S_i(j\omega) * N(j\omega) = \frac{1}{2\pi}\int_{-\infty}^{\infty} S_i(ju) \cdot N(j(\omega - u))du \tag{5.17}$$

Again, the full details of $N_I(j\omega)$ are unavailable to us because of the probabilistic nature of $N(t)$, but it will turn out that we can say something about $|N_I(j\omega)|^2$. To wit:

$$|N_I(j\omega)|^2 = \left|\frac{1}{2\pi}\int_{-\infty}^{\infty} S_i(ju) \cdot N(j(\omega - u))du\right|^2. \tag{5.18}$$

While $N(j\omega)$ is understood to be extremely broad in the frequency domain, we are on safe ground if we insist on our signal $X(j\omega)$ being of finite bandwidth B:

$$|N_I(j\omega)|^2 = \left|\frac{1}{2\pi}\int_{-B}^{B} S_i(ju) \cdot N(j(\omega - u))du\right|^2. \tag{5.19}$$

7 When you multiply in the time domain, you convolve in the frequency domain and vice versa.

We have gone as far as we can without the full details of $N(j\omega)$ and $X(j\omega)$. Boxed into a corner, the conservative thing to do analytically is to establish an *upper bound* on the noise power. We can do this by making use of the Schwarz inequality:

$$\left|\frac{1}{2\pi}\int_{-B}^{B} S_i(ju)\cdot N(j(\omega-u))du\right|^2 \leq \frac{1}{4\pi^2}\int_{-B}^{B}|S_i(ju)|^2 du\cdot\int_{-B}^{B}|N(j(\omega-u))|^2 du. \tag{5.20}$$

Since we expect $|N(j\omega)|^2$ to be constant over the frequency range of interest, we can further simplify to

$$\left|\frac{1}{2\pi}\int_{-B}^{B} S_i(ju)\cdot N(j(\omega-u))du\right|^2 \leq \frac{1}{4\pi^2}\int_{-B}^{B}|S_i(j\omega)|^2 d\omega\cdot\int_{-B}^{B}|N(j\omega)|^2 d\omega, \tag{5.21}$$

and now write

$$|N_I(j\omega)|^2 \leq \frac{1}{4\pi^2}\int_{-B}^{B}|S_i(j\omega)|^2 d\omega\cdot\int_{-B}^{B}|N(j\omega)|^2 d\omega. \tag{5.22}$$

Super interesting. We will use Eq. 5.22 as a *useful, conservative upper bound* on the power spectral density $|N_I(j\omega)|^2$, and note that it scales with the total energy of the $S_i(t)$ symbol. If this does not jump out at you, recall that we can use Parseval's theorem to write

$$\int_{-B}^{B}|S_i(j\omega)|^2 d\omega = 2\pi\int_{-\infty}^{\infty}|S_i(t)|^2 dt. \tag{5.23}$$

Eq. 5.22 now becomes

$$|N_I(j\omega)|^2 \leq \frac{1}{2\pi}\int_{-\infty}^{\infty}|S_i(t)|^2 dt\cdot\int_{-B}^{B}|N(j\omega)|^2 d\omega. \tag{5.24}$$

Going forward, we will take this upper limit as our PSD for N_I, and state firmly that

$$|N_I(j\omega)|^2 = \frac{1}{2\pi}\int_{-\infty}^{\infty}|S_i(t)|^2 dt\cdot\int_{-B}^{B}|N(j\omega)|^2 d\omega. \tag{5.25}$$

Since we are taking our upper bound $|N_\mathrm{I}(j\omega)|^2$ to be a constant, let's capture that in a new, frequency-independent quantity that nevertheless reminds us that we are dealing with a thermal noise source:

$$kT_\mathrm{I} = \frac{1}{2\pi} \int_{-\infty}^{\infty} |S_i(t)|^2 dt \cdot \int_{-B}^{B} |N(j\omega)|^2 d\omega. \tag{5.26}$$

Here, k is the usual Boltzmann constant, and T_I is whatever "noise temperature" is necessary to make good on the equality.

Great. We started this development by trying to figure out how to treat the case of thermal noise being multiplied by one of our symbols. We have our answer now. We redraw our diagram as in Figure 5.13. And then redraw it one more time, with the effect of our noise referred to *after* an integrate-and-dump, as shown in Figure 5.14. The B_N here is the noise bandwidth associated with the integrate-and-dump. Beautiful and straightforward at last. Now we just have to compute that pesky B_N. Recall from Section 4.4 that the noise bandwidth of a linear filter is given by:

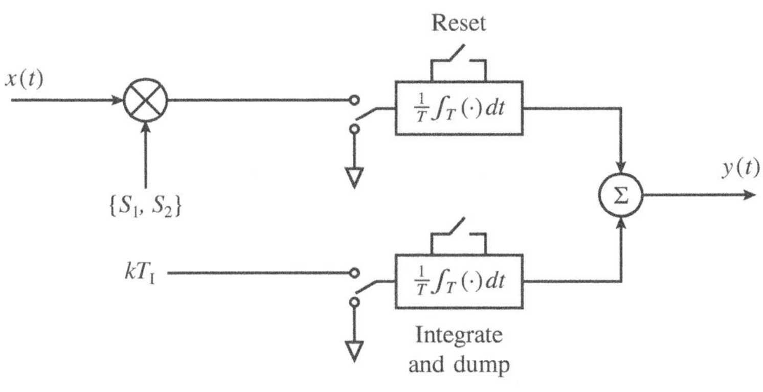

Figure 5.13 Third redrawing of Figure 5.9 for analysis.

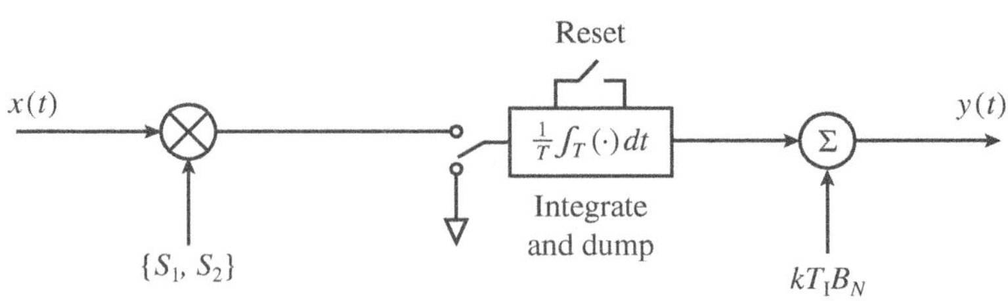

Figure 5.14 Final redrawing of Figure 5.9 for analysis.

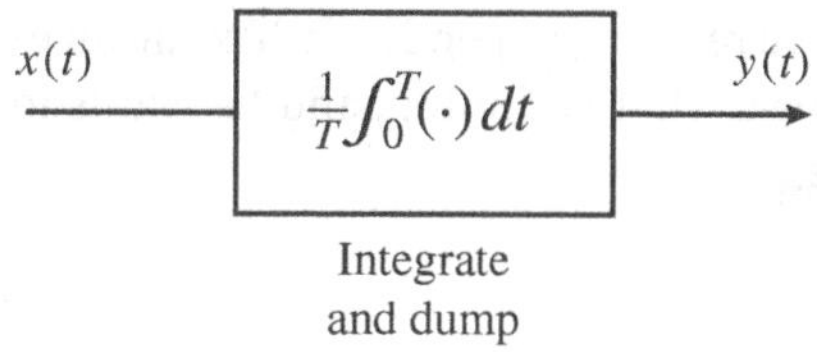

Figure 5.15 We need to determine the transfer function of this filter.

$$B_N = \frac{1}{|H_{\text{peak}}|^2} \int_0^{\infty} |H(f)|^2 df. \tag{5.27}$$

To this end, we must figure out the transfer function $H(j\omega)$, or $H(f)$, for the integration filter given in Figure 5.14, isolated for analysis in Figure 5.15. Those of you with the background may have your own tricks for determining transfer functions, and you are encouraged to dust them off here. For those to whom this is new, all there really is to know is that $H(j\omega)$ is a complex scalar that tells us how an input signal $e^{j\omega t}$ is amplitude scaled and phase shifted before it gets to the output. So... let's make the input $e^{j\omega t}$ and see what happens. Note that when we dive in to do the math, we have to be a little more precise about integration limits,[8] and write that the output at time t is actually

$$y(t) = \frac{1}{T} \int_{t-T}^{t} x(t)dt. \tag{5.28}$$

Now if the input is $x(t) = e^{j\omega t}$, the output is given by

$$\begin{aligned} y(t) &= \frac{1}{T} \int_{t-T}^{t} e^{j\omega t} dt \\ &= \frac{1}{j\omega T} \left[e^{j\omega t} - e^{j\omega t} e^{-j\omega T} \right] \\ &= e^{j\omega t} \cdot \frac{1}{j\omega T} \left[1 - e^{-j\omega T} \right] \end{aligned} \tag{5.29}$$

We therefore identify the transfer function $H(j\omega)$ as

$$H(j\omega) = \frac{1}{j\omega T} \left[1 - e^{-j\omega T} \right]. \tag{5.30}$$

8 Don't sweat the "dump" portion of "integrate-and-dump". It is an implementation detail.

This is a good and valid final answer, but we're going to persist with some algebraic manipulations because the author knows what is coming, and he wants to make *that* integral easier to do. We continue.

$$\begin{aligned}
\frac{1}{j\omega T}\left[1-e^{-j\omega T}\right] &= \frac{1}{j\omega T}e^{-j\omega t/2}\left[e^{j\omega T/2}-e^{-j\omega T/2}\right] \\
&= \frac{1}{j\omega T}e^{-j\omega T/2}\left[2j\sin\omega T/2\right] \\
&= \frac{1}{\omega T/2}e^{-j\omega T/2}\sin\omega T/2 \\
&= e^{-j\omega T/2}\cdot\frac{\sin\omega T/2}{\omega T/2} \\
&= e^{-j\omega T/2}\cdot\mathrm{sinc}\left(\frac{\omega T}{2}\right).
\end{aligned} \tag{5.31}$$

Since $\omega = 2\pi f$, we wind up with

$$H(f) = e^{-j\pi f T}\mathrm{sinc}(\pi f T), \tag{5.32}$$

and the noise bandwidth falls out of the definition.

$$B_N = \frac{1}{1^2}\int_0^\infty \mathrm{sinc}^2(\pi f T)df. \tag{5.33}$$

With the aid of a good integration table, we end at last with a noise bandwidth of

$$B_N = \frac{1}{2T}. \tag{5.34}$$

We finally have all the pieces to compute what we wanted all along, which is the SNR at the output of Figure 5.9. Our noise power is apparently

$$P_N = kT_\mathrm{I}\cdot B_N = kT_\mathrm{I}\cdot\frac{1}{2T}. \tag{5.35}$$

The signal power at the output is the result of integration of the multiplication of signal $A\sin\frac{2\pi}{T}nt$ and $x_i = \sin\frac{2\pi}{T}mt$. When $m = n$ the result of that integration is $A/2$. We may therefore write the signal power as the proportionality

$$P_\mathrm{sig} \propto \frac{1}{4}A^2, \tag{5.36}$$

and our SNR as the proportionality

$$\mathrm{SNR} \propto \frac{A^2}{2kT_\mathrm{I}}\cdot T. \tag{5.37}$$

This is an incredibly important result. It says that, *if you can wait long enough (i.e. make T big enough)*, you can detect the presence of a predetermined signal in an arbitrarily low SNR with an arbitrarily low probability of error. Not only does it say that, but the foregoing treatment even *tells you how to do it*, which is to use correlative sampling.

Herein lies the answer to questions that have dogged some of you since you started to think deeply about the Shannon limit. Recall that according to the strict definition, we are able to have nonzero capacity as long as the SNR is nonzero. Again,

$$C = B \cdot \log_2(\mathrm{SNR} + 1)\,. \tag{5.38}$$

What this equation stubbornly insists is that for an SNR of 0.1 (−10 dB), or even 1×10^{-9} (−90 dB), as long as we are given a channel of nonzero bandwidth and a lot of patience, Nature allows us to get bits through the channel. But if you are practical-minded, you may have wondered how exactly to accomplish this in an environment where your SNR is extremely low. Well, now you know. You must have *long* symbols, which correspond to long integration times in a correlative sampler. We emphasize that the Shannon limit does not prescribe correlative sampling. We note only that, if you are concerned about implementation, correlative sampling is a powerful surrogate that provides much useful insight. Once you have this concept, you will have acquired an understanding that is absolutely foundational to modern digital communications.

5.3 Shannon Warmup Problem: Reading a Capacitor Voltage

First: why a warmup problem? Why not dive right into an understanding of the Shannon limit, the way Shannon himself did and the way information theorists always do?

The answer, of course, is that you are free to do so if you have the background. The right path for you may be to skip this warmup problem and go directly to Shannon's paper (or some other trusted source). For *his* part, the author found these original sources to be a very heavy lift and spent a great deal of time filling in the information theoretic concepts that were taken for granted by those authors. If you find yourself in the same boat, this section is for you.

Very good. We return, as we so often have, to the basic *RC* circuit. Consider a voltage source connected to a capacitor through a resistor, where the voltage source is constrained to assume one of N discrete voltages, from 0 to $(N-1) \cdot \Delta v$. Assume further that the total range is constrained such that $N \cdot \Delta v = V_R$, so that higher N necessarily requires smaller Δv. See Figure 5.16. Our challenge is simple: we want to know how many levels N we can reliably distinguish given the presence of

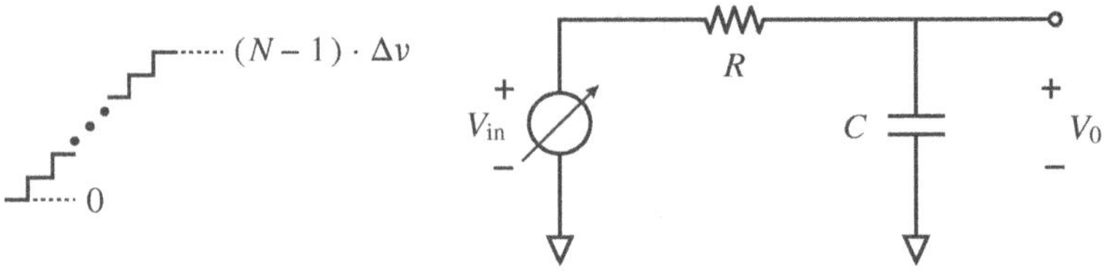

Figure 5.16 Setup for considering how many levels we can reliably distinguish in the presence of thermal noise.

thermal noise *and* given a constraint on our maximum measurement time. Take some time to consider how you might go about answering this question on your own before proceeding.

Remember that even if the noise voltage spread $\sqrt{kT/C}$ is comparable to or even larger than the increment Δv, we are not helpless. We can, for example, use averaging as discussed in Section 4.5 to lower the noise variance. But a key requirement for using this method is that the samples that we average be separated in time long enough so that the noise from sample to sample is uncorrelated. And *that* means that extracting more distinguishable levels from this setup necessarily demands *more time.*

Now perhaps the most exciting insight from Shannon's paper is that you can buy the time you need to get arbitrary accuracy by adding latency, or delay, between the time that you ask for a level determination and the time that you get it. If you can tolerate this delay, the total number of level determinations per time period can still be high, much higher in fact than if you held yourself to a "ask for one result, get one result" sequential cadence. This is called "getting high throughput at the expense of latency, or delay."

We are touching here on another foundational concept in modern digital communications. We will get much more deeply into this in the following development, but now you know where we are headed.

5.3.1 Noise Variance and Error Rate

We're going to start looking at the problem of measuring the voltage off our capacitor by getting very analytical about *how often we expect to make a mistake*. We will quantify the frequency with which we make mistakes as the *error rate*, and we are interested in what we need to do to drive this error rate arbitrarily low.

This is easier to do than it might first appear. We refer back to Section 4.2.1, on Gaussian distributions, and calculate the error rate in two extreme cases. In one case, the standard deviation $\sigma = \sqrt{\frac{kT}{C}}$ is *large* compared to the level spacing Δv. In the second case, we will look at what happens when σ is small compared to Δv.

Assume that in both cases we choose the variable voltage source to be right near midrange, so $V_{\text{in}} \approx \frac{N}{2} \cdot \Delta v$. We will draw a Gaussian distribution of possible voltage measurements, and in each case overlay the voltage levels on the horizontal axis.

For the first case, where σ is large compared to the level spacing, we have Figure 5.17. In the second case, where σ is small compared to the level spacing, we have it graphically as shown in Figure 5.18. Just by looking, you would say that we expect to make errors all the time in the first case, while in the second case errors would be relatively rare.

Can we do better, though, than "all the time" and "relatively rare"? That is to say, can we be quantitative? It turns out that we can. The fraction of time that we will make the *correct* decision is exactly the area under the Gaussian distribution between $\frac{N}{2} \cdot \Delta v - \frac{\Delta v}{2}$ and $\frac{N}{2} \cdot \Delta v + \frac{\Delta v}{2}$, which is the shaded region for the first case.

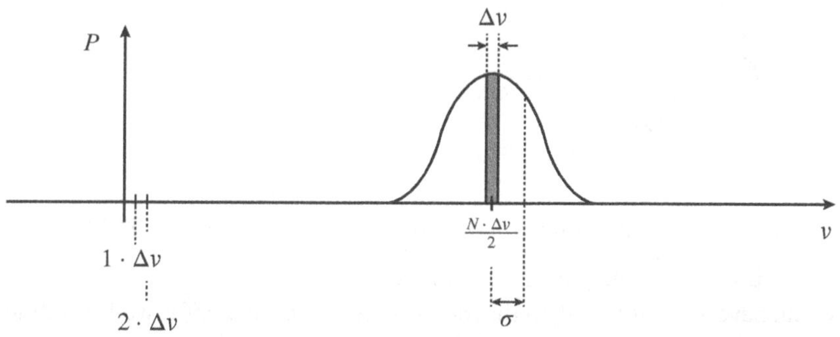

Figure 5.17 The first case, where the standard deviation of the noise voltage is large compared to our level spacing.

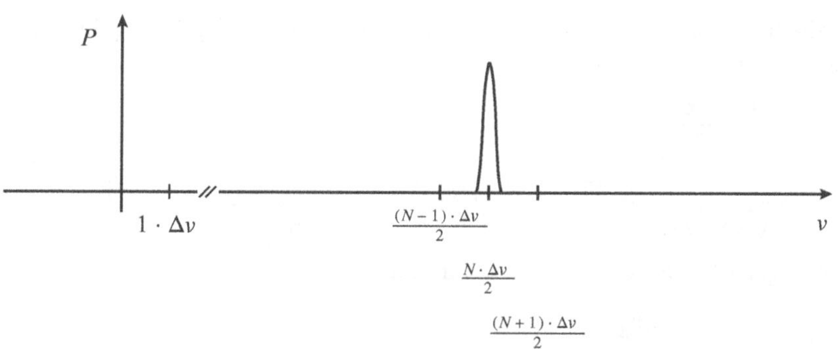

Figure 5.18 The second case, where the standard deviation of the noise voltage is small compared to our level spacing.

If we label this area as P_c, then the probability of error is $P_e = 1 - P_c$. So let's refer back to Section 4.2.1 and devote ourselves to the definite integral

$$P_c = \int_{\frac{N-1}{2}\cdot\Delta v}^{\frac{N+1}{2}\cdot\Delta v} \frac{1}{\sqrt{\frac{kT}{C}}\sqrt{2\pi}} \cdot e^{-\frac{1}{2}\frac{\left(v-\frac{N}{2}\cdot\Delta v\right)^2}{kT/C}}\, dv. \tag{5.39}$$

Correct! But messy. For the purpose of determining an antiderivative, we might make a change of variables

$$\begin{aligned} x &= v - \frac{N}{2} \cdot \Delta v \\ dx &= dv, \end{aligned} \tag{5.40}$$

so now our probability of getting a correct result becomes

$$P_c = \int_{-\frac{\Delta v}{2}}^{+\frac{\Delta v}{2}} \frac{1}{\sqrt{\frac{kT}{C}}\sqrt{2\pi}} \cdot e^{-\frac{1}{2}\frac{x^2}{kT/C}}\, dx. \tag{5.41}$$

Much better.

Or at least it seems, until you dive into your math references and find that the $e^{-\frac{1}{2}\frac{x^2}{kT/C}}$ function has, in fact, no antiderivative.

But we do have a way forward, in that *detailed* numerical tables (as well as online calculators) exist for the so-called "erf" function, defined as

$$\text{erf}(z) = \frac{2}{\sqrt{\pi}} \int_0^z e^{-t^2}\, dt. \tag{5.42}$$

This is the form we are trying to get to, and we are helped along the way by doing one more change of variables:

$$\begin{aligned} p &= \frac{1}{\sqrt{2}} \frac{x}{\sqrt{kT/C}} \\ dp &= \frac{1}{\sqrt{2}} \frac{1}{\sqrt{kT/C}}\, dx. \end{aligned} \tag{5.43}$$

We write Eq. 5.41 in terms of this new variable

$$P_c = \frac{1}{\sqrt{\pi}} \int_{-\frac{1}{2\sqrt{2}}\frac{\Delta v}{\sqrt{kT/C}}}^{+\frac{1}{2\sqrt{2}}\frac{\Delta v}{\sqrt{kT/C}}} e^{-p^2}\, dp. \tag{5.44}$$

Taking advantage of the fact that e^{-p^2} is symmetric about $p = 0$, our match to Eq. 5.42 is complete if we write

$$P_c = \frac{2}{\sqrt{\pi}} \int_0^{+\frac{1}{2\sqrt{2}}\frac{\Delta v}{\sqrt{kT/C}}} e^{-p^2}\, dp. \tag{5.45}$$

If our probability of being correct is

$$P_c = \text{erf}\left(\frac{1}{2\sqrt{2}}\frac{\Delta v}{\sqrt{kT/C}}\right), \tag{5.46}$$

then our probability of error can be expressed using the so-called "erfc" function

$$\begin{aligned} P_e &= 1 - \text{erf}\left(\frac{1}{2\sqrt{2}}\frac{\Delta v}{\sqrt{kT/C}}\right) \\ &= \text{erfc}\left(\frac{1}{2\sqrt{2}}\frac{\Delta v}{\sqrt{kT/C}}\right). \end{aligned} \tag{5.47}$$

We'll do one last manipulation, to focus on the importance of the ratio between Δv and the standard deviation $\sqrt{kT/C}$. Observing that $\frac{1}{2\sqrt{2}} \approx 0.354$, we have at last

$$\begin{aligned} P_c &\approx \text{erf}\left(0.354 \cdot \frac{\Delta v}{\sqrt{kT/C}}\right) \\ P_e &\approx \text{erfc}\left(0.354 \cdot \frac{\Delta v}{\sqrt{kT/C}}\right). \end{aligned} \tag{5.48}$$

From here, it is instructive to use erfc tables or numerical calculators to relate Δv-to-$\sqrt{kT/C}$ ratios to error rates. Referring to Table 5.2, we have that for an error rate below 1%, we must have a ratio $\frac{\Delta v}{\sqrt{kT/C}}$ that exceeds 5.15. If we have a 10 pF capacitor, and we are in thermal equilibrium at 300 K, we have that

$$\sqrt{\frac{kT}{C}} = \sqrt{\frac{1.38 \times 10^{-23}\ \text{J/K} \cdot 300\ \text{K}}{10\ \text{pF}}} = 20.3\ \mu\text{V}. \tag{5.49}$$

An error rate of 1% apparently requires that we separate our levels by $\Delta v = 5.15 \times 20.3\ \mu\text{V} = 105\ \mu\text{V}$. If we say that our total voltage range is 0–5 V, we may say that the total number of levels we can encode and maintain our error rate is therefore $5\ \text{V}/105\ \mu\text{V} = 47,619$. This in turn equates to $\log_2(47619) = 15.5$ bits.

For many of you, this portion of the warmup problem may be the first time you have gotten quantitative about how much information can be represented by an

Table 5.2 Requirements on $\frac{\Delta v}{\sqrt{kT/C}}$ for given error rates.

Error Rate	erfc(·)	$\frac{\Delta v}{\sqrt{kT/C}}$
50%	erfc(0.48)	1.36
10% (10^{-1})	erfc(1.16)	3.28
1% (10^{-2})	erfc(1.82)	5.15
10^{-3}	erfc(2.33)	6.59
10^{-4}	erfc(2.75)	7.78
10^{-5}	erfc(3.12)	8.83
10^{-6}	erfc(3.46)	9.79
10^{-9}	erfc(4.32)	12.2

analog voltage in the presence of thermal noise. Some specific insights that will carry over into a full blown discussion of the Shannon limit:

- All other factors held constant, lowering the temperature increases the number of levels and therefore increases the number of bits.
- If we increase the energy we are willing to store in the capacitor, by increasing C and/or increasing the voltage range, we increase the number of levels and therefore the number of bits.
- The number of bits increases *slowly* with the amount of energy we are willing to store, as the number of bits goes logarithmically with the voltage range. Another way to say that is that for a linear increase in the number of bits, we must increase our maximum energy storage *exponentially*. Not for the first or last time, this is the worst energy trade-off in Nature.

Fine. We have established a baseline of how many levels we can distinguish, but you may rightly object that we have not pulled any of our averaging, filtering, or correlating tricks to help us deal with the noise. We will treat this worthy topic in the next section.

5.3.2 Averaging to Drive the Error Rate Low

How can we do better? That is, how can we squeeze in more levels, represent more bits, without demanding changes to our circuit or the environment?

Recall that in Section 4.5, we showed that if we average M measurements, the rms noise voltage goes down according to

$$\sqrt{\frac{kT}{C}} \rightarrow \frac{1}{\sqrt{M}}\sqrt{\frac{kT}{C}}. \tag{5.50}$$

If we are willing to average aggressively enough, we see that we can distinguish levels that are arbitrarily close together with arbitrarily low error rates.

The cost, of course, is measured in time. To see this, start with a target rms noise voltage σ in order to represent a certain number of bits, and observe that the key proportionality is

$$\sigma \propto \frac{1}{2^{\text{bits}}}. \tag{5.51}$$

But since σ is given by Eq. 5.50, we can substitute and arrive quickly at the proportionality

$$M \propto 2^{2\cdot\text{bits}} \cdot \frac{kT}{C}. \tag{5.52}$$

Now the time required to take our samples is MT_s, where T_s is the minimum time we must wait in order for successive samples to be uncorrelated. From Section 4.3.2, we know that this T_s is inversely proportional to $B = 1/RC$, the bandwidth of the RC filter. We can therefore say that the total time it takes to read enough samples to satisfy our error rate requirements scales according to

$$\begin{aligned} T_\text{tot} &\propto \frac{1}{B} \cdot 2^{2\cdot\text{bits}} \cdot \frac{kT}{C} \\ &\propto kTR \cdot 2^{2\cdot\text{bits}}. \end{aligned} \tag{5.53}$$

We find that we can use the averaging trick to distinguish more levels, but the time we must wait scales exponentially with the number of bits we demand.

If you're hoping that some other fancy filtering method will give us a more favorable trade-off in time, we can extinguish that hope fairly quickly. The setup now is that we follow our RC circuit with a processing filter $H_\text{P}(s)$, and give ourselves unlimited implementation complexity in this filter to achieve our goals.[9]

We know two things from here. The first is that whatever we do for this filter, it is going to have a characteristic width in the frequency domain which we will call B_N. Second, a *valid* use of the Uncertainty Principle from Section 2.5 is to say that if we know the bandwidth B_N, then the time that it takes for the processing filter to settle to its final value is going to be *on the order of* $\frac{1}{\pi B_\text{N}}$.

From here defeat follows swiftly. The number of bits determines the number of levels, and the number of levels together with the total voltage range gives us our required noise variance σ. We have the proportionality $\sigma \propto \sqrt{kTB_\text{N}}$, and therefore

$$B_\text{N} \propto \frac{1}{2^{2\cdot\text{bits}} \cdot kT}. \tag{5.54}$$

---

9 Let us suppose that the filter is linear and time invariant. This is not a terribly restrictive requirement, but it ensures that our analysis is valid.

Taking a cue from the Uncertainty Principle and substituting $B_N \approx \frac{1}{\pi T_s}$, we wind up with

$$T_s \propto \frac{1}{\pi} \cdot 2^{2 \cdot \text{bits}} \cdot kT. \tag{5.55}$$

It is the same exponential dependence of waiting time on bit resolution that we had with averaging. There is no getting around it.

5.3.3 A Way Out: Doing Measurements in Batches

We're stuck, then. It is surely possible to do slightly better, say by optimizing our filters for good settling time or through other tricks. But the bad news from Eq. 5.55 is that the fundamental scaling trends governing measurement time vs. bit resolution are here to stay.

In this section, though, we get to a central insight, arguably *the* central insight, that you need to properly understand and apply the Shannon limit. The insight is as follows. Rather than focusing on getting one capacitor value at a time, averaging/filtering for as long as you need to get the desired bit resolution, *you can instead take a long sequence of capacitor values as a group, and then in your processing untangle the signal from the noise on all the measurements at once.* The delay from the time you start to the time you get your answers is no better than what we get from Eq. 5.55. But a huge consolation is that at the end of your wait, you get a large number of answers at once. In communications parlance, we would say that the latency, or delay, is long, but that is okay because the overall *throughput*, or number of high-resolution capacitor measurements per unit time, is *high.*

This is a great improvement from where we were before, where the delay was high and we only got one capacitor measurement per running of our process. Many times in communications, we are quite tolerant of delay as long as the throughput is high. A great example of this is most video applications, where we need a huge number of bits per second in order to have a pleasing experience, but there is no expectation of synchronicity between events in the video and real time. We can tolerate quite long delays between when we ask for the video to start and when it actually starts; once it gets going, the high throughput is all we care about. This trading of latency for throughput is absolutely foundational to modern digital communications and is very important for you to internalize.[10]

To gain an understanding of trading latency for throughput, we will start a new game with our capacitor circuit.[11] This time rather than hold at one capacitor

10 An important case where latency is completely unacceptable is when the communications link is a means of real-time control of machines like industrial robots and cars.

11 The rules of this game are chosen to make understanding the Shannon limit as easy as possible for readers who, like the author, do not come from a communications theory background.

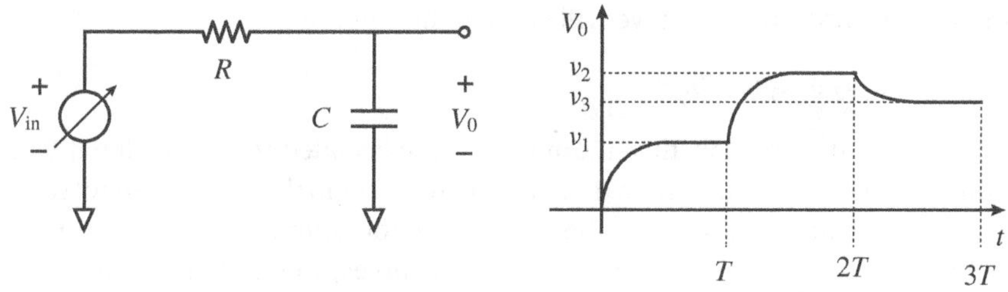

Figure 5.19 A three-symbol sequence through our *RC* channel.

voltage, we set about determining what three-symbol sequence was "transmitted"; see Figure 5.19. The first rule of our game is that we require

$$v_1^2 + v_2^2 + v_3^2 = E. \tag{5.56}$$

Note that eventually we will extend this game to M symbols, where M can be tens, hundreds, thousands, millions, of v_is or more. When M is large, we will talk of the aggregate signal having an average power P_0. By this we mean that we pay attention only to the samples, and expect of the average

$$\frac{1}{M} \cdot \left(\frac{1}{R_L} \sum_{i=1}^{M} v_i^2 \right) = P_0. \tag{5.57}$$

The resistance R_L is a load resistance that is large enough to not significantly load the *RC* circuit, and is only really there for practical minded circuit people who can't let go of the fact that an unloaded *RC* will deliver *no* power. Then to reckon with communications theory literature, we will throw these constants aside and write

$$\sum_{i=1}^{M} v_i^2 = E, \tag{5.58}$$

focusing only on the fact that the sum of the squares of the v_is is a constant. Again, for M large enough we can relax and just know that this will be true, but at this point in our game with a small number of samples, $M = 3$, we have to be strict and enforce the equality.

Next, we have that each sample v_i will be addled by additive noise values labeled x_i. Our second rule will be that

$$x_1^2 + x_2^2 + x_3^2 = N. \tag{5.59}$$

That is to say, we will be definitive about the additive noise power in the same way that we were definitive about the signal power.

Finally, our third rule is that we will enforce the equality

$$x_1 v_1 + x_2 v_2 + x_3 v_3 = 0. \tag{5.60}$$

This is simply an assertion that the noise is uncorrelated with the signal (see Section 4.2.2). Geometrically, note that this is equivalent to the noise vector of samples (x_1, x_2, x_3) being *orthogonal* to the signal vector of samples (v_1, v_2, v_3). As with the other rules, when there is a large number of samples, and when the underlying noise mechanism is thermal, we get this equality for free. But when the number of samples is as small as three, we must enforce the equality as a rule in our game.

Now we get to the key question, whose answer is foundational to understanding the Shannon limit. And that question is: How many signal sequences (v_1, v_2, v_3) can we reliably distinguish, given noise energy N and signal energy E?

This is a significant shift in thinking from previous sections, where we were focused on determining one value on the capacitor at a time. Instead, we now collect three-sample sequences, and set ourselves the task of determining how many distinct three-sample sequences we can read given the signal-to-noise ratio E/N. It takes us MT seconds to collect our samples ($M = 3$ here), which time we now call the "latency."

Note also that by choosing $M = 3$, we make an appeal to geometrical arguments practically irresistible, indeed the only compelling path toward understanding. Hold on to this warm and cozy feeling as long as you can. It will be seriously challenged as we go to even four samples and beyond. It turns out that geometrical reasoning will still be the best path forward, but we *will* have to lean, with gratitude, on the work of our mathematician friends, who long ago worked out the geometry of high dimensional spaces that we are helpless to visualize. It will require attention and discipline to see where our understanding of three dimensional (3D) spaces carries over, and *where it does not*. Keep an eye on this.

Having said all of that, we set about visualizing the space of possibilities for our three-sample sequences, which have in common that $v_1^2 + v_2^2 + v_3^2 = E$. This is, of course, a sphere in three-space, shown in Figure 5.20. How do we treat the presence of additive noise in this geometric picture? Suppose first that the noise vector is nominally of length $\sqrt{N}$, but let us suppose for a moment that we are not so strict about the orthogonality that we required in Eq. 5.60 (we will come back to this). The effect of adding a noise vector can be captured by appending a noise sphere, or "noise ball," whose center is the original signal vector. See Figure 5.21. From here it is instructive to zoom in on three special cases for the noise: (1) when the noise is completely correlated with the signal; (2) when the noise is completely *anti*-correlated with signal; (3) the nominal case, when the noise is uncorrelated with the signal. We look at these cases in Figure 5.22. When we draw pictures like this, we are saying, "If we step our capacitor voltage

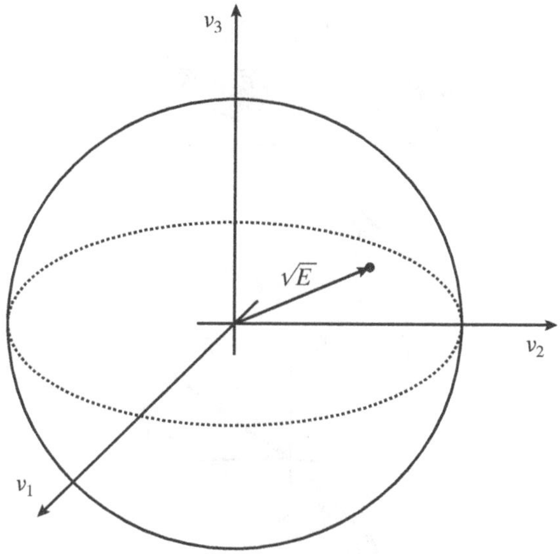

Figure 5.20 We can visualize all the possibilities for which $v_1^2 + v_2^2 + v_3^2 = E$ by imagining a sphere of radius $\sqrt{E}$.

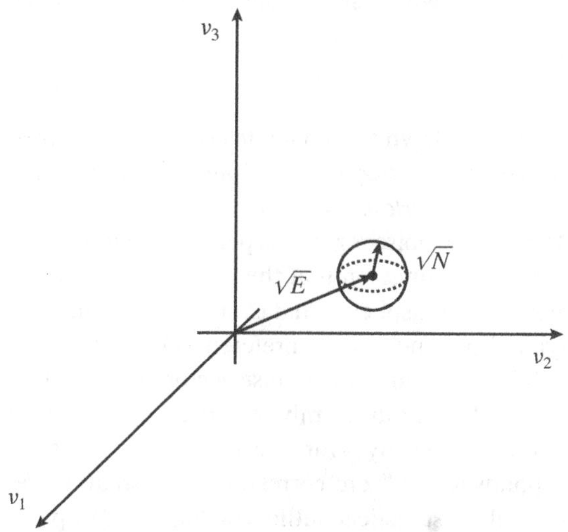

Figure 5.21 The concept of a noise sphere.

through the three-voltage sequence trying to arrive at the vector $\vec{v}$ in our signal space, the presence of noise means we instead wind up *somewhere* in a sphere defined by the noise ball centered on the intended $\vec{v}$." So when we consider our key question, and think about what it means to "reliably distinguish" different

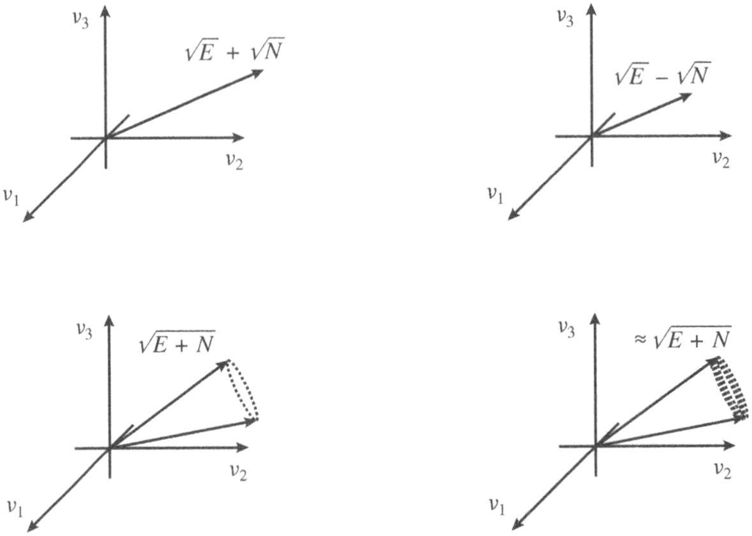

Figure 5.22 Different cases of correlation and how the $\sqrt{E}$ and $\sqrt{N}$ vectors combine. On the top two axes, we have the noise and signal being completely correlated on the left, and being completely anti-correlated on the right. The bottom two axes are what we normally expect, where the noise has exactly zero correlation (left) or a small correlation coefficient (right).

three-voltage sequences, we can boil this down to a concrete criterion: *two three-voltage sequences are reliably distinguishable if they are far enough apart in the signal space such that their noise balls do not overlap.* See Figure 5.23.

A quick housekeeping note. Drawing the noise regions as perfectly solid spheres is a *little* misleading, in that it gives the impression that the noise-addled symbol will routinely wind up anywhere in the noise sphere. That is, strictly speaking, not what we expect. We *expect* the symbol-plus-noise to be preferentially scattered on the *surface* of the noise ball, which is consistent with a noise vector length of $\sqrt{N}$. Moreover, we do not expect it to be scattered uniformly over the surface of said ball. We instead expect the distribution to heavily favor a narrow band circling the equator of the ball, which corresponds to the "zero correlation" or "small correlation" cases of Figure 5.22. Nevertheless, statistical outliers do happen. For probabilistic signals, then, the mental model of solid noise spheres builds in a useful kind of conservatism when it comes to questions of distinguishing symbols with arbitrarily small chances of error.

We think now of committing to a library of three-sample symbols, and these will be the only symbols that are allowed. The next important question for us is, how many such symbols can we accommodate? We know from Figure 5.20 that if

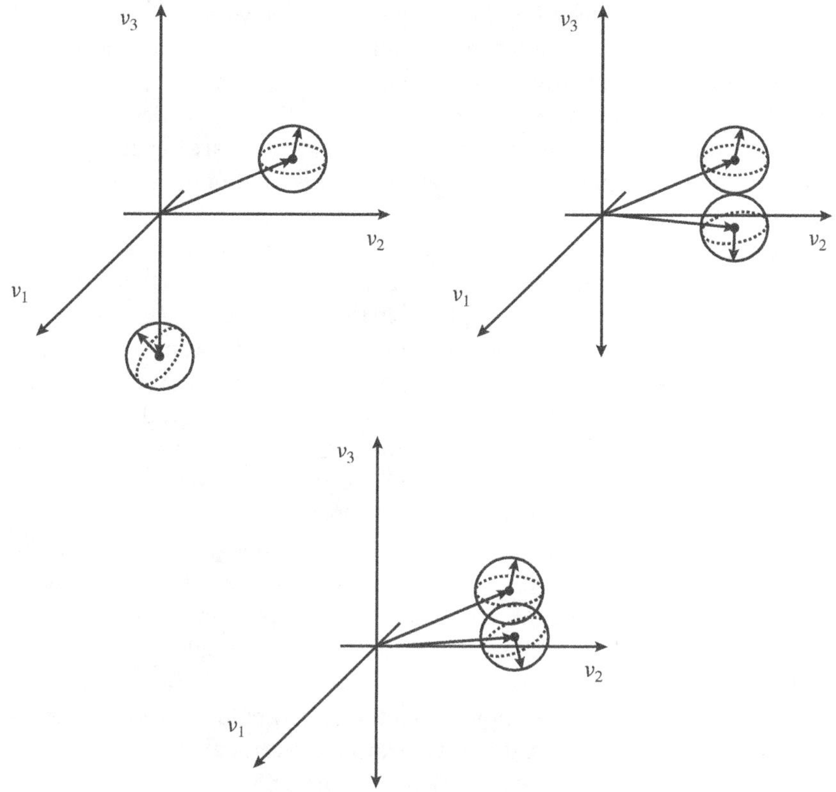

Figure 5.23 Different choices for our symbol library. If we choose symbols according to the top two sets of axes, then even in the presence of noise we are safe from making an error. The bottom set of axes shows overlap of the noise spheres, and therefore a nonzero error probability.

we choose our symbols to be of energy E, then they will all lie on the surface of a sphere or radius $\sqrt{E}$. When noise comes into the picture, we make the assumption that the noise is uncorrelated with the signal and therefore that the noise vector is orthogonal to the signal vector. Geometrically, this means that once we have noise added, any received signals at the other end of the communications channel will all lie on a sphere of radius $\sqrt{E+N}$.

We can now tackle the real question, which is, how many points can we accommodate on the surface of the $\sqrt{E}$ sphere *if we insist on non-overlapping noise spheres of diameter* $\sqrt{N}$? This is a complicated question of high dimensional geometry, one that we approach carefully because our intuition from three-space is not entirely reliable. It turns out that one thing we do to make our lives easier is to focus on getting an *upper bound* on the number of points we can get. If we build inaccuracies into our calculation for the sake of simplicity, it is then okay as long as those mistakes cause us to overcount instead of undercount.

A candidate upper bound is the volume of the $\sqrt{E+N}$ sphere divided by volume of the $\sqrt{N}$ sphere. This is obviously overcounting, because we are including in our count a bunch of noise spheres in the interior. That is kind of okay, because we agreed ahead of time that we're looking for an upper bound. But it actually gets better for two reasons. The first is a curiosity of high dimensional spheres, which is that almost all of their volume is near their surface. This means that the cases that you might worry about, the noise spheres in the interior of the $\sqrt{E+N}$ sphere, wind up being less and less consequential, numerically, as the dimensionality of the space increases. The second reason to keep your eye on is that when we go from the number of points to the number of bits, we do so through the argument of a base 2 logarithm. If you are going to be making approximations, you cannot ask for more favorable circumstances than to have your approximation processed through the soft focus of a compressive nonlinearity.

If we're going to go with this ratio of volumes idea, though, an aspect of the problem that we must finesse is illustrated in Figure 5.24. We speak of how many noise spheres can fit inside the $\sqrt{E+N}$ sphere, but we see from the top of Figure 5.24 that when we commit to symbol on the surface of the $\sqrt{E}$ sphere, a good fraction of the ordinary, centered noise sphere lies just outside of the $\sqrt{E+N}$ sphere. Now we agreed early in our development that the noise would always be orthogonal to the signal, so in practice we don't ever expect to receive a signal with energy $\sqrt{E}+\sqrt{N}$. But we also insist on non-overlapping noise spheres when considering how many symbols we can distinguish with arbitrarily low error. This means that for every symbol in our library that is anywhere on the surface or within the interior of the $\sqrt{E}$ sphere, *we must associate with that symbol the full volume of a $\sqrt{N}$ noise sphere that lies entirely within the $\sqrt{E+N}$ sphere.* In our mental accounting, then, it is useful and important to associate noise spheres with *all* of our symbols in the offset way illustrated in the bottom of Figure 5.24.

Fine. Let's firm up the notion of a high-dimensional sphere having little volume in its interior. Consider two concentric spheres, as shown in Figure 5.25. To demonstrate that the volume of the interior becomes less important as the dimensionality increases, we look at the ratio of the volume of sphere 1 to the volume of sphere 2. What we expect is that for a given ratio R_1/R_2, the corresponding volume ratio should drop dramatically with increasing dimensionality. Suppose that we make $R_1/R_2 = 0.9$. For two-space:

$$\frac{\pi(0.9R)^2}{\pi R^2} = 0.81. \tag{5.61}$$

Figure 5.24 To every chosen symbol on the surface of or interior to the $\sqrt{E}$ sphere, we must attribute the volume of a full noise sphere that falls entirely within the $\sqrt{E+N}$ sphere. The bottom figure illustrates one possibility.

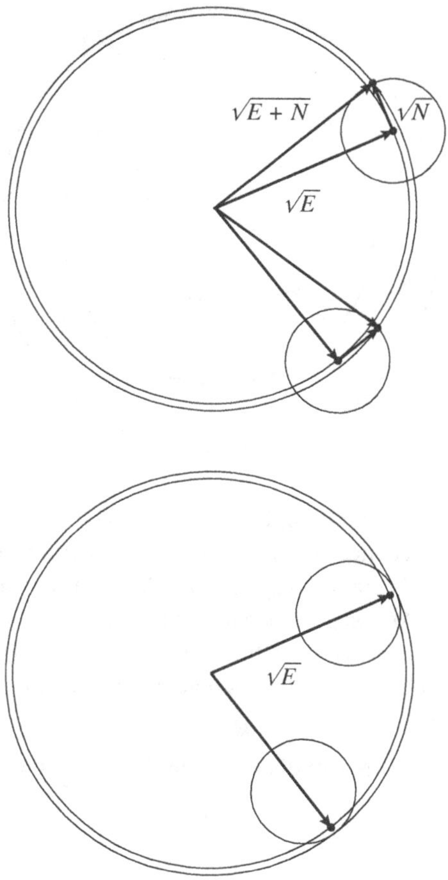

For three-space:

$$\frac{\frac{4}{3}\pi(0.9R)^3}{\frac{4}{3}\pi R^3} = 0.73. \tag{5.62}$$

In 10-space:

$$\frac{\frac{\pi^5}{\Gamma(6)}}{\frac{\pi^5}{\Gamma(6)}} \cdot \frac{(0.9R)^{10}}{R^{10}} = 0.35. \tag{5.63}$$

And in 100-space:

$$\frac{\frac{\pi^{50}}{\Gamma(51)}}{\frac{\pi^{50}}{\Gamma(51)}} \cdot \frac{(0.9R)^{100}}{R^{100}} = 2.7 \times 10^{-5}. \tag{5.64}$$

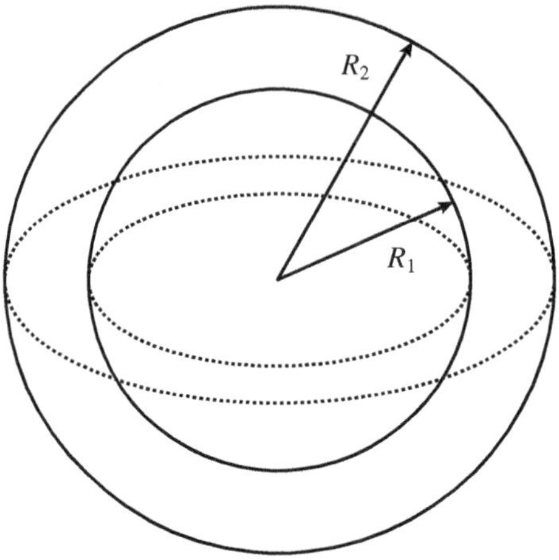

Figure 5.25 Two concentric, high dimensional spheres.

The Γ function is given to us by our mathematician friends, and the details need not concern us here since this function cancels out in the ratio. The critical fact for us is that for a sphere in n-space, the volume of that sphere goes as r^n.

At this point we are knocking on the door of deriving the Shannon limit. Once you are satisfied that we can get a useful upper limit on the number of distinguishable symbols that is reachable by dividing volumes of hyperspheres, we can write

$$\begin{aligned}\text{No. of distinguishable symbols} &= \left(\frac{\sqrt{E+N}}{\sqrt{N}}\right)^D \\ &= \left(\sqrt{\frac{E}{N}+1}\right)^D \\ &= (\text{SNR}+1)^{D/2}, \end{aligned} \tag{5.65}$$

where D is the dimensionality of the symbol space, or the number of samples that we consider at any one time for our batch processing.

We'll close this section by noting that if we have a total of $(\text{SNR} + 1)^{D/2}$ distinguishable symbols, then the number of bits represented by any given symbol is simply

$$\begin{aligned}\text{No. of bits} &= \log_2(\text{SNR}+1)^{D/2} \\ &= \frac{D}{2}\log_2(\text{SNR}+1). \end{aligned} \tag{5.66}$$

In the next section, we'll relate the dimensionality D to the channel bandwidth, and complete our journey to deriving the Shannon limit.

5.4 Arriving at the Shannon Limit

Looks like we are almost there. We have the logarithm of the SNR (plus 1!) that we need, but on the left side we have bits where we want bits per second. And instead of bandwidth on the right side, we have the dimensionality of the signal space. So we have a little work to do.

5.4.1 How Many Samples Do We Need for a Bandlimited Signal?

When we talk about the Shannon limit, we always start by asserting, or assuming, that our communications channel has a well-defined, finite bandwidth. In real systems, the finiteness of this bandwidth is enforced by legal regulation (as in, a government entity enforcing a limit on the amount of spectral real estate you are allowed to occupy), and/or by the physics of the systems involved.

So let's start with this. Suppose that our communications channel is rigorously enforced to be of bandwidth B, and we want to figure out how many bits per second we can send. If we know our SNR, and we know the number of samples we are going to end in a time interval T, then Section 5.3.3 tells us how many bits we can get in that time interval. The important question, therefore, is: how many samples do we need to take in interval T to completely characterize the signal? If we take fewer samples than this, we will miss information. If we take more samples than this, then the excess samples are more work for us with no payoff.

The answer is given to us in Section 2.2.3, where we observed that once we start sampling, we can sample without any aliasing if we require that our sampling frequency be equal to twice the highest frequency component B. If we adhere strictly to this rule, then for a signal of bandwidth B, we will sample at a rate of of $2B$ samples per second. In the time interval T, then, we have

$$\text{No. of Samples} = 2B \cdot T. \tag{5.67}$$

And now we have D, for the purposes of Eq. 5.66.

5.4.2 The Final Result

Now let's close the deal. From Eq. 5.66, we have that the number of bits we can extract from D samples is given by

$$\text{No. of bits} = \frac{D}{2}\log_2(\text{SNR} + 1). \tag{5.68}$$

But with $D = 2B \cdot T$, where T is the time interval over which we collect our samples, we have

$$\text{No. of bits} = BT \cdot \log_2(\text{SNR} + 1). \tag{5.69}$$

Dividing through both sides by T, on the right we get channel capacity C in bits per second, at last:

$$C = B \cdot \log_2(\text{SNR} + 1). \tag{5.70}$$

Quite a lot of work for an astonishingly simple result, but there it is! We got there.

5.4.3 A Note on the Geometry of Low SNRs

We kicked off this chapter putting the Shannon limit to work, and did so without fear of SNRs that were lower than unity. As the SNR gets smaller than unity, the argument of the logarithm, SNR + 1, gets closer to 1, and therefore the logarithm itself gets closer to zero...but not *quite* zero. And if is not zero, then in principle we can squeeze bits through the channel.

This was all well and good when we were just looking at a formula. But after spending Section 5.3.3 slogging through the geometry of high dimensional spheres, we chose examples where the $\sqrt{N}$ spheres were clearly small compared to the $\sqrt{E+N}$ sphere. Does the geometrical picture fall apart for us when the signal is buried in noise?

In the case of very low SNRs, our geometrical picture becomes what we see in Figure 5.26. No way we get more than one noise sphere in there, right?

There are two keys to moving forward from this point. The first is to remind ourselves that our intuition that comes from visualizing 2D and 3D spheres continues to *not* be useful. We are quite firmly in the realm of high dimensional spheres now. Then we look back, for example, at Eq. 5.64, for a reminder of how dramatically small differences in radius can make for huge differences in the volumes of

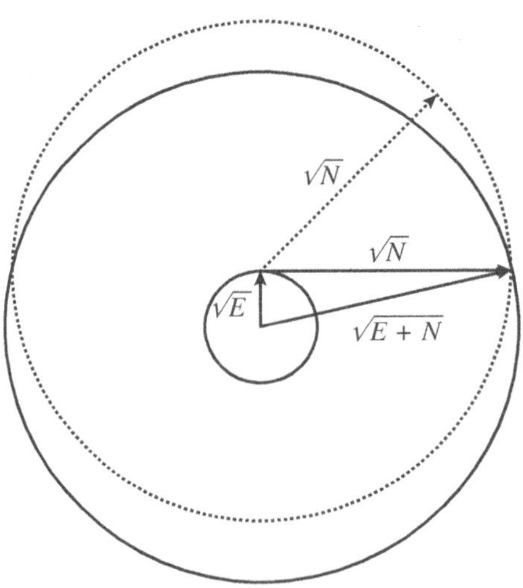

Figure 5.26 The geometry of low SNRs, looking hopeless in two dimensions. (As it should.)

spheres in high dimensions. As strange as Figure 5.26 looks, it remains true that $\sqrt{E+N} > \sqrt{N}$. And that means, for a sufficiently high dimensional space, we can in fact fit multiple noise spheres inside the $\sqrt{E+N}$ sphere. Despite 2D and 3D appearances, the underpinnings of the Shannon limit derivation still hold up for low SNRs.

5.5 Chapter Summary

There is one more thing that we need to justify, which is all the fuss over coming up with a fundamental, upper bound on the bit rate that we can achieve communicating across a channel. Why not just try your hardest, and see what happens?

The answer is that, in a deep way, there is all the difference in the world between trying to hit a goal that you know Nature allows, versus flailing at a goal that may or may not be achievable. Our greatest sprinters will strive their whole careers to run 100 m in under 9.5 seconds; tell them to try go under 6.5 seconds, and they will not try at all.[12] For a long time, running the mile in under four minutes was considered impossible. The feat was achieved for the first time in 1954, and has now been achieved by more than 1700 athletes.

In engineering, then, do not neglect to consider what Nature fundamentally allows. Many, many research challenges are launched with the mandate to advance the field by "10×" in one direction or another, with the idea that even if it is impossible, there will still be useful results that fall out along the way. This is not, respectfully, the author's experience. In the case of the Shannon limit, a tremendous amount of creative and useful work was accomplished over decades precisely because those pioneers knew, or thought they knew, what Nature's limit is. We are not always so fortunate to have fundamental limits come out as simply and beautifully as the Shannon limit does. And even when we think we have such a neat answer, humility is always called for in the form of openness to checking assumptions, since these limits may take decades to interrogate by experiment. But whatever the challenge, careful, diligent, creative searching for performance limits based on fundamental principles is never wasted effort.

---

12 At the time of this writing, the world record is 9.58 seconds.

6

Digital Modulation Basics

Radio communication starts with a remarkable fact, which is that if we drive a transmitting antenna with a sinusoid, we can detect that sinusoid in a suitably built receiver (see Figure 6.1 for a cartoon rendering). Now, suppose we turn on both the transmitter and receiver, as well as the sinusoidal drive. If we just leave it on like this...no information is transmitted. Or maybe minimal information: detecting a sinusoid in the receiver indicates that everything is in working order, whereas detecting no sinusoid tells us that things are not in working order. So that is one bit of information at most.

This traveling electromagnetic (EM) wave is called a "carrier," and the whole trick once we have the basic hardware in place is to modify this carrier in such a way that we encode our information on this far-traveling EM wave. We call this encoding process "modulation," or "modulating the carrier." In Section 1.5, we took a tentative stab at this by doing simple, on–off keying of the carrier. When it comes to using spectrum efficiently, we shall see that we can do much better than this.

6.1 Modulating a Carrier

We jump straight to the modulation technique that is most widely used in high-capacity networks today. This technique is called "quadrature amplitude modulation," or QAM for short and it is illustrated in Figure 6.2. I and Q are just normal, serial bit streams, shown in Figure 6.3. Now, recall from Section 2.5.2 that the spread of our signal in the frequency domain is determined by our basic pulse shape. Figure 6.4 shows the rectangular pulse shape and its corresponding shape in the frequency domain.

Energy Fundamentals of Radio, First Edition. Joel L. Dawson.

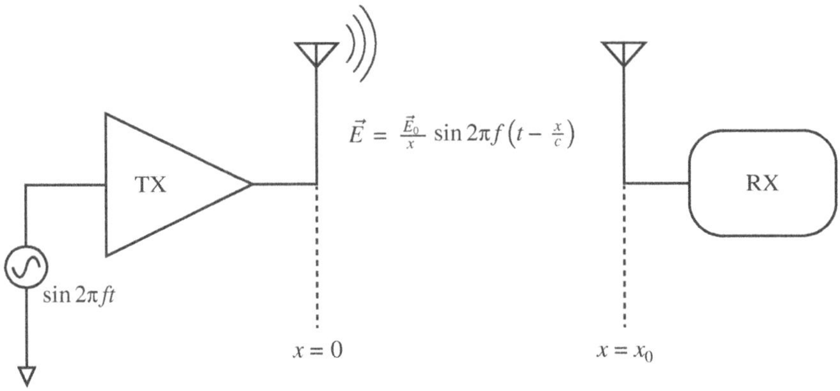

Figure 6.1 Transmitting and receiving an unmodulated carrier.

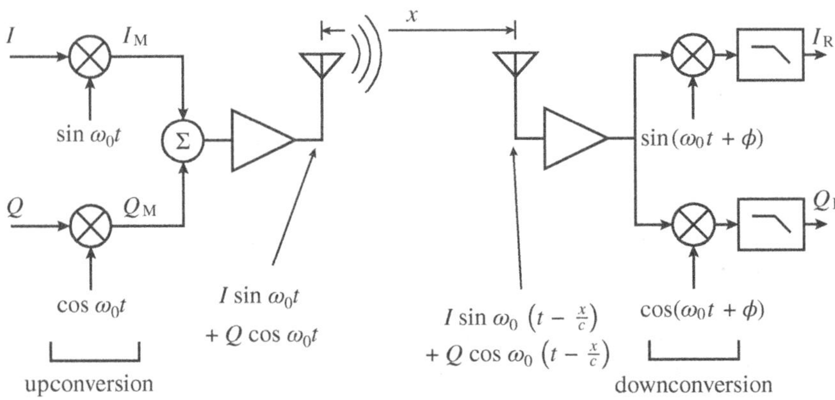

Figure 6.2 Illustration of QAM.

One of the things we love about QAM is that we have the ability to shape our basic pulses in exquisite detail. Fast digital-to-analog converters, which enable high degrees of oversampling of our digital signals, make this possible. Given that spectrum is a finite resource, we are particularly interested in pulse shapes that take up as little spectral real estate as possible, while also allowing us to pack our pulses as closely together in time without interfering with one another. We will have a great deal to say about this in Section 6.2.1.

Now, what is with the whole "multiply by $\sin \omega_0 t$ or $\cos \omega_0 t$" that we show in Figure 6.2? Suppose that we didn't, and that instead, we fed $I(t)$ and/or $Q(t)$ directly into an antenna. Figure 6.5 shows what $I(j\omega)$ and $Q(j\omega)$ might look like in the frequency domain.

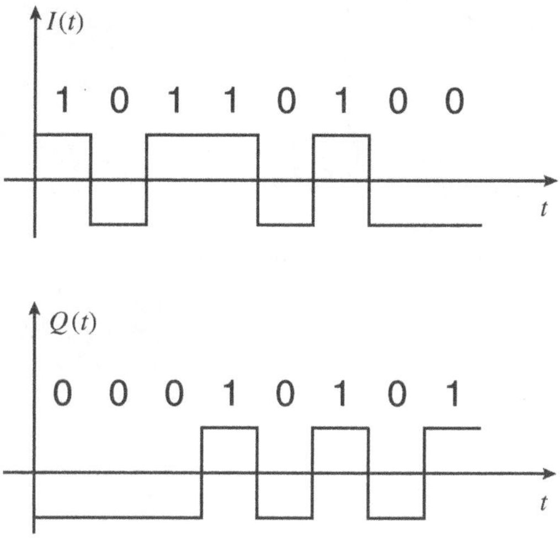

Figure 6.3 Example *I* and *Q* bit streams.

The practical problem with feeding signals like this to a physical antenna becomes clear when you sit down to, well, physically dimension the antenna. It helps to know that antennas are sized in proportion to the free-space wavelength of the sinusoids they are designed to transmit or receive. The wavelength of an EM wave in free space of frequency f is given by

$$\lambda = \frac{c}{f}, \tag{6.1}$$

where $c = 3.0 \times 10^8$ m/s and f is in hertz. The first problem we have, then, is that our low-frequency content requires huge antennas. If we assume a quarter-wave monopole as our baseline antenna structure, then for any content around, say, 100 kHz, we would need an antenna roughly of length

$$\frac{\lambda}{4} = \frac{1}{4} \cdot \frac{3.0 \times 10^8 \text{ m/s}}{1 \times 10^5 \text{ Hz}} = 750 \text{ m}. \tag{6.2}$$

The second problem is that the content just a short way away, at 50 kHz, would need a dramatically different antenna length: 1500 m. This issue is really only manageable if we keep our signals incredibly narrowband, where by "narrowband" we mean that the lower part of the band and the upper part of the band are separated by a small number of absolute hertz.

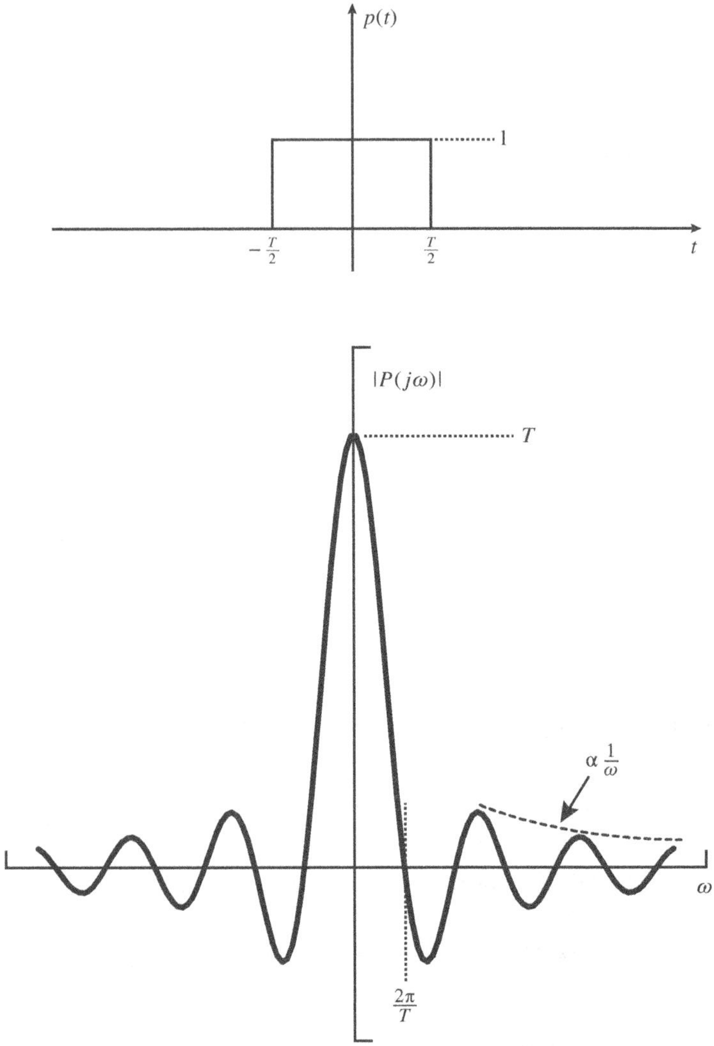

Figure 6.4 A rectangular pulse in time and its corresponding Fourier transform.

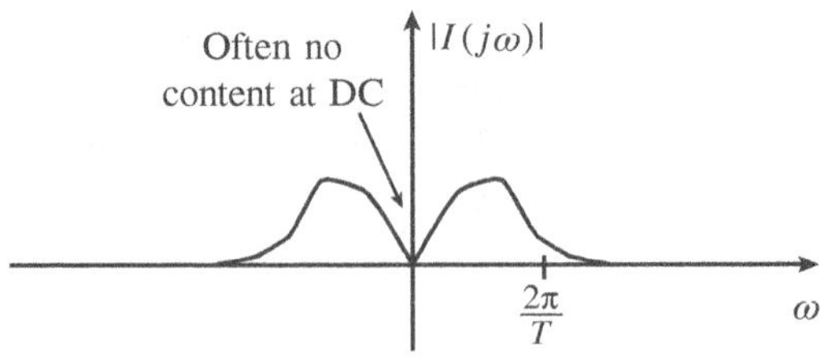

Figure 6.5 Fourier transform of a typical, unmodulated bit stream.

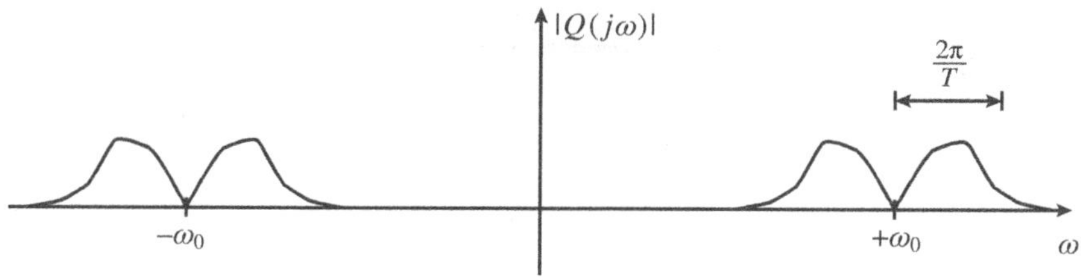

Figure 6.6 Fourier transform of a modulated bit stream.

We are now in position to appreciate the magic of modulators. Consider the product $Q_M = Q(t)\cos\omega_0 t$, and let $Q(j\omega)$ be the Fourier transform of $Q(t)$. We ask for $Q_M(j\omega)$:

$$\begin{aligned} Q_M(j\omega) &= \frac{1}{2}\int_{-\infty}^{\infty} Q(t)e^{j\omega_0 t}e^{-j\omega t}dt + \frac{1}{2}\int_{-\infty}^{\infty} Q(t)e^{-j\omega_0 t}e^{-j\omega t}dt \\ &= \frac{1}{2}\int_{-\infty}^{\infty} Q(t)e^{-j(\omega-\omega_0)t}dt + \frac{1}{2}\int_{-\infty}^{\infty} Q(t)e^{-j(\omega+\omega_0)t}dt \\ &= \frac{1}{2}Q(j(\omega-\omega_0)) + \frac{1}{2}Q(j(\omega+\omega_0)). \end{aligned} \tag{6.3}$$

In pictures, we have the new spectrum after modulating in Figure 6.6.

Let's redo our antenna sizing exercise now, with a carrier frequency of $\omega_0 = 2\pi \cdot 1$ GHz. The rough dimension of our quarter-wave monopole becomes

$$\frac{\lambda}{4} = \frac{1}{4} \cdot \frac{3.0 \times 10^8 \text{ m/s}}{1 \times 10^9 \text{ Hz}} = 7.5 \text{ cm}. \tag{6.4}$$

Much better! And what if we now consider the ideal monopole length for content that is 50 kHz away:

$$\frac{\lambda}{4} = \frac{1}{4} \cdot \frac{3.0 \times 10^8 \text{ m/s}}{(1 \times 10^9 + 50 \times 10^3) \text{ Hz}} = 7.499625 \text{ cm}. \tag{6.5}$$

That is to say: 7.5 cm is close enough! The beauty of modulating up to a high carrier frequency is that we can accommodate a large absolute bandwidth Δf while using resonant structures that have a small relative bandwidth $\Delta f/f_0$. This narrowband assumption is built into the reflexes of RF engineers and is absolutely fundamental to signal chains for wireless systems.

Now let's get to what happens on the receive side. The carrier has a phase shift associated with the distance z between the transmitter and the receiver. In addition, the local oscillator on the receive side will have some phase shift relative to the local oscillator in the transmitter. Mathematically, we can lump these two phase

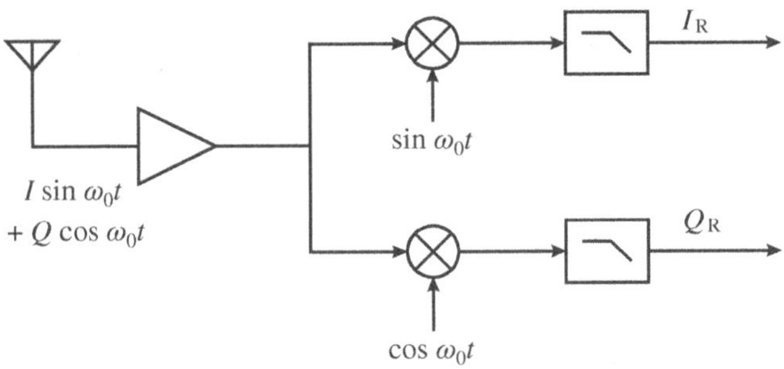

Figure 6.7 Coherent demodulation of a QAM signal.

shifts together in the parameter ϕ. So rather than have $\sin \omega_0 t$ and $\cos \omega_0 t$ as we had in the transmitter, we now have $\sin(\omega_0 t + \phi)$ and $\cos(\omega_0 t + \phi)$. And then in coherent receivers, which we will represent as the norm here, ϕ is chosen specifically to zero out phase shifts due to propagation and due to phase offsets between the local oscillators.[1] The effect of all of these careful parameter settings is that we can, with a clear conscience, draw our receiver as shown in Figure 6.7.

It is about to become clear why this quadrature modulation business is universally beloved. If we derive the output signals I_R and Q_R, we get

$$\begin{aligned} I_R &= (I \sin \omega_0 t + Q \cos \omega_0 t) \sin \omega_0 t \\ &= I\sin^2 \omega_0 t + Q \cos \omega_0 t \sin \omega_0 t \\ &= I\left(\frac{1}{2} - \frac{1}{2}\cos 2\omega_0 t\right) + \frac{1}{2}\sin 2\omega_0 t. \end{aligned} \tag{6.6}$$

The low pass filter cuts out the $2\omega_0$ terms, and we are left with

$$I_R = \frac{1}{2}I. \tag{6.7}$$

Similarly,

$$\begin{aligned} Q_R &= (I \sin \omega_0 t + Q \cos \omega_0 t) \cos \omega_0 t \\ &= I \sin \omega_0 t \cos \omega_0 t + Q\cos^2 \omega_0 t \\ &= \frac{1}{2}I \sin 2\omega_0 t + Q\left(\frac{1}{2} + \frac{1}{2}\cos 2\omega_0 t\right). \end{aligned} \tag{6.8}$$

So,

$$Q_R = \frac{1}{2}Q. \tag{6.9}$$

1 In mobile applications, the necessary phase shift is actively tracked.

Don't take the factors of 1/2 too seriously: we include them here to be algebraically correct, but there are so many other gain factors in these complicated signal chains that the absolute values of the 1/2s can pretty much be dismissed. What *is* meaningful, and exciting, is that with this quadrature modulator setup, *we have two parallel, noninterfering channels that nevertheless occupy the same spectral real estate!* We call them "orthogonal channels."

We never get tired of this trick.

6.2 Spectral Efficiency Is Important

We emphasized at the end of Section 6.1 that we love quadrature modulation. Because sin $\omega_0 t$ and cos $\omega_0 t$ are orthogonal, QAM affords us a particularly straightforward way to pack more bits/sec into a given slice of spectrum than we might have done with a less sophisticated modulation strategy. Why so much affection?

The answer is that bandwidth, or spectrum, is a finite resource. We spent all of Chapter 5 examining the fact that the upper limit on data capacity in a communications link is proportional to B, to how much spectrum is available. In some settings, such as very dense, urban areas, or at a busy event like an outdoor concert, the sheer number of users makes spectrum downright scarce. The prime directive in these settings is to absolutely maximize the number of bits per second flowing through the network (which is not separate from maximizing the number of users) for the given available B. Underneath that righteous upper limit $B\log_2(\text{SNR}+1)$ lies a world of implementations that fall short. What we find is that QAM figures prominently in the family of implementations that meet or get very close to the Shannon capacity limit. A network that features a high ratio of data capacity to spectral bandwidth, or C/B, is said to have *high spectral efficiency*.

So QAM is important, but there are more details to fill in. Let's look at some of those.

6.2.1 Pulse Shaping to Stay Within Our Spectral Boundaries

Now when we put QAM to work, there are a few guiding principles to keep in mind:

- Knowing that our channel will be corrupted by noise, our overall modulation strategy should have some way of dealing with this reality. A powerful concept we explored earlier is correlative sampling, covered in Section 5.2. It would be good to leverage the insights we gathered there.
- "Leveraging that insight," to be concrete, necessarily means forging an agreement between transmitter and receiver on a library of possible symbols. This way, our detector can use correlation to see through the fog of the noise floor.

In practice, we will see that we commit to pulse shapes $p(t)$ that are somewhat more nuanced than the rectangular shapes that one might write down at first.

- Being careful about pulse shape $p(t)$ allows us to keep our modulated signal band limited, as discussed in Section 2.5.2. However, the narrower our bandwidth B, the wider our pulse will be in time. And the wider the pulse in time, the fewer pulses we can fit in a given time window...and therefore, the fewer the bits per second we can send. A desirable pulse shape will be as narrow as it can be in time without being too broadband as required by the uncertainty principle, so we can squeeze these pulses as close together in time as we can.[2]

We have given voice to some important concerns and guiding principles. From here, the faster we get to implementation details, the better.

Start by agreeing that at the end of any given symbol period T, our goal is to deliver either a "1" or a "0." Our basic pulse shape will be $p[n]$, as shown in Figure 6.8, and the idea is that N sample intervals constitute a time period T. In this scheme, when we want to send a "1" we send

$$b[n] = +1 \cdot p[n]. \tag{6.10}$$

And when we want to send a "0," we send

$$b[n] = -1 \cdot p[n]. \tag{6.11}$$

The right picture to have in your mind for our transmit machine is given in Figure 6.9. We chose the transmit filter to have the "impulse response" equal to $p[t]$,[3] something that is easy to do when the filter is implemented digitally. The shape of $p[t]$ allows us to choose the bandwidth occupied by our pulse train,

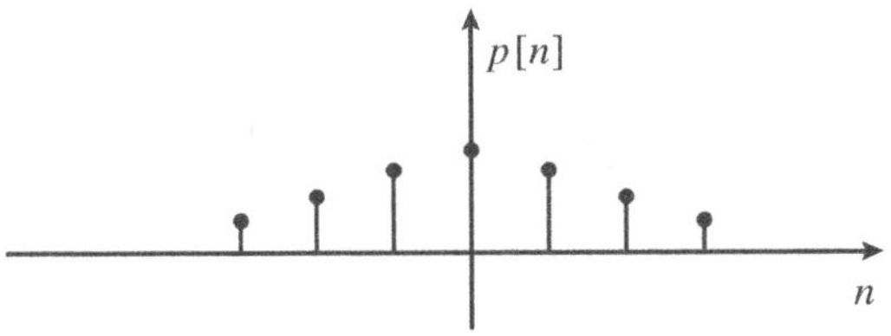

Figure 6.8 An example pulse shape.

2 A little lookahead for you modulation jockeys out there. Defining our pulse shapes using Nyquist filters is an incredibly clever way of allowing us to pack pulses in time much more closely than you might ordinarily imagine.

3 See Sections 11.2.1 and 11.2.2 for a detailed discussion of the impulse response for linear, time-invariant systems.

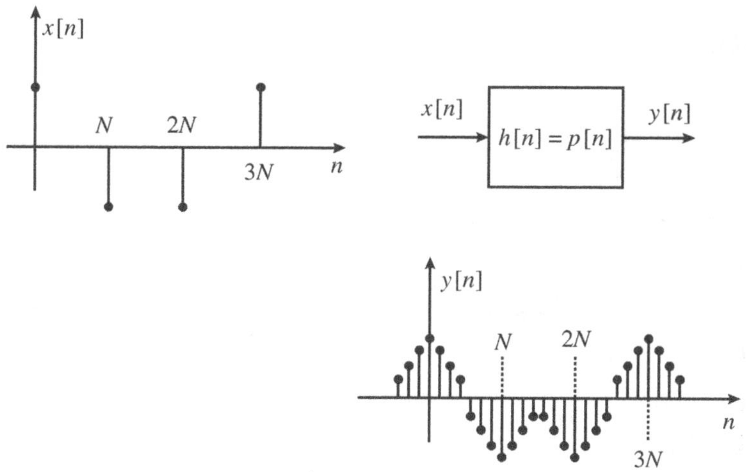

Figure 6.9 Using a filter with impulse response $p[n]$ to generate a pulse train.

and moreover the sample-by-sample specification sets us up well to use correlative sampling on the receiving end. There is yet more cleverness to be mined in picking a pulse shape, as we shall see.

Let's jump to the receiver, now, and suppose that we greet the received pulse by putting it through another filter whose impulse response is $h_r[n]$. It turns out that we can write the output from this receive filter as the convolution sum:[4]

$$y[n] = \sum_{k=-\infty}^{\infty} h_r[n-k] \cdot (\pm 1) \cdot p[k]. \tag{6.12}$$

The first trick we are going to play here is to choose both $h_r[n]$ and $p[n]$ to be symmetric, such that $h_r[n] = h_r[-n]$ and $p[n] = p[-n]$. If we do this, it becomes valid to rewrite Eq. 6.12 as:

$$y[n] = \sum_{k=-\infty}^{\infty} h_r[k-n] \cdot (\pm 1) \cdot p[k]. \tag{6.13}$$

And now the fun starts. Suppose we choose the impulse response $h_r[n]$ of the receive filter to *match* the pulse shape $p[n]$. Then, we have

$$y[n] = \sum_{k=-\infty}^{\infty} p[k-n] \cdot (\pm 1) \cdot p[k]. \tag{6.14}$$

Our overall system now looks like Figure 6.10.

4 See Section 11.2.2 for an explanation of the convolution sum.

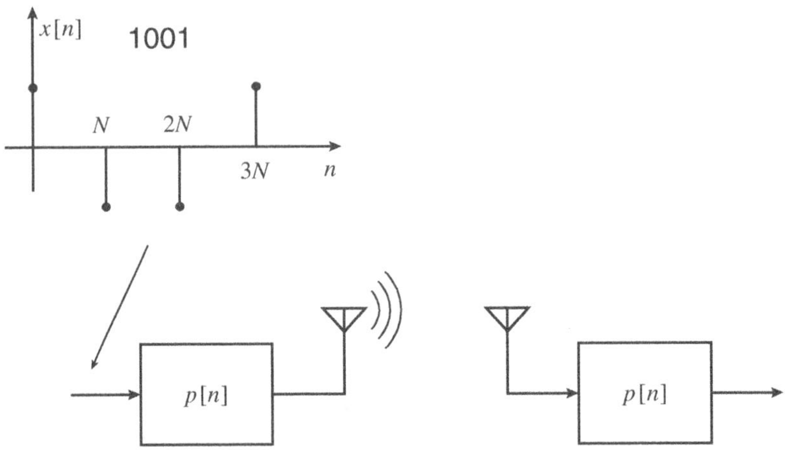

Figure 6.10 Simplified end-to-end system, signaling with our symmetric pulses $p[n]$.

Finally, we remind ourselves that we agreed on a symbol period of N, and that we are only interested in a final answer or decision every N samples. This means that at the output of the receive filter, a series of three pulses will look like

$$y[n] = \sum_{k=-\infty}^{\infty} p[k-n] \cdot ((\pm 1) \cdot p[k] + (\pm 1) \cdot p[k-N] + (\pm 1) \cdot p[k-2N]) . \tag{6.15}$$

Well, look what happens at the output of the receive side every N samples:

$$\begin{aligned} y[0] &= \sum_{k=-\infty}^{\infty} p[k] \cdot (\pm 1) \cdot p[k] \\ y[N] &= \sum_{k=-\infty}^{\infty} p[k-N] \cdot (\pm 1) \cdot p[k-N] \\ y[2N] &= \sum_{k=-\infty}^{\infty} p[k-2N] \cdot (\pm 1) \cdot p[k-2N]. \end{aligned} \tag{6.16}$$

Look familiar? It should! By choosing the impulse response of our receive filter to *match* our pulse shape, we wind up mathematically with the exact, magical, correlate-and-dump strategy that we explored in Section 5.2.2. We call this strategy "using matched filtering." It turns out that this is the optimal linear filtering strategy in the presence of noise.

There is only one other aspect of our discussion that we need to touch on, and that is the detailed shape of $p[n]$. We have already agreed that it should be symmetrical about $p[0]$. We also have committed to evaluating our receive output every

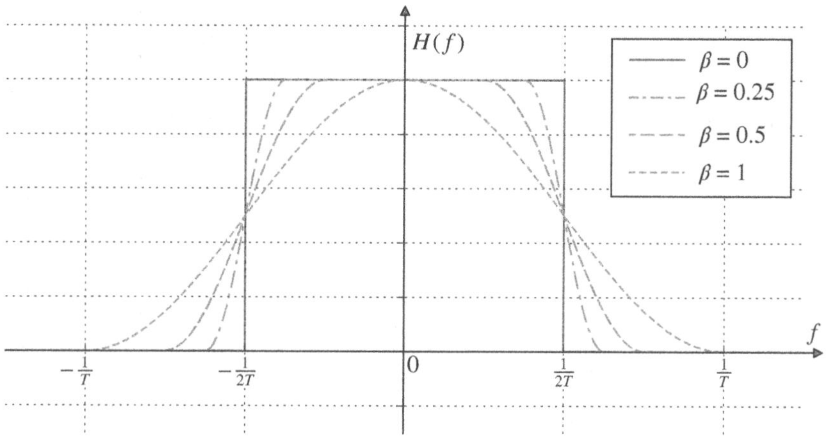

Figure 6.11 The frequency response of a raised-cosine filter. Credit: Krishnavedala/ wikimedia Commons/CC BY 3.0/https://en.wikipedia.org/wiki/Raised-cosine_filter.

N samples, at which time we make a decision about whether we received a "1" or a "0." It would be...really nice if consecutive symbols could be made invisible to each other in the time domain at the critical n values 0, N, $2N$, and so on.

Well!

It turns out that the "raised-cosine filter" is an example of a filter that checks *all* the boxes. The frequency response of a raised-cosine filter is given in Eq. 6.17, and illustrated in Figure 6.11.

$$H(f) = \begin{cases} 1 & |f| \leq \dfrac{1-\beta}{2T} \\ \dfrac{1}{2}\left[1 + \cos\left(\dfrac{\pi T}{\beta}\left[|f| - \dfrac{1-\beta}{2T}\right]\right)\right] & \dfrac{1-\beta}{2T} < |f| \leq \dfrac{1+\beta}{2T} \\ 0 & \text{otherwise.} \end{cases} \tag{6.17}$$

Here the symbol period T is of course equal to $N \cdot \Delta t$, where Δt is our sample period.

The impulse response is given by Eq. 6.18.

$$h(t) = \frac{1}{T} \cdot \frac{\cos\left(\pi\beta\frac{t}{T}\right)}{1 - \left(2\beta\frac{t}{T}\right)^2} \cdot \frac{\sin\left(\frac{\pi t}{T}\right)}{\frac{\pi t}{T}}. \tag{6.18}$$

(Sharp-eyed readers will have some concern about what happens when t approaches $\pm T/2\beta$, where both the numerator and the denominator vanish and

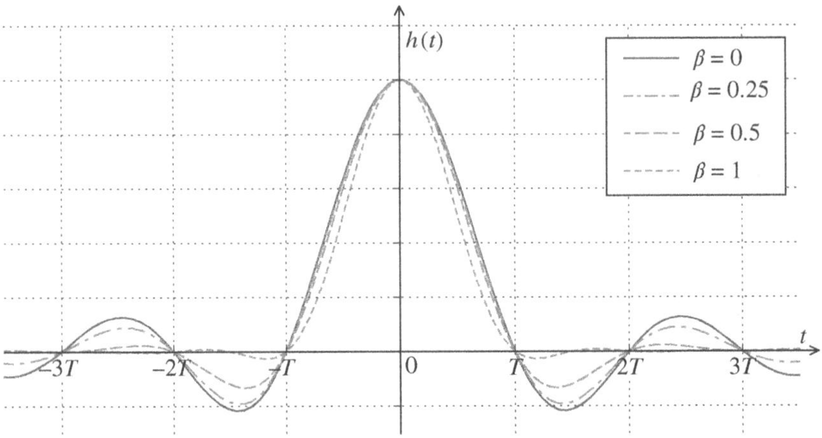

Figure 6.12 The impulse response of a raised cosine filter. Credit: Krishnavedala/wikimedia Commons/CC BY 3.0/https://en.wikipedia.org/wiki/Raised-cosine_filter.

then you have to figure out what happens. This turns out to be the perfect time to dust off L'Hôpital's Rule, which states that under suitable conditions,

$$\lim_{x\to c}\frac{f(x)}{g(x)}=\lim_{x\to c}\frac{f'(x)}{g'(x)}. \tag{6.19}$$

An eye-rolling piece of calculus trivia when you're first introduced to it, but it is amazing how many times it winds up saving the day.) We illustrate the impulse response in Figure 6.12.

Now, take a series of these impulse responses, and space them T (equals $N \cdot \Delta t$ seconds) apart, and behold: at the key instants where we make a decision about whether we have received a "1" or a "0," the contributions of prior pulses are exactly zero! (see Figure 6.13). We speak of individual pulses "being invisible to each other" at decision boundaries. In the field, this is called "having zero intersymbol interference." Because we never, ever let an opportunity to spawn a new acronym simply pass us by, you will hear this discussed as "having zero ISI." Pulse shapes that have this magical property are generated by what we call Nyquist filters.

If we put it all together, our signaling strategy looks as depicted in Figure 6.14. Now a final bit of cleverness. In Figure 6.14, we actually require the filter cascade $h_t(t) * h_r(t)$ to be a raised-cosine filter. By splitting the raised-cosine filter function across the transmitter and the receiver this way, we get both no ISI and the benefits of correlative sampling to combat noise. You would be forgiven for falling over at how nicely this all comes together. The net effect is that we are able to push our signaling pulses very closely together in time without requiring them to be highly

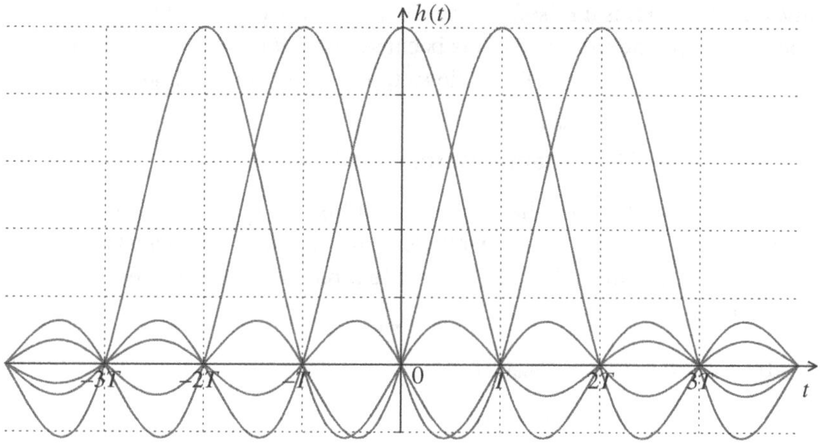

Figure 6.13 A series of raised-cosine filter impulse responses, spaced properly every T seconds to ensure zero ISI. Credit: Krishnavedala/wikimedia Commons/CC BY 3.0/https://en.wikipedia.org/wiki/Raised-cosine_filter.

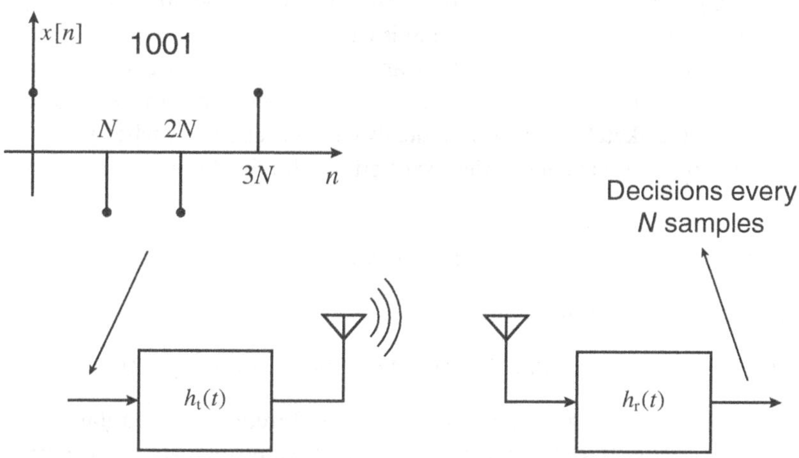

Figure 6.14 Simplified end-to-end system, now using our zero ISI filter strategy.

localized in time (and therefore very broadband). We thereby pack more pulses into a given time interval, which means more bits per second, and at the very least feel that we have made a strenuous effort to get close in performance to the Shannon limit. Raised-cosine filters are a great example of the kind of creativity that our communications theory friends have produced in the decades since Shannon elucidated Nature's limit on capacity.

Finally, if $h_t(t) * h_r(t)$ is a raised-cosine filter, it is conventional to call $h_t(t)$ and $h_r(t)$ "root raised-cosine filters." This is because since $h_t(t) = h_r(t) = h(t)$ and we now require $|H(j\omega)|^2 = |H_{rcf}(j\omega)|$, it follows that $|H(j\omega)| = \sqrt{|H_{rcf}(j\omega)|}$.

6.2.2 We Often Avoid Constant-envelope Modulation...

Constant-envelope modulation has a pretty rich history, but is not so commonly used today because it is not spectrally efficient. It is worth seeing why this is.

In the most abstract form, we might encode a message as a phase-modulated carrier as follows

$$x_{pm}(t) = A \sin\left(2\pi f_0 t + m_{pm}(t)\right) \tag{6.20}$$

and a frequency-modulated carrier as

$$x_{fm}(t) = A \sin\left(2\pi f_0 t + m_{fm}(t) \cdot t\right), \tag{6.21}$$

with $m_{pm}(t)$ having units of radians, and $m_{fm}(t)$ having units of radians per second. Sometimes we favor constant-envelope modulation because it leads to simplifications in the design of our circuits. At other times, we may explicitly favor energy efficiency over spectral efficiency when that is called for (see Section 6.2.3).

Now, frequency and phase modulation are famously awkward for analysis. The clearest analytical path is to step back, for a moment, from thinking about implementation, and sketch out mathematically what is happening when we modulate this way. Trigonometric identities, so often helpful, lead the way. For phase modulation, we have

$$x_{pm}(t) = A \sin(2\pi f_0 t) \cos(m_{pm}(t)) + A \cos(2\pi f_0 t) \sin(m_{pm}(t)), \tag{6.22}$$

and for frequency modulation, we have

$$x_{fm}(t) = A \sin(2\pi f_0 t) \cos(m_{fm}(t) \cdot t) + A \cos(2\pi f_0 t) \sin(m_{fm}(t) \cdot t). \tag{6.23}$$

One more piece, which is that phase modulation and frequency modulation are linearly related cousins. If we want to, we can send the same message two different ways if we equate[5]

$$m_{fm}(t) = \frac{d}{dt} m_{pm}(t). \tag{6.24}$$

Mathematically, we can depict these two constant-envelope modulation strategies as shown in Figure 6.15.

5 This equivalence is well known to those familiar with phase-locked loops. See Section 11.3 for details.

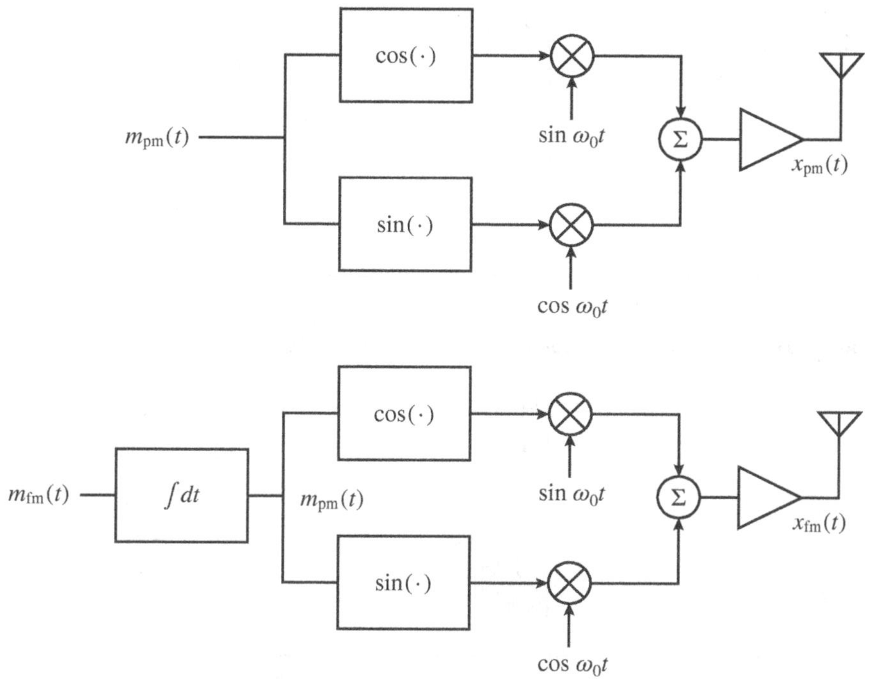

Figure 6.15 Abstract representations of phase modulation and frequency modulation. The emphasis here is on "abstract": note that with frequency modulation in particular we can, and often do in practice, replace the integrator, sine and cosine blocks, mixers, and summation with a voltage-controlled oscillator. This abstract structure is mathematically equivalent, though, and easier for hand analysis.

Now the subject here is spectral efficiency. We are powerfully aided by reminding ourselves of the polynomial expansions of $\sin x$

$$\sin x = x - \frac{x^3}{3!} + \frac{x^5}{5!} - \cdots \tag{6.25}$$

and $\cos x$

$$\cos x = 1 - \frac{x^2}{2!} + \frac{x^4}{4!} - \cdots \tag{6.26}$$

It turns out that higher order polynomial terms wreak havoc on the bandwidth that our modulated signal occupies. This is well-known among power amplifier designers, and indeed, we will spend some time on this in Section 11.4.2. Let's look, for example, at the distortion byproducts caused by the sine block of Figure 6.15. For now, suppose that $m_{pm}(t)$ is a signal of bandwidth B; we might draw $M_{pm}(j\omega)$ as in Figure 6.16. From our discussion in Section 2.3.3, we know that in the privacy of our minds, we can think in terms of a Riemann sum as depicted in Figure 6.17. To see how nonlinearity impacts how our signal is distributed in the frequency

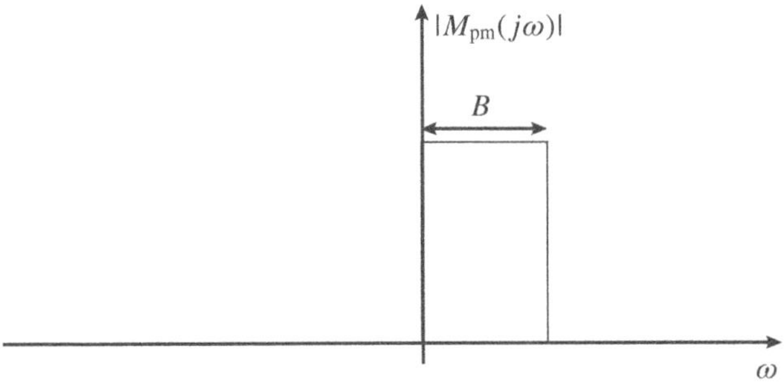

Figure 6.16 A low-pass signal containing our message $m(t)$.

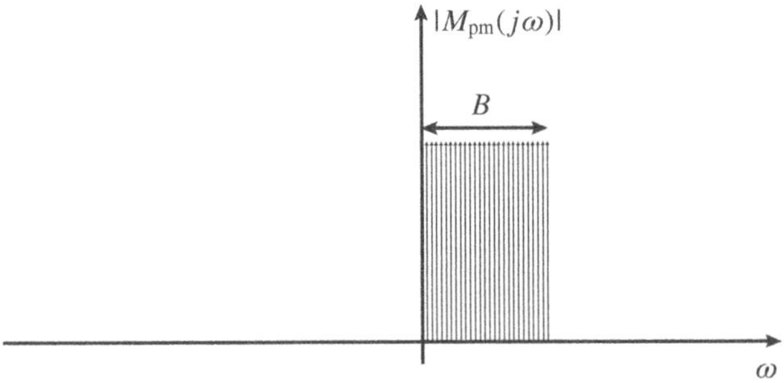

Figure 6.17 Our low-pass signal, discretized in the frequency domain for purposes of analysis.

domain, we need only examine how a sum of closely spaced sinusoids is treated by an individual polynomial term. Start with the sum itself, written out for clarity in terms of, say, N closely spaced frequency components:

$$m_{pm}(t) = \sum_{i=0}^{N-1} A_i \cos \omega_i t + B_i \sin \omega_i t. \tag{6.27}$$

Next, pick a nonlinear term (x raised to the second power or higher) from Eq. 6.25 or 6.26. Easiest to start, actually, with x raised to the power of 2; we will leave that to the reader to try out, and choose instead here the slightly more complicated x^3 term from the expansion of the sine in Eq. 6.25:

$$\begin{aligned} \frac{m_{pm}^3(t)}{3!} =& \frac{1}{3!}\left(\sum_{i=0}^{N-1} A_i \cos \omega_i t + B_i \sin \omega_i t\right) \\ &\times \left(\sum_{j=0}^{N-1} A_j \cos \omega_j t + B_j \sin \omega_j t\right) \\ &\times \left(\sum_{k=0}^{N-1} A_k \cos \omega_k t + B_k \sin \omega_k t\right). \end{aligned} \tag{6.28}$$

The right thing to do with hand analysis is to leave the detailed calculations to computer aids, and see what we can glean instead here in terms of insight. We know that when the dust settles, we will have the summation of a bunch of terms of the form:

$$A_i \cos \omega_i t \cdot B_j \sin \omega_j t \cdot A_k \cos \omega_k t \tag{6.29}$$

with all the combinations of As, Bs, and sines and cosines eventually represented. While the tedium of computing all the terms is best left to a computer, we can gain insight by noticing that there will be a group of terms involving the highest frequency component, ω_{N-1}. One such term will be

$$B^3_{N-1}\sin^3\omega_{N-1}t, \tag{6.30}$$

for example, and another will be

$$B^3_{N-1}\cos^3\omega_{N-1}t. \tag{6.31}$$

If you work your trigonometric identities, what you will find is that such terms contribute frequency components at $3\omega_{N-1}$, or three times the highest frequency component in our original signal!

We will give an example of working out more details of such a calculation in Section 11.4.2. For now, the key insight is that a low-pass $m_{\mathrm{pm}}(t)$ of bandwidth B, when processed by a cubic nonlinearity, morphs into low-pass signal of bandwidth $3B$. If you do enough calculations, you will be able to convince yourself that Figure 6.18 is a serviceable cartoon for this bandwidth expansion. The figure gives an idea of what would happen if we passed $m_{\mathrm{pm}}(t)$ through $x - \frac{x^3}{3!}$, which are the first two terms of the polynomial expansion of sin(x). We see that the third-order term has expanded our spectral footprint by 3× without doing us the courtesy of

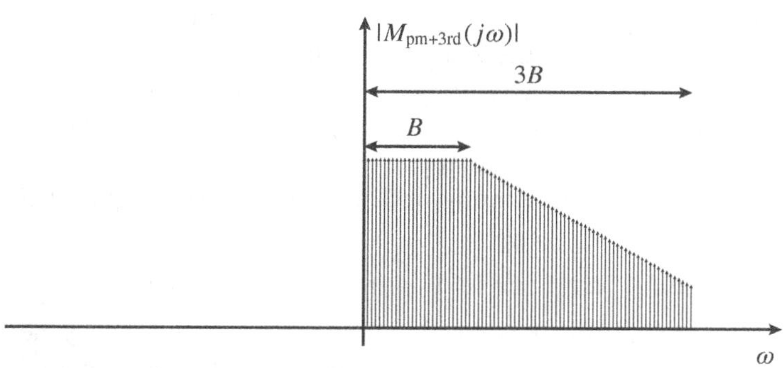

Figure 6.18 Conceptual drawing of the low-pass signal after processing by the sine function, but truncated at just two terms: $x - \frac{x^3}{3!}$.

increasing our information content by 3×. Awful. Worse, the pattern repeats: the second-order term (from the cosine in Eq. 6.26) doubles the bandwidth, the fifth-order term commits the same crime to the tune of 5×, the seventh-order term to the tune of 7×, and so on. Now, the steadily increasing factorial denominators do help ($5! = 120$, so down by 42 dB). Still, from the standpoint of spectral efficiency, this is not what we like to see.

6.2.3 ...Except When Energy Is More Scarce Than Spectrum

If you are not careful, the sheer analytical might of the Shannon limit can lead you to a blind spot. This has certainly happened to a number of authors over the last few decades, some of whom will casually use words like "awful" and "crime" in connection with modulation strategies that are not spectrally efficient.

But there are things to love about a constant envelope. For one thing, it is possible to design radio transmitters that are *much* more energy efficient if we know ahead of time that we are using a constant envelope. Sometimes, in the real world, things like parts count, or the convenience of a legacy design that speeds time to market, etc., are more important than spectral efficiency. This is as good a time as any to remind yourself that "to a hammer, everything looks like a nail." Always make sure you are putting your solution in service of the problem in front of you, and not the other way around.

6.2.4 Navigating the Modulation Alphabet Soup

There are many, many modulation strategies in circulation at the time of this writing. Most have two things in common: (1) an impenetrable acronym that leaves the newcomer bewildered and overwhelmed[6]; (2) mathematically, they can usually be regarded as a special case of QAM. For this reason, and in the interest of brevity where brevity is possible in this complex field, we drill down on QAM and leave most of these other techniques for the reader's exploration. You will be helped in your exploration by having studied the fundamentals here.

That said, at this point in the evolution of wireless systems there is one modulation strategy that deserves special mention. It is built on the foundation of QAM, and is a powerful way to organize network traffic among multiple users. It is called orthogonal frequency division multiplexing, or OFDM (naturally). Let's get into it.

6.2.5 OFDM: A Powerful Technique for Sharing Network Resources

OFDM is currently the dominant modulation strategy for high-capacity networks. It is so good, in fact, that it has achieved the rare feat of being both an

---

6 For phase shift keying *alone*, an incomplete list: BPSK, QPSK, OQPSK, SOQPSK, $\pi/4$-QPSK, DPQPSK...

implementation focal point for 4G networks and of continuing in this role for 5G. In order to properly understand OFDM, it helps to understand why we love it and what problems it solves. In brief:

1. In an OFDM system, the modulator and demodulator functions can be performed using the Inverse fast Fourier transform (IFFT) and fast Fourier transform (FFT) algorithms, respectively. We will get more into why this is, but it is impossible to overstate the practical and emotional impact of having the FFT at the heart of your strategy, almost no matter what you are doing in a signal chain. The FFT is actually an implementation strategy for computing the discrete Fourier transform (DFT; see Section 2.3.4) that is hugely more computationally efficient than a straightforward mapping of Eq. 2.96. In his book on linear algebra, Gilbert Strang described the FFT as "the most important numerical algorithm of our lifetime."[7] It also happens to be a tremendous fit for wireless communications employing OFDM.
2. A major, major challenge of high data-rate communications is that frequency-dependent characteristics of the channel tend to smear our symbols out in time, kind of like what we see in Figure 5.19. In that figure, you can see that we changed our signal slowly enough so as not to have a problem. But it is clear that if we made T much smaller, we could say that adjacent symbols would interfere with each other. There are two basic approaches to dealing with this problem. One is to keep squeezing T smaller and smaller, and then use an equalization filter that undoes the dispersion of the resistor–capacitor (*RC*) filter. If that sounds hard, that's because it *is* hard and involves complicated filters that consume a lot of computational resources. The other approach is to live with a nice, long, T, and use *parallel channels* to allow us to get a high aggregate symbol rate. OFDM is effectively a realization of the second approach.
3. Take a moment to think about how *you* would share wireless spectrum among a large number of users. It makes it easier if you free yourself with the admission that there are many ways to do this.[8] You could do it by allocating the whole swath of spectrum to every user, but restricting them to use nonoverlapping time slots. Depending on their need, you could assign users proportionally more or fewer time slots as appropriate. Another approach is to divide your spectrum up into small individual slices, parallel channels if you will; users with heavier needs get more slices, users with lighter needs get fewer. Whatever scheme you come up with....the only technique currently that lets you exploit the power of the FFT is OFDM. So here we are.

---

7 Strang. *Introduction to Linear Algebra.* Wellesley, MA: Wellesley-Cambridge Press, 1993.
8 In fact, it is a good exercise in creativity to come up with as many ways on your own as you can.

The DFT for Modulation and Demodulation

Consider, again, Eq. 2.96. We are going to invert the matrix using Eq. 2.103 and write the *C*s in terms of the *x*s:

$$\begin{bmatrix} C_0 \\ C_1 \\ \vdots \\ C_{N-1} \end{bmatrix} = \frac{1}{N} \begin{bmatrix} 1 & 1 & \cdots & 1 \\ 1 & e^{-j\left(\frac{2\pi}{N}\cdot 1\right)\cdot 1} & \cdots & e^{-j\left(\frac{2\pi}{N}\cdot(N-1)\right)\cdot 1} \\ \vdots & \vdots & \vdots & \vdots \\ 1 & e^{-j\left(\frac{2\pi}{N}\cdot 1\right)\cdot(N-1)} & \cdots & e^{-j\left(\frac{2\pi}{N}\cdot(N-1)\right)(N-1)} \end{bmatrix} \begin{bmatrix} x[0] \\ x[1] \\ \vdots \\ x[N-1] \end{bmatrix} \quad (6.32)$$

The matrix on the right with all the exponentials is the conjugate transpose of the matrix in Eq. 2.96 (you need the factor of $1/N$ to make it the inverse, in that the product of the matrices results in a matrix with all ones on the diagonal and zeros everywhere else). This matrix, when applied to a series of time samples (arranged as a column vector as shown), gives us the DFT of that sequence. When we talk specifically of the *fast* Fourier transform, or FFT, we mean implementation details that allow for its computation with minimal hardware resources and maximum speed. It helps, for example, for N to be an integer power of 2.

Let's pay attention to the left side of Eq. 6.32. Suppose that we have a buffer of N time samples. Our computational engine operates by grabbing those N samples, performing an FFT, and giving a new buffer consisting of the C_is. Let us further index each complete buffer with the index m. The mth collection of C_is would be given by

$$\begin{bmatrix} C_0[m] \\ C_1[m] \\ \vdots \\ C_{N-1}[m] \end{bmatrix} = \frac{1}{N} \begin{bmatrix} 1 & 1 & \cdots & 1 \\ 1 & e^{-j\left(\frac{2\pi}{N}\cdot 1\right)\cdot 1} & \cdots & e^{-j\left(\frac{2\pi}{N}\cdot(N-1)\right)\cdot 1} \\ \vdots & \vdots & \vdots & \vdots \\ 1 & e^{-j\left(\frac{2\pi}{N}\cdot 1\right)\cdot(N-1)} & \cdots & e^{-j\left(\frac{2\pi}{N}\cdot(N-1)\right)(N-1)} \end{bmatrix} \begin{bmatrix} x_{\mathrm{m}}[0] \\ x_{\mathrm{m}}[1] \\ \vdots \\ x_{\mathrm{m}}[N-1] \end{bmatrix}. \quad (6.33)$$

In *words*, then, successive values of $C_0[m]$ tell the story of how the DC content of the original signal $x[n]$ evolves from buffer to buffer; the (complex) $C_1[m]$ tells us the amplitude and phase of $e^{j\frac{2\pi}{N}\cdot n}$ or, equivalently, the mix of $\cos\left(\frac{2\pi}{N}\cdot n\right)$ and $\sin\left(\frac{2\pi}{N}\cdot n\right)$ that are in the mth buffer of $x[n]$; $C_2[m]$ tells us the mix of $\cos\left(\frac{2\pi}{N}\cdot 2\cdot n\right)$ and $\sin\left(\frac{2\pi}{N}\cdot 2\cdot n\right)$ that are in the mth buffer of $x[n]$; and so on.[9]

9 Remember that the *C*s are complex, and so, for example, $C_1 = C_{1R} + jC_{1I}$. When we multiply C_1 by $e^{j\left(\frac{2\pi}{N}\cdot 1\right)\cdot n}$, we get

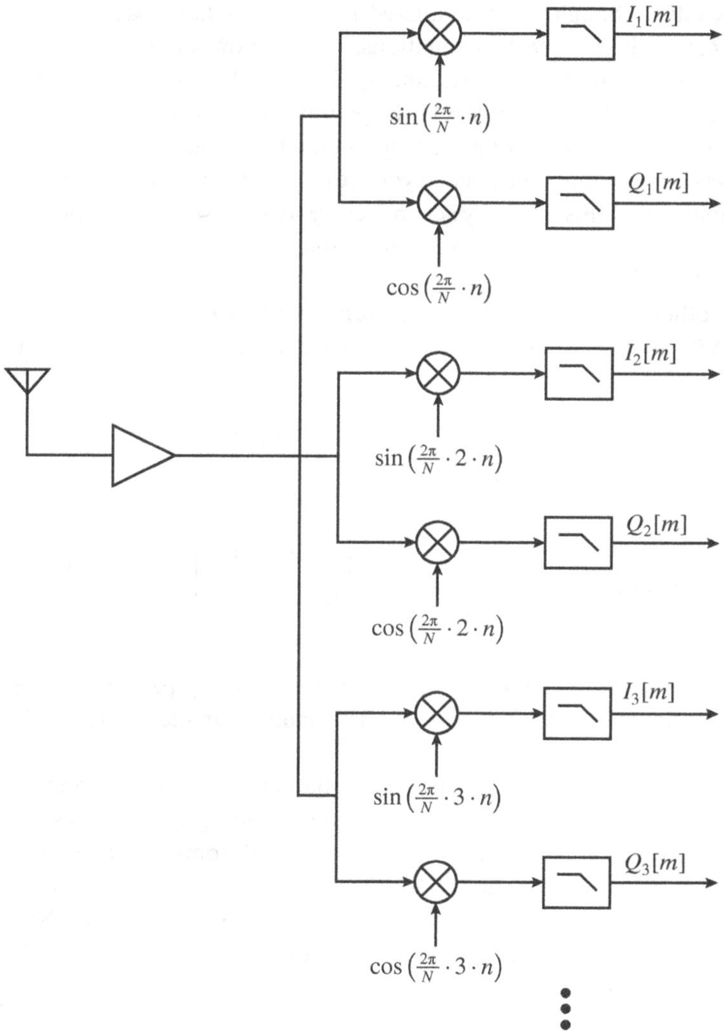

Figure 6.19 The DFT as a demodulator.

$$
\begin{aligned}
C_1 e^{j\left(\frac{2\pi}{N}\cdot 1\right)\cdot n} = &\left(C_{1R}\cos\left(\frac{2\pi}{N}\cdot 1\cdot n\right) - C_{1I}\sin\left(\frac{2\pi}{N}\cdot 1\cdot n\right)\right) \\
&+ j\left(C_{1R}\sin\left(\frac{2\pi}{N}\cdot 1\cdot n\right) + C_{1I}\cos\left(\frac{2\pi}{N}\cdot 1\cdot n\right)\right)
\end{aligned}
\tag{6.34}
$$

In our construction, $x[n]$ is purely real, so the imaginary part in Eq. 6.34 is ultimately canceled in the inverse DFT sum because of conjugate symmetry among the Cs (as we saw in our first

Which is...exactly as though we demodulated a bunch of equally spaced carriers all at once! What we just showed in equations, we will now show in a picture. When you take an N-sample time buffer, and apply the FFT, it is exactly like you have done what is depicted in Figure 6.19. Hopefully, now you see what OFDM really is. We choose to transmit a tight group of equally spaced subcarriers and apply QAM to each individual subcarrier. If you are oriented toward analog circuits and imagine doing all of this with physical mixers, you are quickly overwhelmed by the sheer parts count. But what we see here is that the DFT makes it possible, and the FFT makes it easy.

It works the other way as well. On the modulator side, we use Eq. 2.96 in its original form. We call the matrix, in this case, the inverse DFT or, more commonly, the IFFT.

$$\begin{bmatrix} 1 & 1 & \cdots & 1 \\ 1 & e^{j\left(\frac{2\pi}{N}\cdot 1\right)\cdot 1} & \cdots & e^{j\left(\frac{2\pi}{N}\cdot(N-1)\right)\cdot 1} \\ \vdots & \vdots & \vdots & \vdots \\ 1 & e^{j\left(\frac{2\pi}{N}\cdot 1\right)\cdot(N-1)} & \cdots & e^{j\left(\frac{2\pi}{N}\cdot(N-1)\right)(N-1)} \end{bmatrix} \begin{bmatrix} C_0[m] \\ C_1[m] \\ \vdots \\ C_{N-1}[m] \end{bmatrix} = \begin{bmatrix} x_{\mathrm{m}}[0] \\ x_{\mathrm{m}}[1] \\ \vdots \\ x_{\mathrm{m}}[N-1] \end{bmatrix} \tag{6.35}$$

The Cs are complex, and by choosing the real and imaginary parts we choose the I and Q amplitudes for each subcarrier. The *modulator* picture looks like Figure 6.20.

Finally, when you have mastered the details, the hardest part about understanding OFDM can be getting the Big Picture in your head (see Figure 6.21). It's a lot to take in, but it helps understanding if we flesh it out with some concrete implementation details. In real systems, the DFT subcarriers are going to relate to analog frequencies through an internal sample interval Δt. That is, subcarrier No. 1 will modulate an analog-world complex sinusoid of the form

$$e^{j\left(\frac{2\pi}{N\cdot\Delta t}\cdot 1\right)n\Delta t}, \tag{6.36}$$

subcarrier No. 2 will modulate

$$e^{j\left(\frac{2\pi}{N\cdot\Delta t}\cdot 2\right)n\Delta t}, \tag{6.37}$$

and so on. We may properly say that our sampled-data signal chain discretizes real-world time in chunks of sample period Δt. Our subcarrier spacing is then

$$\Delta f = \frac{1}{N\cdot\Delta t}. \tag{6.38}$$

---

discussion of the DFT in Section 2.3.4). We see here that $C_{1R}[m]$ and $C_{1I}[m]$ allow us to read off $Q_1[m]$s and $I_1[m]$.

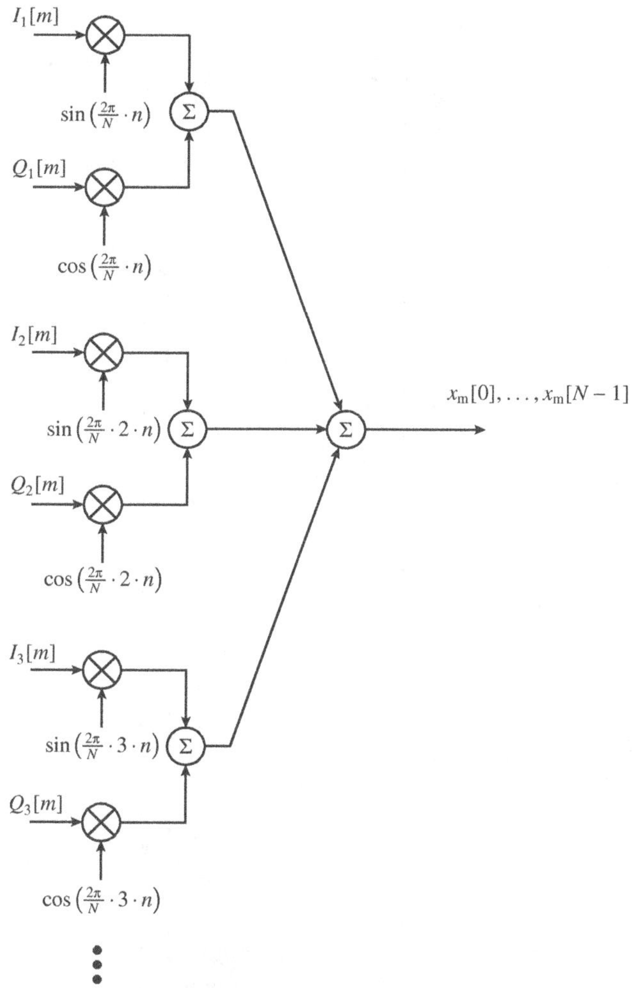

Figure 6.20 The IDFT as a modulator.

Here we get to an important piece of OFDM arithmetic that can be quite mystifying the first few times one approaches this implementation strategy. The parameter N tells us the dimensions of the FFT and IFFT matrices, and therefore the number of subcarriers in our OFDM signal. Notice also in the demodulator that we process our receive time domain samples in blocks of N samples ($x_m[0] \ldots x_m[N-1]$ in Figure 6.21). By committing to processing our time samples in batches of N, we necessarily mean that the $C_i[m]$ is constant for this entire N-sample interval of time. Each subcarrier only changes its mix of I and Q every N samples, which leads to $N \cdot \Delta t$ being called the "symbol period."

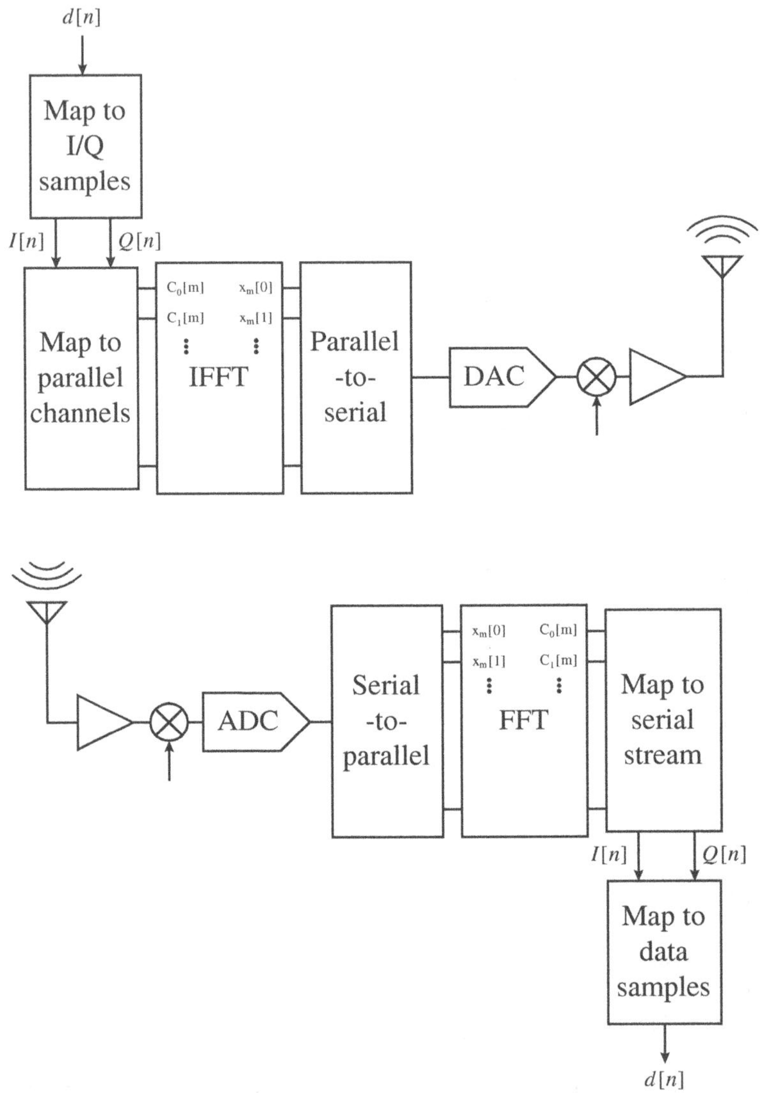

Figure 6.21 Overall picture of a link utilizing OFDM.

And what of Δt, the sample rate? In real-world situations, we sometimes have to deduce this, but it turns out to be easy enough. When you talk to OFDM people, they almost always lead with the carrier spacing (in hertz) and the number of carriers, and may or may not say what the symbol period is. All you need is N, the number of subcarriers, and *either* the symbol period *or* the carrier spacing since the last two are reciprocals of each other.

As an example, an 802.11 n/ac WiFi channel at the time of writing may have 64 subcarriers (notice 64 is a power of 2 ($= 2^6$)), and with a subcarrier spacing of 312.5 kHz. The means that the symbol period is $1/(312.5\ \text{kHz}) = 3.2\ \mu\text{s}$. Our sample period for the purposes of the FFT is therefore

$$\begin{aligned} N \cdot \Delta t &= 3.2\ \mu\text{s} \\ \Delta t &= \frac{3.2\ \mu\text{s}}{64} = 50\ \text{ns}. \end{aligned} \tag{6.39}$$

Once you understand the foregoing, you have the conceptual essence of OFDM. There are always important nuances as you get closer to implementation, but what is presented here has served the author well in terms of staying conceptually grounded.

What Is "Orthogonal" in OFDM?

If you wade into the literature on OFDM, the one concept you could never miss is that the subcarriers are *orthogonal.* You can see now that by having the DFT as the foundational signal processing operation, the subcarriers are in fact orthogonal according to the definition we worked out in Eq. 2.103. You could argue that nothing more really needs to be said.

However, *much* more is almost always said, usually along the lines of: "The symbol period is $1/\Delta f$ to make the subcarriers orthogonal." You can see now the sense in which this is certainly true. We just showed that the I and Q values, which are determined by the complex Cs, must indeed be constant for the duration of the N-sample time buffer in order for our FFTs and IFFTs to make sense. If we accept that, then yes, the symbol period $N \cdot \Delta t$ does indeed equal $1/\Delta f$ according to Eq. 6.38.

To prepare you for what's out there, then, let's look at an IFFT input in Figure 6.21, say the complex $C_1[m]$, and further focus on just the $I_1[m]$ part. When we insist that $I_1[m]$ can only change on the symbol boundaries, what mean is shown in Figure 6.22. Referring back to Section 2.5.2, we can see that for OFDM, we have gone with a rectangular pulse shape as shown in Figure 6.23. The continuous-time (CT) Fourier transform of the pulse is also shown in Figure 6.23. What we see is that by using rectangular time pulses of width $N \cdot \Delta t$, the CT Fourier transform happens have nulls right in the middle of all the neighboring subcarriers. Since the DFT-based modulator can be seen as sampling in the frequency domain right at those same frequency points, we may say, with no small satisfaction, that we have chosen our pulse widths to result in "zero intercarrier interference." The parallels with the zero intersymbol inteference project described in Section 6.2.1 are irresistible.

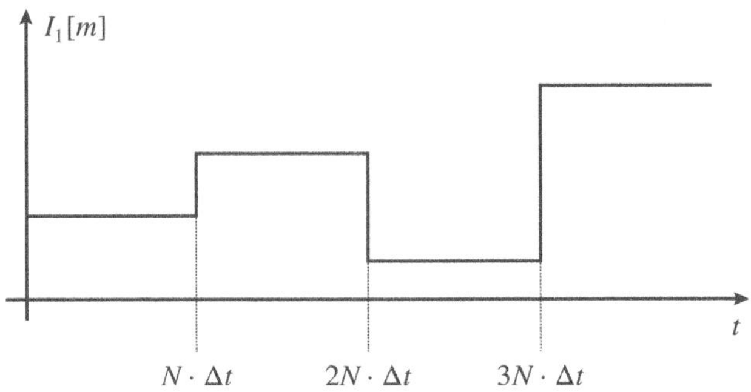

Figure 6.22 In OFDM, we typically only change the *C*s at the symbol rate, which is every *N* samples.

How OFDM Simplifies the Problem of Equalization

It is widely asserted that OFDM signaling "greatly simplifies" the challenge of channel equalization. A great way to understand this is to consider the challenge of pushing a signal of bandwidth *B* through our familiar stand-in for all of analog electronics, the *RC* filter. The setup is that, regardless of the bandwidth *B*, we succeed in imposing the desired spectrum on the output capacitor. The equalization filter/amplifier is realized in whatever extreme way is needed to make this happen (see Figure 6.24). We have from Eq. 3.26 that the average power that must be injected into the *RC* is given by

$$P_{\text{avg}} = \frac{1}{6}\left(\frac{1}{\omega_{-3\text{dB}}}\right)^2 \frac{A^2}{R} B^3. \tag{6.40}$$

If we assume that we are in the positive signal-to-noise ratio (SNR) region ("Region 2") described in Section 5.1.2, what we can see is that for every doubling of the channel capacity (which, in this region, goes 1:1 with bandwidth), the power output required by the equalization filter goes up by a factor of 8.

Now let's look at a different concept, where we use two channels in parallel, and where we use only half of the bandwidth in each channel (see Figure 6.25). Let's look at the average power required to get our signal through now:

$$\begin{aligned} P_{\text{avg}} &= P_{\text{avg1}} + P_{\text{avg2}} \\ &= \frac{1}{6}\left(\frac{1}{\omega_{-3\text{dB}}}\right)^2 \frac{A^2}{R}\left(\frac{B^3}{8} + \frac{B^3}{8}\right) \\ &= \frac{1}{4}\left[\frac{1}{6}\left(\frac{1}{\omega_{-3\text{dB}}}\right)^2 \frac{A^2}{R} B^3\right]. \end{aligned} \tag{6.41}$$

So breaking things up this way lowers the power output demanded of our equalization filter by a factor of 4. And as we'll argue in Section 12.1.3, lowering the

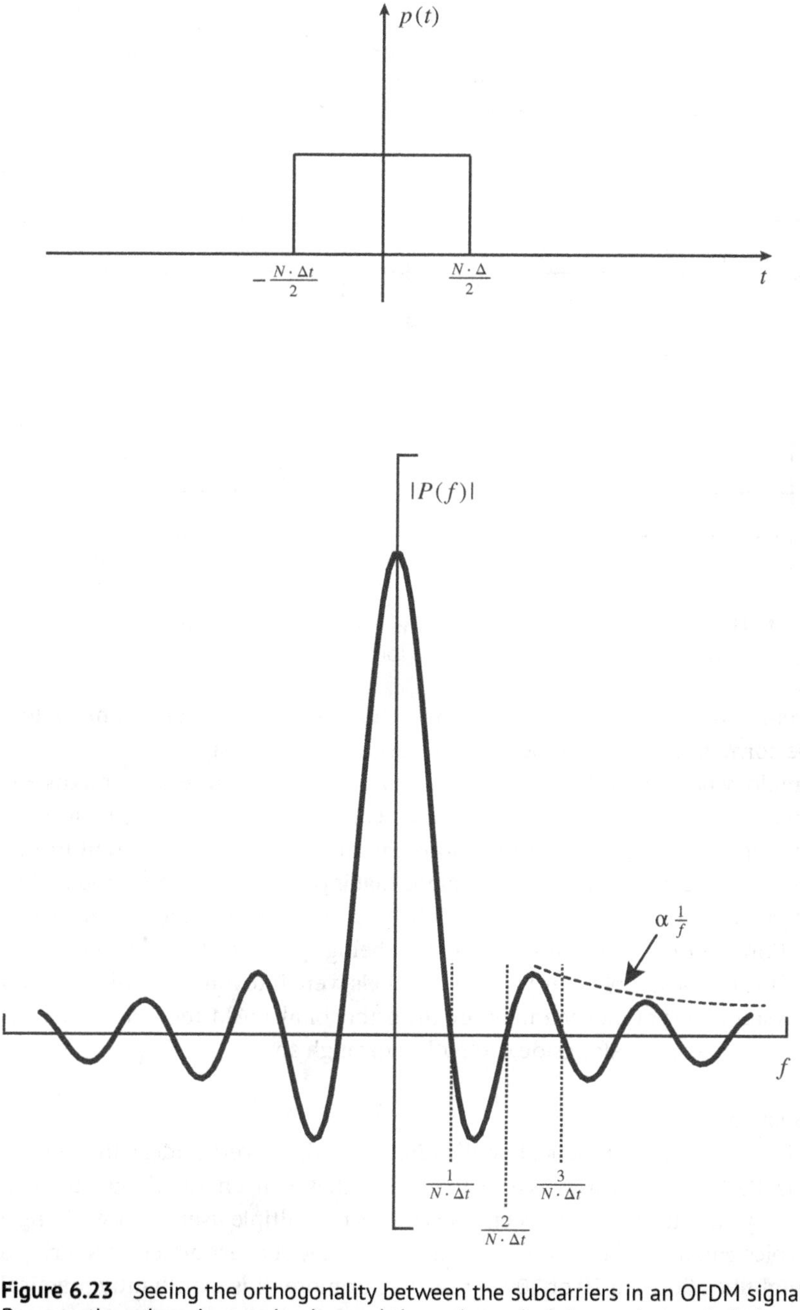

Figure 6.23 Seeing the orthogonality between the subcarriers in an OFDM signal. Because the subcarrier spacing is equal the reciprocal of the symbol period, the nulls in the Fourier transform of the rectangular pulse land right in the center of all neighboring frequency bins.

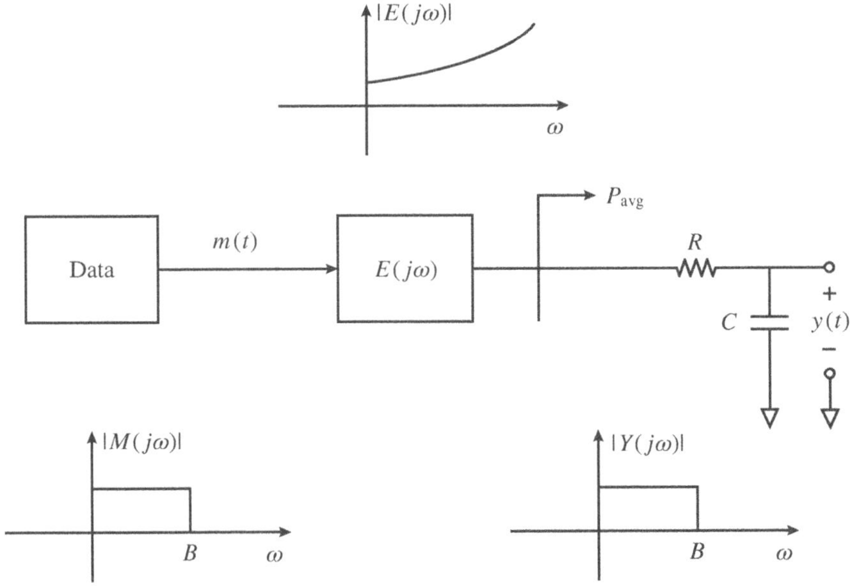

Figure 6.24 Using an equalizer to counteract the low pass filter characteristic of an *RC* filter. If performed properly, $Y(j\omega)$ is restored to looking like $M(j\omega)$.

power required by this filter moves it from a high-performance circuit block to a lower performance block and therefore makes it easier to design.

The analogy between this *RC* channel and an OFDM signal sent over a dispersive propagation path is actually pretty exact, conceptually speaking. The form of dispersion in the wireless communications channel is, of course, different from a simple single-pole filter. But the fact remains that if you push for higher and higher B through a dispersive channel, you will need a higher and higher performance equalization filter to keep your message from being splattered beyond all recognition. In the above example, our parallel channels were individual *RC* filters. In an OFDM system, the parallel channels are the orthogonal, QAM-modulated subcarriers, each propagating alongside each other through space.

OFDM Summary

OFDM is a tremendous practical application of many powerful ideas in communications. Perhaps its greatest asset is the ease with which a network operator can dynamically allocate precious spectral resources to multiple users. A user doing a simple voice call might be allocated the use of a single subcarrier; a user sharing a video might be allocated 10 or 100 subcarriers or more. It is a brilliantly practical strategy, built on the computational efficiency of the FFT and IFFT algorithms. Nothing is forever in engineering, but for high-capacity communications, OFDM

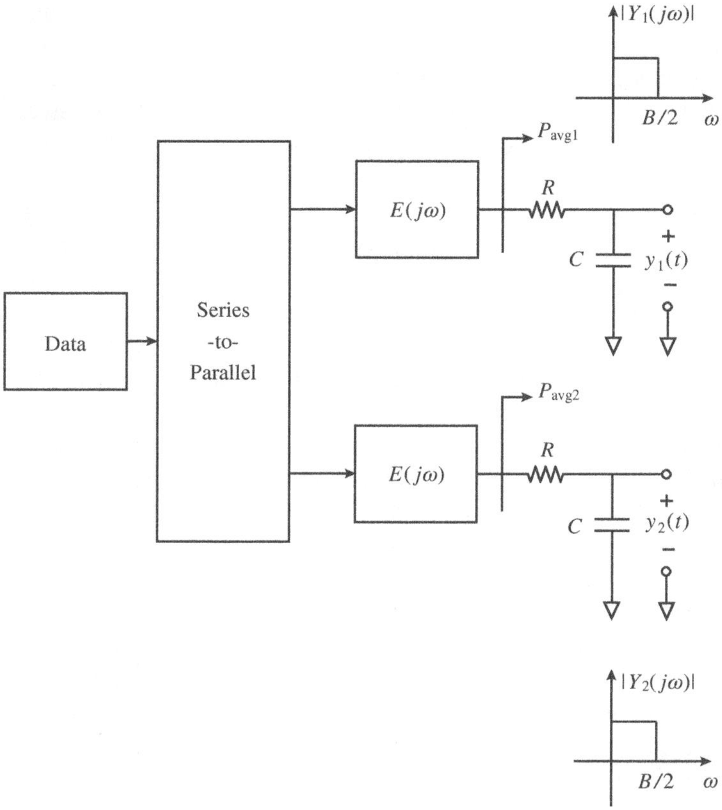

Figure 6.25 Making the equalizer's job easier by splitting the data stream into two parallel paths.

will be tough to beat. And when something *does* come along to beat it, a good understanding of OFDM will help you to appreciate the beauty of the next thing, whatever it is.

6.3 Error Correction Codes

The Shannon limit is remarkable in its simplicity but, as we are discovering, it provides no practical guidance as to how to implement systems that get to that magical limit of throughput. We have already discussed one extremely powerful set of techniques, consisting of correlative sampling and matched filtering. There is another broad class of techniques to discuss briefly, which is called error correction codes.

This is a pretty deep hole to go down, but one can get a sense of it by looking at an extremely simple, relatively inefficient, easy-to-understand technique called a repetition code. Very simply, in a repetition code, you transmit each bit multiple times (the number of repetitions is agreed upon by both transmitter and receiver). Then, on the receive side, you make decisions after each received bit. The "true" bit is decided by majority vote.

Let's say that the bit sequence we want to transmit is a single bit, and that that bit is a 1. We might transmit 111. The receiver would do its best in the presence of noise, and if it perceived a sequence of 011, 101, 110, or 111, it would deduce that the transmitted bit was a 1. The strength of the code comes from the observation that the probability of an error on any one repetition is small(ish); the probability of making the wrong call on two of the repetitions in the group is correspondingly diminished.

This is, of course, nowhere near the sophistication of error correction codes in current circulation. But it does have the benefit of being easy to understand, and also illustrative of an important fundamental trade-off. We call the ratio of the information bits to the total bits the "code rate," and in our example that ratio would be $1/3 \approx 0.33$. Generally speaking, we refer to codes with low code rates as "strong," and obviously, the noisier the channel, the stronger the code needs to be. What we see is a concrete illustration of the Shannon limit's assertion that a lower SNR *means* a lower effective bit rate. For a given bandwidth, there are only so many bit pulses that we can transmit in a given time window. The higher the noise floor, the more bits of redundancy we need, and therefore the fewer information bits we can send in that window.

6.4 Are Our Modulation Techniques Optimal?

The answer to this question is simple. We work as hard as we can with our modulation strategy, matched filters, error correction codes, and whatever else, and then go out and measure the bit rate we are able to achieve. We then compare that number to $B\ \log_2(\mathrm{SNR}+1)$, filling in the measured SNR and B of our actual channel. If these two numbers match, we are done!

Okay, that is a little simplistic. One could, and *should always*, ask, even in the face of successfully hitting the limit, whether we have done so at the cost of excessive power consumption, or implementation complexity, or cost of parts, or any number of things. If you keep these ideas in mind, you have a pretty good idea of how the last several decades of development in communications theory played out. A very exciting time.

7

Selected Topics in Electromagnetic Theory

If you hadn't already before you picked up this book, after reading a chapter or two you can appreciate the difficulty facing a newcomer who seeks to understand wireless links. There are just so many topics: linear system theory, circuit theory, thermodynamics, probability and statistics, information theory, digital signal processing...and that's just so far. Well, now we arrive at another big one: electricity and magnetism (E&M). There is no getting around this.

So, what to do? There are many long books on E&M. It is fine to argue that the following is a carefully curated subset of topics, a minimum needed to understand wireless links. The author certainly believes that to be true. But for the purists, the physicists, wouldn't they be better served first by a fuller study of E&M, untainted by application? And even for the more pragmatic engineers, wouldn't they too be more versatile with a more general grounding in this critical topic?

The answer here almost certainly varies from person to person. For the author, the understanding of general physical principles has always been enhanced and enriched by an intense focus on performance-based application. The resulting depth of understanding naturally leads to versatility. If you are like this, this chapter is written for you.

If you are *not* like this, then you are invited to an experiment. Give both approaches a try. Dust off your old E&M textbook, or find an online course, and get an understanding that you find personally satisfying. Then come back here and compare. Or reverse the order. At minimum, you will benefit from increased knowledge of your learning style. You may even surprise yourself.

Energy Fundamentals of Radio, First Edition. Joel L. Dawson.

7.1 Action at a Distance Is Weird

Imagine being an early observer of two charged particles. (We'll skip over the parts where we decide what it means for a particle to be "charged," and further how we decided to rate the degree of chargedness on a linear scale that we call "coulombs.") In our experiment, we take two charged particles and fix one in space according to Figure 7.1. We move the other particle around manually and observe...that they repel each other! That is, the charge on the right resists our attempts to move it closer to the fixed charge. The closer the charges get to each other, the greater the repelling force. Moreover, that force gets stronger *faster* as we close the distance. What in the world is going on? Relatedly, how would you describe this phenomenon to a friend?

Part of the answer is that you would keep it to yourself while you tried to figure out if there was a hidden mechanical connection between the charges. In vain, you wave your hand between them with your senses on high alert, trying to feel something. It's almost, almost, like there is a hidden spring between them, but something feels off about that too.[1]

If you rule out hidden mechanical interaction, you're left with only a couple of classes of explanations. You can go with Magic, which is appealing because you can neither see nor feel anything and because the behavior is like nothing you've seen before. Or you can invoke a kind of Force Field, which at first seems similar to Magic and therefore leaves you vulnerable to mockery from your more serious-minded friends. But at least with a force field, you can lean on the following facts that you glean from careful observation:

1. Careful measurements reveal that the magnitude of the force is *exactly* proportional to the product of the two charge magnitudes.
2. You were right to feel that behavior does not quite mimic a linear spring. The repelling force falls off *exactly* as $1/r^2$, where r is the distance between the charges.

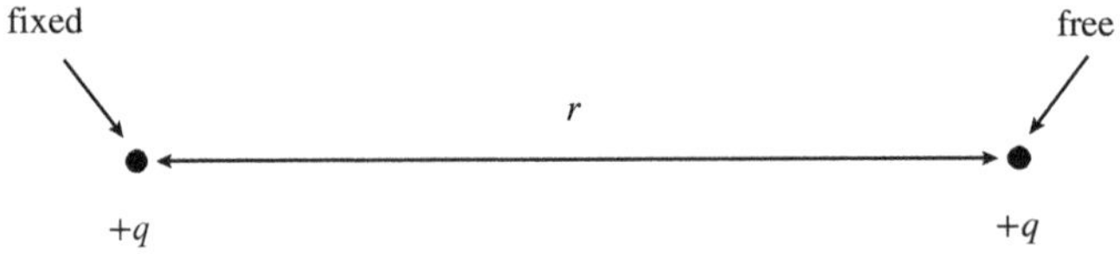

Figure 7.1 Two charges in space, one fixed in place and one free to move.

1 A classic spring increases its repelling force in a way that varies linearly with distance. When you cut the distance in half, the force doubles. But here, as they get closer the charges start to repel each other with hysterical, nonlinear urgency.

3. There appear to be two, and only two, types of charges. Opposite charges attract, similar charges repel. Let's call one flavor positive, and the other flavor negative, foreshadowing the superposition discussion to follow.
4. No matter how long you observe, *these rules never seem to change.* They are ironclad and fixed.
5. The mathematical relationships are *simple.* We don't get Bessel functions, we don't get complicated time dependencies, we don't even get noninteger powers like $1/r^{2.009}$. We get $1/r^2$ exactly.

This is quite an impressive lineup in favor of the force field hypothesis. Because now not only are you *not* invoking Magic, but also you *are* invoking Math. And Math always makes people take your observations more seriously.

It gets better. Suppose we revise the experimental setup to include three charges, and fix two charges in place with one charge still free to move. It is a more complicated system, to be sure. But the joyous good news is that we can attack this problem exactly via superposition. That is, we could first remove the second charge and measure the force vector $\vec{F}_1$ due to the first. We could then remove the first, place the second fixed charge and measure the force vector $\vec{F}_2$ due to it. It turns out that the force vector due to the combination is exactly the vector sum of $\vec{F}_{\text{tot}} = \vec{F}_1 + \vec{F}_2$ (see Figure 7.2). This straightforward superposition of forces will bore some readers and surprise others. In this case, it is better to be in the latter camp. If you are bored, it may be because you have a deep background in physics and have seen superpositions so many times that you now take it for granted.

Don't take it for granted. The fact that the forces due to charged particles combine so simply is remarkable, wonderful, and a tremendous boon to both analysis and conceptual insight. Our favorite tricks from Chapter 2 have versions that apply here, like analyzing continuous charge distributions by discretizing them into

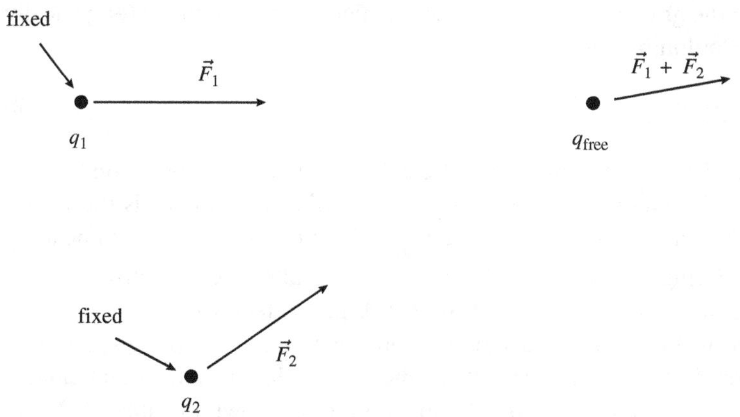

Figure 7.2 Three charges in space, two fixed in place, and one free to move.

Riemann sums. And many times before we get that far, we can spot and exploit geometrical symmetries that give exact answers almost immediately.

In this section, we look at the electric fields of point charges and the magnetic fields of current elements. As humble as these examples are, we will come to appreciate that all the electric and magnetic fields that we rely on for wireless communications can be broken down into superposition sums of these primitives.

7.1.1 Forces on Charges: How We Know There Is an Electric Field

Time to get more precise about what we mean by an "electric field." It is simple to state: an electric field is a map in three-dimensional space. At any point in that space, the field tells you the *magnitude* and the *direction* of the force that a unit charge placed at that point will experience. We tend to say that an electric field is an example of a "vector" field.

We specify the electric field mathematically as a vector function of three spatial coordinates x, y, and z. For every point in space (x_0, y_0, z_0), we associate a vector $\vec{E}(x_0, y_0, z_0)$. The force experienced by a charge of quantity q_0, placed at point (x_0, y_0, z_0), is given by

$$\vec{F}(x_0, y_0, z_0) = q_0 \cdot \vec{E}(x_0, y_0, z_0). \tag{7.1}$$

It bears emphasis: the way that we know that there are electric fields in a given space is because charged particles placed in that space experience forces.

7.1.2 The Electric Field of a Point Charge at Rest

Now we can quickly get an understanding of the electric field associated with a point charge. Suppose that we have a single point charge sitting in space. For illustration purposes, we are going to place a charge q on a rectangular grid in space. At each point on the grid, we evaluate the electric field according to the (empirically determined) Coulomb's law:

$$\vec{E} = \frac{q}{4\pi\epsilon_0 r^2}\hat{r}. \tag{7.2}$$

In words, Eq. 7.2 states that the electric field due to a point charge lies on the line between the point charge and the point in space in question (this is the meaning of the $\hat{r}$). The magnitude of the field is $\frac{q}{4\pi\epsilon_0}\frac{1}{r^2}$, and the direction will be away from the originating point charge. The constant ϵ_0 is called the permittivity of free space and has the value 8.85×10^{-12} F/m (farads per meter). We graphically illustrate this field in Figure 7.3. We suppose here that we are sampling space on a two-dimensional grid in increments of one meter. The electric field at each hollow circle is in the direction illustrated, with magnitudes as shown in Table 7.1. If you

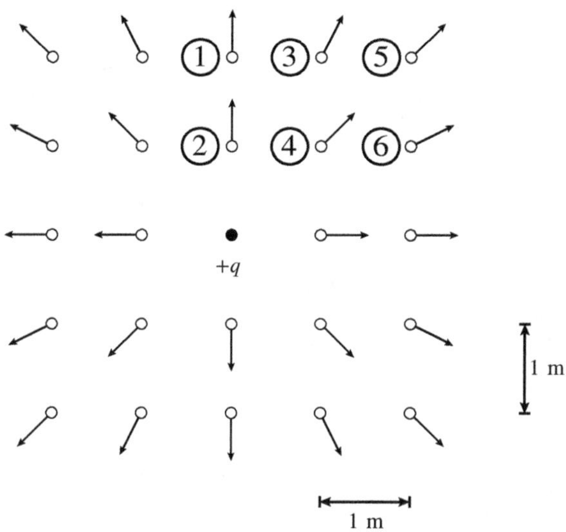

Figure 7.3 Graphically illustrating the electric field due to charge $+q$ on a grid.

Table 7.1 Example electric field magnitudes for various points on the grid.

Point	$\lvert\vec{E}\rvert$
1	$\frac{q}{4\pi\varepsilon_0}\frac{1}{(2\,\text{m})^2}$
2	$\frac{q}{4\pi\varepsilon_0}\frac{1}{(1\,\text{m})^2}$
3	$\frac{q}{4\pi\varepsilon_0}\frac{1}{\sqrt{(2\,\text{m})^2+(1\,\text{m})^2}^2}$
4	$\frac{q}{4\pi\varepsilon_0}\frac{1}{\sqrt{(1\,\text{m})^2+(1\,\text{m})^2}^2}$
5	$\frac{q}{4\pi\varepsilon_0}\frac{1}{\sqrt{(2\,\text{m})^2+(2\,\text{m})^2}^2}$
6	$\frac{q}{4\pi\varepsilon_0}\frac{1}{\sqrt{(2\,\text{m})^2+(1\,\text{m})^2}^2}$

imagine now sampling space on a grid so fine that it looks continuous, you get the figure that you probably expected to see as shown in Figure 7.4.

You can imagine now extending this accounting to more complicated charge distributions by just lowering our heads and using superposition. In practice, once you get more than a few point charges the accounting gets too unwieldy to tackle without computational aid. Instead, in hand analysis it is common to formulate problems such that symmetry saves you from being so loud about adding up

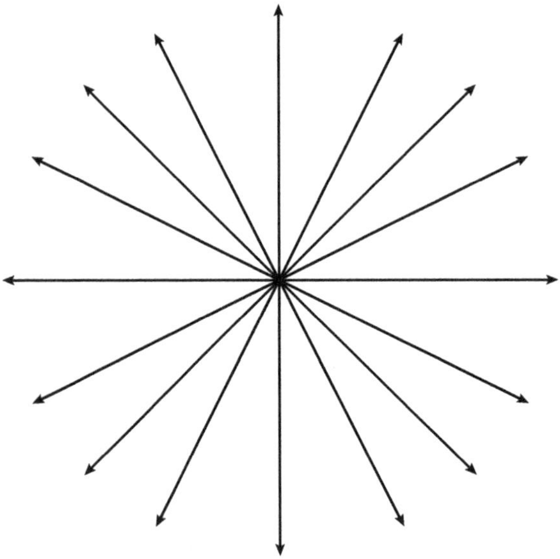

Figure 7.4 Conventional representation of the electric field for a point charge.

individual charge contributions. This is the reason that textbooks on E&M so commonly dwell on "infinite planes," "infinitely long lines," or "perfect spheres" of charge. Superposition is still the underlying principle, but symmetry will often make the math visual.

7.1.3 Forces on Charges: How We Know There Is a Magnetic Field

It turns out that we are not done when it comes to forces on charges. There is one other phenomenon that we must explain and document mathematically. We find that if we place two current-carrying wires next to each other, with currents going in the same direction, the two wires will be attracted to each other (see Figure 7.5). As savvy veterans of earlier subsections, and being aware that an electrical current is composed, in fact, of charged particles, we might suppose that something funny is going on with regard to electric fields.

So we reach for our pencils, eager to apply superposition to an admittedly large number of charged particles. Or not: we are stopped in our tracks when we remember that the material in wires, as is true with most matter, is electrically neutral. That is, it contains equal numbers of protons and electrons. Superposition leads us to understand immediately, without writing any equations, that this large collection of charged particles in a small space, with half at $+q$ and the other half at $-q$, *must* result in almost no electric field when we get outside the wires.

Yet the forces between the wires are substantial and easily observable. If we stick with the idea that this force is tied in some way to charged particles, then we need a new field, totally different from that described by Eq. 7.2.

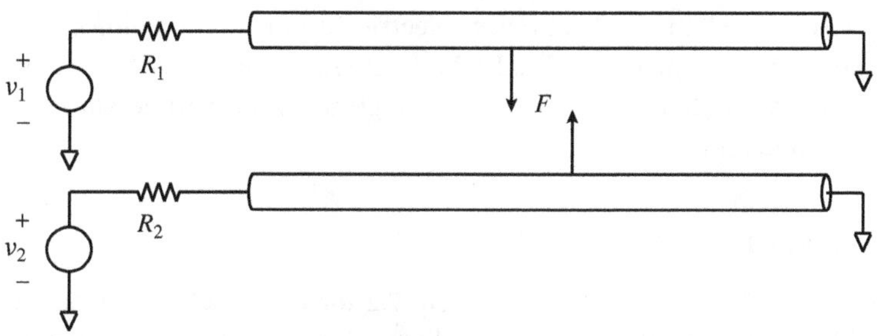

Figure 7.5 Force between two current-carrying wires.

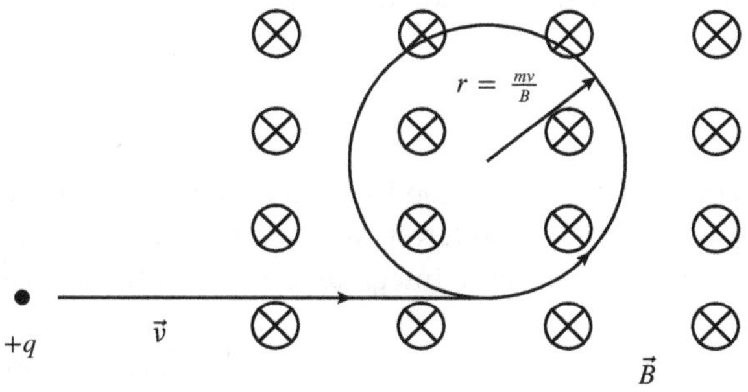

Figure 7.6 A particle of charge $+q$ and mass m entering a region with magnetic field B pointed into the page.

We call this new field a "magnetic field," which in most treatments is unhelpfully labeled $\vec{B}$.[2] We may very reasonably define both $\vec{E}$ and $\vec{B}$ through the forces that they exert on a theoretical point charge q. That force, called the Lorentz force, is given in full by

$$\vec{F} = q\left(\vec{F} + \vec{v} \times \vec{B}\right), \tag{7.3}$$

where $\vec{v}$ is the velocity of the charged particle in the relevant frame of reference.[3] The part due to the E field we now know about from Section 7.1.2, but the $q\vec{v} \times \vec{B}$ part is just a comparatively bizarre fact of Nature that we capture using the vector cross product. Note that the cross-product means, for example, that if the charge is traveling in the direction of the B field, it experiences *no* force. Figure 7.6 captures what would happen if we shot a charged particle of mass m and charge $+q$ into a static magnetic field pointed into the page. (We also assume no other forces acting on the particle, like gravity.)

2 See also I for current, Q for charge, p for momentum, x for position, L for inductance...
3 N.B. keep an eye on this frame of reference caveat. We'll have more to say about this in Section 7.1.5.

We close by re-emphasizing that we *define* electric and magnetic fields by the forces that they exert on charged particles. E&M theory has a lot to say about these fields. For the author, it helps tremendously if one grounds one's understanding in this basic, easy-to-keep-track-of fact.

7.1.4 The Magnetic Field of a Current Element

We need the equivalent of Coulomb's law in Eq. 7.2 for the B field. It turns out that the primitive for this field is not a point charge, but rather a steady current flowing through a wire of infinitesimal length $d\vec{l}$. The expression for the field due to a current element is given by the Biot–Savart Law, and is as follows:

$$d\vec{B} = \frac{\mu_0}{4\pi} \frac{Id\vec{l} \times \hat{r}}{|\vec{r}|^2}. \tag{7.4}$$

So if we have a current carrying wire, we first chop it up into segments that are short compared to the physical dimension of interest. We then get a picture like that in Figure 7.7. To get the total B field, we sum (integrate) over the entire length of the wire.

As with the electric field and point charges, we rarely, in hand analysis, resort to actually calculating integrals to get the magnetic field. Instead, we tend to focus on structures such as the infinitely long wire in Figure 7.8. And even in the special case of the infinite wire, we have an analytical alternative called Ampere's Law that gives the same answer and that we will discuss with the rest of Maxwell's equations in Section 7.2. Regardless, it is worth getting straight in your mind this idea of the current element being the primitive for the B field.

7.1.5 Frame of Reference and *E* Versus *B* Fields

You will notice that the Lorentz force, $\vec{F} = q\left(\vec{F} + \vec{v} \times \vec{B}\right)$, is dependent upon the velocity of the charged particle. Now suppose that we have a charged particle which is moving with velocity $\vec{v}$ in some agreed-upon frame of reference. And then we give ourselves two observers, one who is *stationary* in that frame, and a second

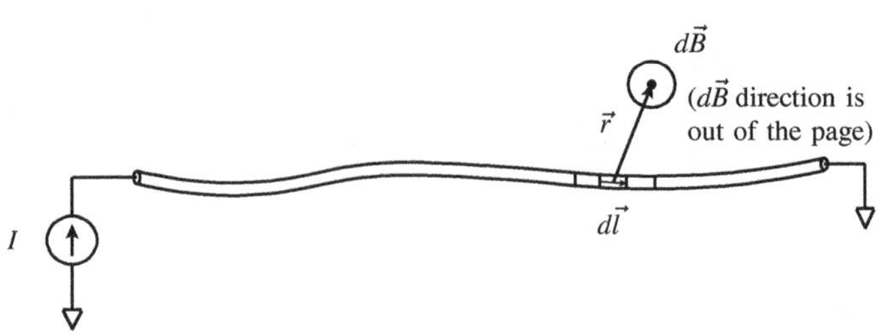

Figure 7.7 Magnetic field due to an infinitesimal current element.

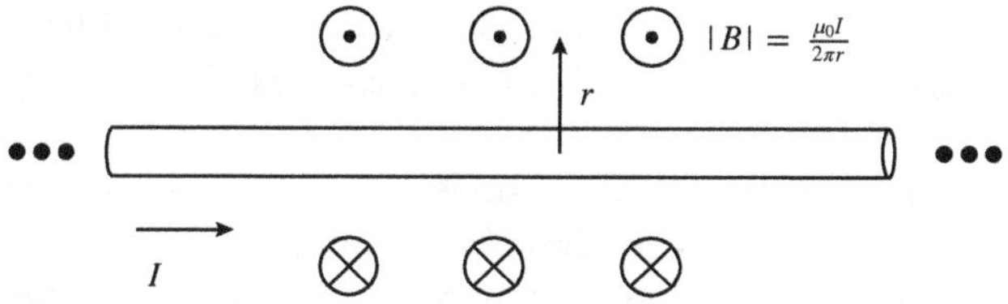

Figure 7.8 Magnetic field of an infinite wire.

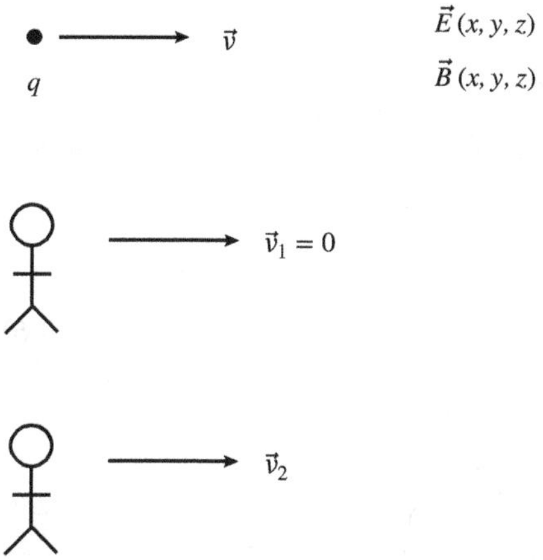

Figure 7.9 When it comes to forces on charges, what we attribute to magnetic fields versus electric fields depends on our frame of reference.

observer who is moving with constant velocity $\vec{v}_2$. We illustrate in Figure 7.9. Suppose further that each observer is able to measure the force on the charged particle as it moves through space. The important implication of the innocuous-looking $\vec{v}$ in the Lorentz force is that the different observers will report different E and B fields.

This little tidbit is not so important for wireless links. However, unwinding this scenario in detail figures prominently in Albert Einstein's theory of special relativity. We digress briefly to acknowledge the reader who might be wrong-footed by the inclusion of a velocity without specifying a frame of reference. It leads down a rabbit hole toward one of the great theoretical results of the early twentieth century.

7.2 We Use Maxwell's Equations When Fields Are Not Static

We will not spend a great deal of time on Maxwell's equations, foundational though they are to E&M theory. We maintain our focus on wireless communications, and so for us, Maxwell's equations are a means to an end. Ultimately, they show us that electromagnetic (EM) waves can propagate without loss and without dispersion through free space. We focus on getting to that conclusion as quickly as possible.

Step one is Maxwell's equations in integral form. This is the easiest form for most people to grasp, and for the effort yields a considerable jump in understanding of EM fields. Step two will be to reintroduce these equations in differential form. We do step two only to enable step three, which is to use our new differential equations to derive the famous *wave solutions* to Maxwell's equations.

7.2.1 Maxwell's Equations in Integral Form

It turns out that each of the four modern Maxwell's equations has their own names. In short order:

- Gauss's Law:

$$\oint\!\!\!\oint_S \vec{E} \cdot d\vec{A} = \frac{1}{\epsilon_0} Q_{\text{encl}}. \tag{7.5}$$

- Gauss's Law for Magnetism:

$$\oint\!\!\!\oint_S \vec{B} \cdot d\vec{A} = 0. \tag{7.6}$$

- Maxwell–Faraday equation:

$$\oint \vec{E} \cdot d\vec{l} = -\frac{d}{dt} \iint \vec{B} \cdot d\vec{A}. \tag{7.7}$$

- Ampere–Maxwell equation:

$$\oint \vec{B} \cdot d\vec{l} = \mu_0 I_{\text{encl}} + \mu_0 \epsilon_0 \frac{d}{dt} \iint \vec{E} \cdot d\vec{A}. \tag{7.8}$$

Let's take these one at a time.

Gauss's law says in words, "The total flux of the electric field through a closed surface is proportional to the charge enclosed." To get this, we need to first understand what we mean by "flux." When we compute the flux of the electric field through a surface, we first break the surface into infinitesimal area fragments. For each fragment, we compute the component of the E field normal to that surface. If the normal E field points to the interior of a closed surface, we consider that negative flux. If it points to the exterior, we consider that positive flux. Fragment by fragment, we compute the product of the normal E field and the magnitude of the area element dA. The total flux is the sum or the integral of all of the $\vec{E} \cdot d\vec{A}$s (see Figure 7.10). The important concept that Gauss's law captures is that the result of the integral is the same regardless of the shape of the surface, as long as the

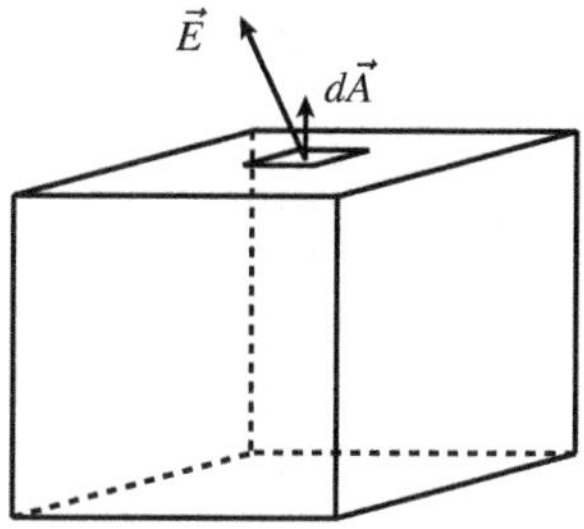

Figure 7.10 When we compute flux through a surface, we sum at every point the product of the normal component of the electric field and the magnitude of the area element.

quantity of the enclosed charge is the same. That is to say, if we performed the flux integral over all of the surfaces in Figure 7.11, the result would be the same. In most texts, the case of the sphere in Figure 7.11 is deservedly central. If you think about it, it must be true that the electric field due to a point charge has spherical symmetry: for a point charge at the center of the sphere, your first guess would be that the electric field would be normal to the surface of the sphere at any point, and also that it would have the same magnitude at any point on the surface of a given sphere. This is the kind of symmetry argument that spares us from actually performing the integral we threaten to do in Gauss's law. Instead, we write down the answer:

$$\oiint_S \vec{E} \cdot d\vec{A} = |\vec{E}| 4\pi r^2 = \frac{1}{\epsilon_0} Q_{\text{encl}}. \tag{7.9}$$

This gives us

$$|\vec{E}| = \frac{1}{4\pi r^2} \frac{1}{\epsilon_0} q, \tag{7.10}$$

where we understand that for a positive charge $+q$ we have the E field pointing outward, normal to the surface, at all points on the sphere. This is none other than Coulomb's law from Eq. 7.2.

Gauss's law tells us, again, that the total flux through a closed surface is proportional only to the total charge enclosed. This means that in the examples shown in Figure 7.12, properly performing a detailed flux integral would result in exactly zero. Any electric flux into the surface, according to Gauss's Law, is exactly balanced by electric flux out of the surface.

Once you internalize Gauss's law in the foregoing, you get Gauss's Law for magnetism for free. This law says that the total magnetic flux through any closed surface is *always* zero. You will sometimes hear people talk about this law as an assertion that "there is no such thing as a magnetic monopole." This is a fancy way

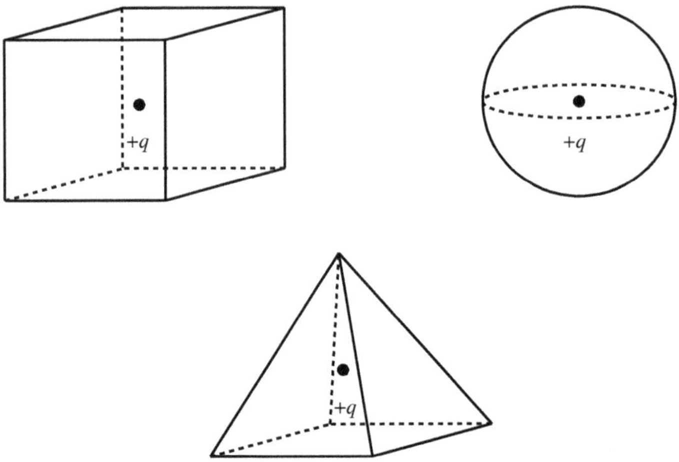

Figure 7.11 Gauss's law says that the flux integrals for all of these closed surfaces will be the same.

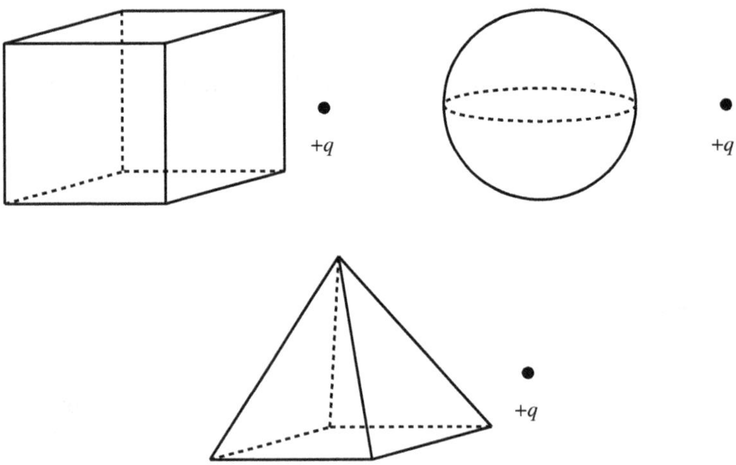

Figure 7.12 Gauss's law says that the flux integrals for all of these closed surfaces will be zero.

of saying that there is no magnetic analog in Nature to the electric point charge $+q$ of Figure 7.4.

The Maxwell–Faraday equation is actually best illustrated with a concrete example. Our setup is shown in Figure 7.13. If we start with the line integral $\oint \vec{E} \cdot d\vec{l}$, we choose in this example a loop that follows the wire from the starting point all

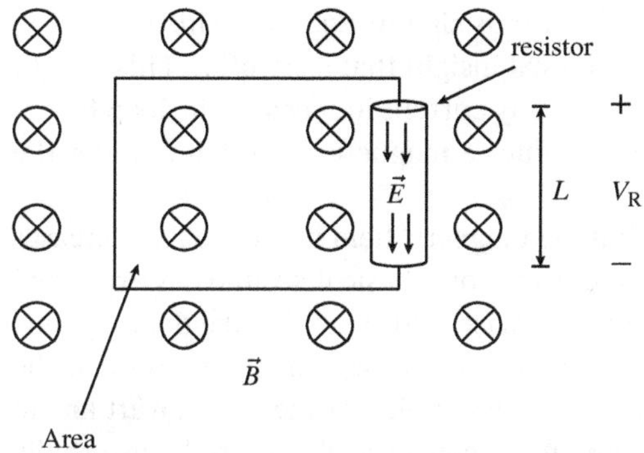

Figure 7.13 A time-varying magnetic flux induces electrical current in a loop.

the way around in a clockwise direction. We stipulate that the wires are perfectly conducting, and therefore the electric field within the wire is zero.[4] Eventually, we get to the part of the path that traverses the resistor, and we have

$$\oint \vec{E} \cdot d\vec{l} = \int_{\text{resistor}} \vec{E} \cdot d\vec{l}. \tag{7.11}$$

If the electric field is uniform within the resistor, then $\int_{\text{resistor}} \vec{E} \cdot d\vec{l} = E \cdot L$, which winds up being the formal definition of voltage across the resistor.[5]

You may know this phenomenon as EM induction, whereby a time-varying magnetic field is known to induce voltages in electrical circuits. Filling in the other side of the Maxwell–Faraday equation, we have that

$$V_R = -\frac{d}{dt} \iint \vec{B} \cdot d\vec{A} \tag{7.12}$$

where V_R, the voltage across the resistor, is $E \cdot L$. If the area of the circuit loop is A, and we suppose that the magnetic field is uniform, the voltage across the resistor can be written

$$V_R = -A\frac{dB}{dt}. \tag{7.13}$$

4 If the electric field were not zero, the electrical current in the conductor would have to be infinite.

5 The formal definition of electric potential, measured in volts and properly understood as the work done per unit charge to move a positive charge from p_1 to p_2, is given by $\phi_{21} = -\int_{p_1}^{p_2} \vec{E} \cdot d\vec{l}$.

Finally, there is the little matter of the minus sign in Eq. 7.13, which is easiest to remember if you bear in mind the physical insight that it signifies. This insight is called Lenz's law, and its essence is that any current induced in the loop by the time-varying magnetic field will itself produce a magnetic field that opposes the change in the magnetic field.

In this present example, suppose that the magnetic field into the page increases over time in magnitude. The induced current, on physical grounds as expressed by Lenz's law, must need to be counterclockwise. This would make the magnetic field associated with the induced current out of the page in the interior of the loop. (Now is a good time to check your understanding of the Biot–Savart law in Section 7.1.4.) This induced field, therefore, opposes the *change* in the externally applied B field, which is growing in the direction that is into the page.

There are plenty of details to get into when it comes to EM induction. What is important for understanding wave propagation of EM fields is that a time-varying magnetic field gives rise to an electric field (and, it will turn out shortly, vice versa). This is big news. Look for its implications when we get to wave solutions of Maxwell's equations in Section 7.3.

Finally, we get to the Ampere–Maxwell equation. Those of you with the background may recall using this law, without the second, time-dependent term, and in place of the Biot–Savart law, to calculate the strength of the magnetic field surrounding a wire of infinite length as shown in Figure 7.14. This is yet another example of using symmetry and, well, gigantic structures to get around doing messy integrals. With the infinite wire in particular, you can see on symmetry grounds that the magnetic field must be uniform all along the wire, and that it can only reasonably vary with distance r from the wire. Biot–Savart gives us an assist, allowing us to further deduce that the field lines of the magnetic field must themselves be circles around the wire, with the wire passing through the center.

This last insight allows us to use the first term of the Ampere–Maxwell equation directly and easily. Choosing our contour to be a circle, and choosing the wire to pass straight through the center of this circle, we write

$$B \cdot 2\pi r = \mu_0 I. \tag{7.14}$$

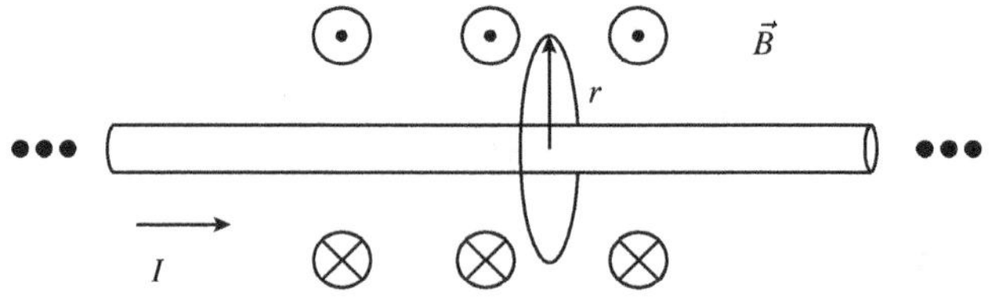

Figure 7.14 Magnetic field of an infinite wire.

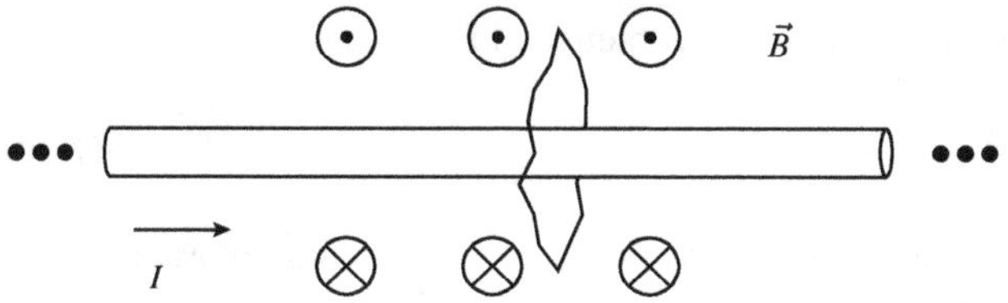

Figure 7.15 Magnetic field of an infinite wire, choosing an arbitrary contour of integration.

It follows that

$$B = \frac{\mu_0 I}{2\pi r}. \tag{7.15}$$

Note that if we chose contours other than perfect circles, or even circles whose centers were offset from the wire, the net result of the contour integral would still be determined solely by $\mu_0 I_{\text{encl}}$ (see Figure 7.15).

The second piece of the Ampere–Maxwell equation, the $\mu_0 \epsilon_0 \frac{d}{dt} \iint \vec{E} \cdot d\vec{A}$ piece, tells us that just as time-varying magnetic fields give rise to swirling electric fields, so too do time-varying electric fields give rise to swirling magnetic fields.

7.2.2 Divergence and Curl

Moving on. We need to introduce Maxwell's equations in differential form, because that allows us to describe mathematically the propagation of the EM waves upon which everything in this book depends.

Before we do that, there are a couple of operations from multivariable calculus that we will need. The equations can get big and complicated looking, even if the underlying concepts are not so difficult to grasp. If you have had multivariable calculus, and remember it making sense to you, now is a good time to dig up the reference that you are used to and remind yourself of what divergence and curl are all about. If these concepts are new to you, or if you only remember a red fog of confusion, pull up a chair!

The easiest way to understand these operations is to look at them directly in their form for Cartesian coordinates. Suppose that $\vec{F}$ is a vector field in three dimensions, by which we mean

$$\vec{F} = F_x(x, y, z)\hat{x} + F_y(x, y, z)\hat{y} + F_z(x, y, z)\hat{z}, \tag{7.16}$$

or when we are not in the mood to write everything out so explicitly,

$$\vec{F} = F_x\hat{x} + F_y\hat{y} + F_z\hat{z}. \tag{7.17}$$

In Cartesian coordinates, we write the "divergence of $\vec{F}$" as

$$\nabla \cdot \vec{F} = \frac{\partial F_x}{\partial x} + \frac{\partial F_y}{\partial y} + \frac{\partial F_z}{\partial z}. \tag{7.18}$$

From here, actually, the clearest path forward is slow and steady plowing using the formal definitions of the derivative. We have

$$\begin{aligned}\nabla \cdot \vec{F} = &\lim_{\Delta x \to 0} \frac{F_x(x_0 + \Delta x, y_0, z_0) - F_x(x_0, y_0, z_0)}{\Delta x} \\ &+ \lim_{\Delta y \to 0} \frac{F_y(x_0, y_0 + \Delta y, z_0) - F_y(x_0, y_0, z_0)}{\Delta y} \\ &+ \lim_{\Delta z \to 0} \frac{F_z(x_0, y_0, z_0 + \Delta z) - F_z(x_0, y_0, z_0)}{\Delta z}.\end{aligned} \tag{7.19}$$

Often authors will refer to $\nabla \cdot \vec{F}$ as a "flux density," in the sense that $\nabla \cdot \vec{F}\,\Delta V = \nabla \cdot \vec{F}\,\Delta x \Delta y \Delta z$ is the next flux through a volume element of length Δx, width Δy, and height Δz (see Figure 7.16). The word choice "flux density" turns out to be suggestive. If we mean "density" in the usual sense of the word, then integrating this flux density (divergence) throughout a closed volume ought to get us the *net flux of that vector field through the surface that encloses the volume.* Let's see if we can verify this.

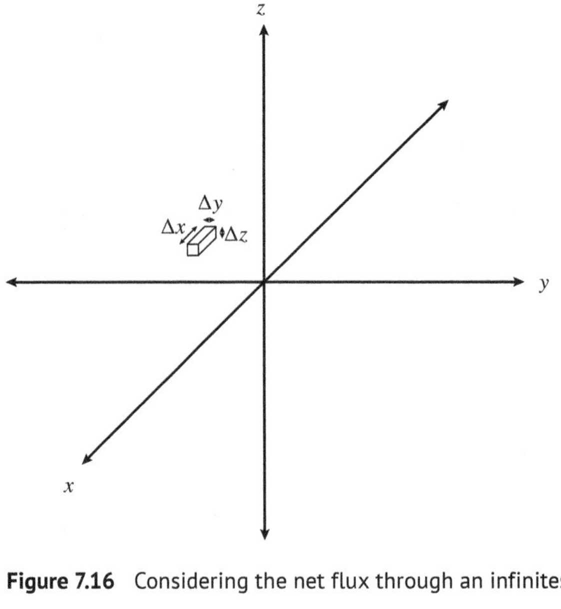

Figure 7.16 Considering the net flux through an infinitesimal volume element.

Start by writing down $\nabla \cdot \vec{F}\,\Delta V = \nabla \cdot \vec{F}\,\Delta x\Delta y\Delta z$, and let's drop the whole limit as $\Delta x \to 0$ notation and just keep in mind that the Δx, etc. are so small as to make our discretization of space indistinguishable from a truly continuous representation. We now write

$$\begin{aligned}\nabla \cdot \vec{F}\,\Delta x\Delta y\Delta z = {} & [F_x(x_0 + \Delta x, y_0, z_0) - F_x(x_0, y_0, z_0)] \cdot \Delta y\Delta z \\ & [F_y(x_0, y_0 + \Delta y, z_0) - F_y(x_0, y_0, z_0)] \cdot \Delta x\Delta z \\ & [F_z(x_0, y_0, z_0 + \Delta z) - F_z(x_0, y_0, z_0)] \cdot \Delta x\Delta y \end{aligned} \tag{7.20}$$

A picture helps greatly. Take the first term of Eq. 7.20

$$[F_x(x_0 + \Delta x, y_0, z_0) - F_x(x_0, y_0, z_0)] \cdot \Delta y\Delta z. \tag{7.21}$$

We represent this term pictorially in Figure 7.17. As with any good discrete approximation, we assert that *to the desired degree of accuracy*, the x-component of the vector field has a constant value $F_x(x_0, y_0, z_0)$ all over surface no. ①, and value $F_x(x_0 + \Delta x, y_0, z_0)$ on surface no. ②. With this setup, it is clear that the net flux leaving this closed, infinitesimal prism through the yz surfaces is nothing other than Eq. 7.21. If you are comfortable with that, then it follows that the net flux leaving the entire (infinitesimal) prism is given by Eq. 7.20.

Now we need to go one step further, and consider the net flux through a larger box that consists of two of these infinitesimal boxes that abut one another

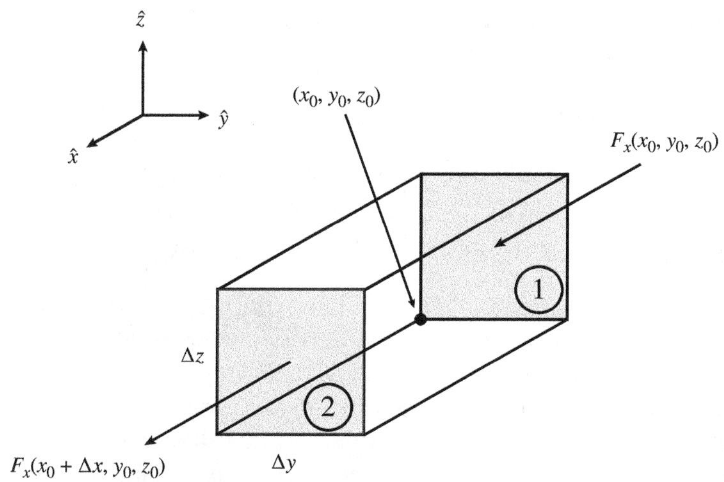

Figure 7.17 Detail of divergence calculation.

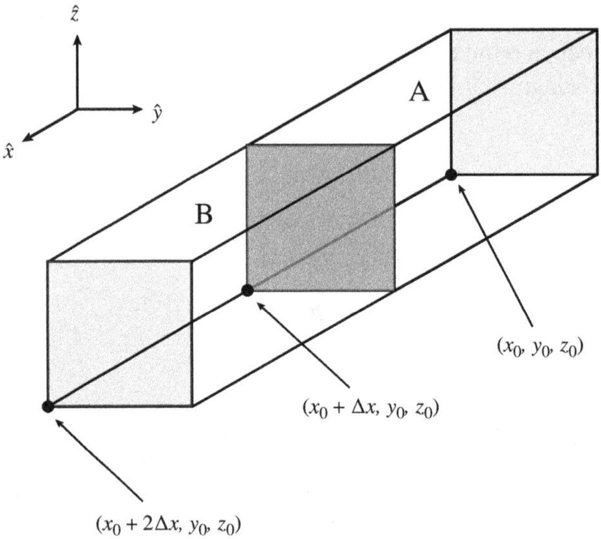

Figure 7.18 Two abutting infinitesimal boxes for a net flux calculation.

(see Figure 7.18). We are interested in the *sum* of the net flux for both boxes. Borrowing from Eq. 7.20, the net flux for Box A is

$$\begin{aligned} \text{Flux}_A = & [F_x(x_0 + \Delta x, y_0, z_0) - F_x(x_0, y_0, z_0)] \cdot \Delta y \Delta z \\ & [F_y(x_0, y_0 + \Delta y, z_0) - F_y(x_0, y_0, z_0)] \cdot \Delta x \Delta z \\ & [F_z(x_0, y_0, z_0 + \Delta z) - F_z(x_0, y_0, z_0)] \cdot \Delta x \Delta y. \end{aligned} \tag{7.22}$$

Similarly, for Box B we have

$$\begin{aligned} \text{Flux}_B = & [F_x(x_0 + 2\Delta x, y_0, z_0) - F_x(x_0 + \Delta x, y_0, z_0)] \cdot \Delta y \Delta z \\ & [F_y(x_0 + \Delta x, y_0 + \Delta y, z_0) - F_y(x_0 + \Delta x, y_0, z_0)] \cdot \Delta x \Delta z \\ & [F_z(x_0 + \Delta x, y_0, z_0 + \Delta z) - F_z(x_0 + \Delta x, y_0, z_0)] \cdot \Delta x \Delta y. \end{aligned} \tag{7.23}$$

Looking carefully, we see that if we add $\text{Flux}_A + \text{Flux}_B$, the $[F_x(x_0 + \Delta x, y_0, z_0)] \cdot \Delta y \Delta z$ term, which deals with flux through the *internal* boundary of our two-segment prism, cancels! This is exactly what we were hoping for, if we want to validate the use of the divergence as a flux density. All flux terms that arise from internal boundaries *should* cancel in our volume integral, and all we should be left with is contributions due to flux through the outer surfaces *only*. This is exactly what we wind up with.

The logical end of this thinking is the divergence theorem, or Gauss's theorem, which now you are in a position to appreciate:

$$\iiint_V \nabla \cdot \vec{F}\, dV = \oiint_S \vec{F} \cdot d\vec{A}. \tag{7.24}$$

Remember that the whole game is to get you even a little bit comfortable with the concept of divergence, because the form of Maxwell's equations that give rise to propagating wave solutions are most clearly written using this ∇ operator. It turns out that we need to look into one more operation, which is the curl of $\vec{F}$, or $\nabla \times \vec{F}$.

The basic approach is the same as for divergence. It's not going to look the same, because the cross-product gives us a lot of terms, but try not to let that be a distraction. Again in Cartesian coordinates, the curl of vector field $\vec{F}$ is given by

$$\nabla \times \vec{F} = \left(\frac{\partial F_z}{\partial y} - \frac{\partial F_y}{\partial z}\right)\hat{x} + \left(\frac{\partial F_x}{\partial z} - \frac{\partial F_z}{\partial x}\right)\hat{y} + \left(\frac{\partial F_y}{\partial x} - \frac{\partial F_x}{\partial y}\right)\hat{z}. \tag{7.25}$$

Notice that, unlike the divergence $\nabla \cdot \vec{F}$, the curl $\nabla \times \vec{F}$ is itself a vector quantity. It has an x-component

$$\begin{aligned}\left(\nabla \times \vec{F}\right)_x = \lim_{\Delta y \to 0} &\frac{F_z(x_0, y_0 + \Delta y, z_0) - F_z(x_0, y_0, z_0)}{\Delta y} \\ - \lim_{\Delta z \to 0} &\frac{F_y(x_0, y_0, z_0 + \Delta z) - F_y(x_0, y_0, z_0)}{\Delta z},\end{aligned} \tag{7.26}$$

a y-component

$$\begin{aligned}\left(\nabla \times \vec{F}\right)_y = \lim_{\Delta z \to 0} &\frac{F_x(x_0, y_0, z_0 + \Delta z) - F_x(x_0, y_0, z_0)}{\Delta z} \\ - \lim_{\Delta x \to 0} &\frac{F_z(x_0 + \Delta x, y_0, z_0) - F_z(x_0, y_0, z_0)}{\Delta x},\end{aligned} \tag{7.27}$$

and a z-component

$$\begin{aligned}\left(\nabla \times \vec{F}\right)_z = \lim_{\Delta x \to 0} &\frac{F_y(x_0 + \Delta x, y_0, z_0) - F_y(x_0, y_0, z_0)}{\Delta x} \\ - \lim_{\Delta y \to 0} &\frac{F_x(x_0, y_0 + \Delta y, z_0) - F_x(x_0, y_0, z_0)}{\Delta y}.\end{aligned} \tag{7.28}$$

Now, we refer to $\nabla \times \vec{F}$ as the "circulation density" of a vector field. What we mean by that is that if we integrate the curl of a vector field over a surface, the result of that integral is equal to a line integral of the vector field around the boundary of that surface. In equations:

$$\begin{aligned}\oint \vec{F} \cdot d\vec{l} &= \iint_A \left(\nabla \times \vec{F}\right) \cdot d\vec{A} \\ &= \iint_A \left(\nabla \times \vec{F}\right) \cdot (dy\, dz\, \hat{x} + dx\, dz\, \hat{y} + dx\, dy\, \hat{z}) \\ &= \iint_A \left(\nabla \times \vec{F}\right)_x dy\, dz + \iint_A \left(\nabla \times \vec{F}\right)_y dx\, dz + \iint_A \left(\nabla \times \vec{F}\right)_z dx\, dy.\end{aligned} \tag{7.29}$$

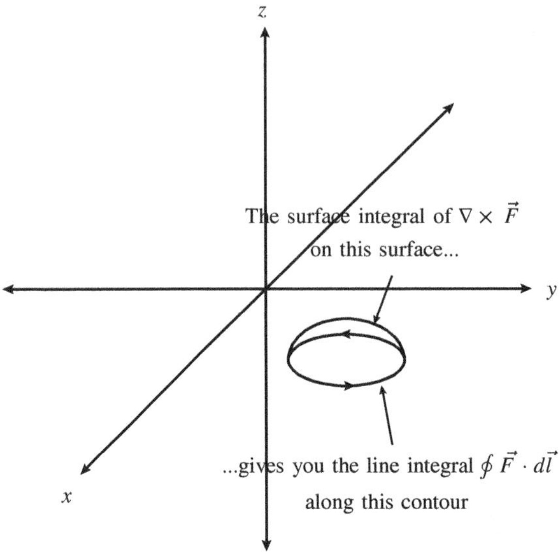

Figure 7.19 Illustration of Stokes' theorem.

To see this in pictures see Figure 7.19.

A couple of notes. First, the equivalence between the surface integral of the curl and the line integral around the border is a result from multivariable calculus called Stokes' theorem. Second, why do we even care about any of this? We care, because we are interested in as detailed a picture as we can get of the EM fields. A beautiful thing about writing in these easily discretizable forms, as opposed to the integral forms that we first discussed when introducing Maxwell's equations, is that a sophisticated computer does not share our human difficulty of keeping careful track of thousands, millions, or billions of terms. If we can get convinced of the foundations of our approach for tiny fragments of space, then macro-scale fields can be analyzed by computational engines of sufficient strength.

To that end, let's look at the detailed calculation of the curl. We start by examining $\left(\nabla \times \vec{F}\right)_x$'s contribution to the surface integral, which is $\left(\nabla \times \vec{F}\right)_x \Delta y \, \Delta z$. We write the approximation

$$\begin{aligned}\left(\nabla \times \vec{F}\right)_x \Delta y \, \Delta z = &\left(\frac{F_z(x_0, y_0 + \Delta y, z_0) - F_z(x_0, y_0, z_0)}{\Delta y}\right.\\ &\left. - \frac{F_y(x_0, y_0, z_0 + \Delta z) - F_y(x_0, y_0, z_0)}{\Delta z}\right) \Delta y \Delta z \\ = &\left[F_z(x_0, y_0 + \Delta y, z_0) - F_z(x_0, y_0, z_0)\right] \Delta z \\ &- \left[F_y(x_0, y_0, z_0 + \Delta z) - F_y(x_0, y_0, z_0)\right] \Delta y.\end{aligned} \tag{7.30}$$

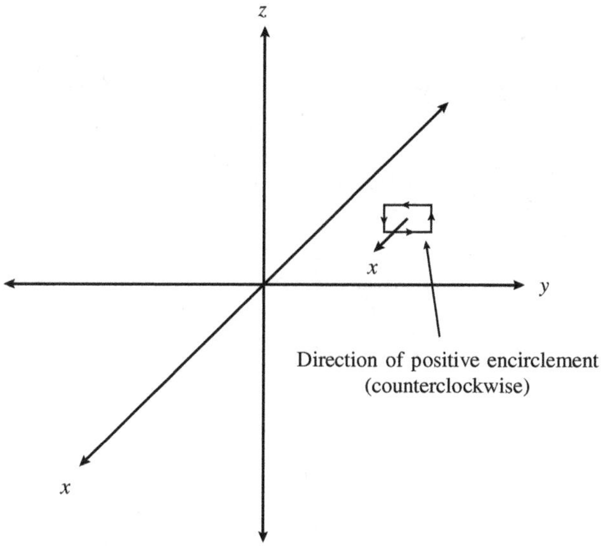

Figure 7.20 The curl at a particular point is the line integral around an infinitesimal loop.

A key detail in this calculation is proper understanding of the $d\vec{A}$ vector: the x-component of the $d\vec{A}$ vector has magnitude $dy\,dz$ ($\Delta y\,\Delta z$ in our discretized approximation), and the $\hat{x}$ is orthogonal to the yz plane. We choose the sign of the $\hat{x}$ component to set the convention for positive encirclement of the area fragment. A picture helps greatly again (see Figure 7.20). This sign convention is almost universally called the "right-hand rule": if you take the fingers of the right hand, and point them in the indicated direction around this tiny loop, the thumb points along the normal in the direction that we consider positive.

Referring to Eq. 7.30, we illustrate in Figure 7.21 the meaning of $\left(\nabla \times \vec{F}\right)_x$ evaluated at the point (x_0, y_0, z_0). You can see that as we allow Δy and Δz to get smaller and smaller, we do a fair job of calculating the x-component of the circulation "at" the point (x_0, y_0, z_0). Taking the next step of imagining the net contribution of a bunch of these little squares, take a look at Figure 7.22 and compare it to Figure 7.18. As we saw with the divergence, we get the net circulation of the larger loop by summing the contributions of the two smaller loops. This is what we expect from a quantity the purports to represent a "circulation density."

Finally, if you really want to understand the curl of a vector field, the last step is to examine an area fragment that does not lie neatly in the xy, xz, or yz plane (see Figure 7.23). Worth pondering. Meantime, note that while we have gestured here toward a proof of Stokes' theorem, we have not gone all the way there because the details are more than what we need for this text. If you are interested, any quality book on multivariable calculus will satisfy your curiosity.

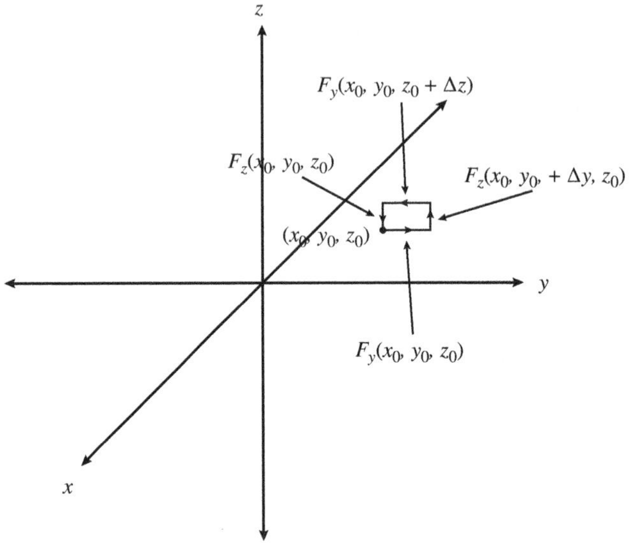

Figure 7.21 Detailed calculation of the curl at a particular point (x_0, y_0, z_0).

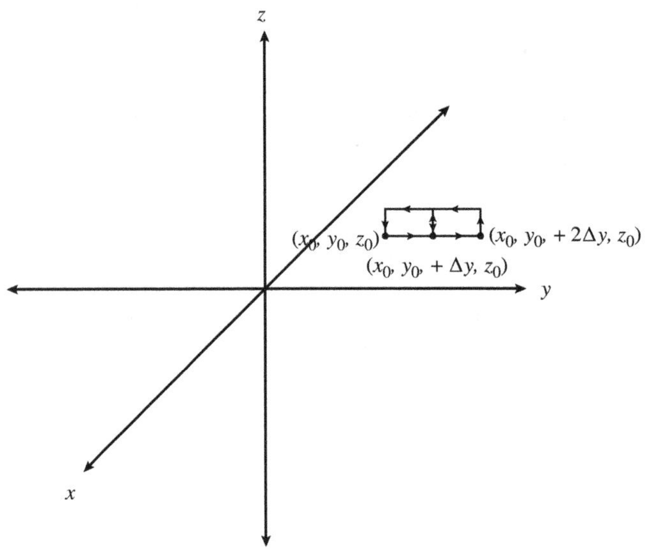

Figure 7.22 As we integrate the curl over a surface, we see that the contributions due to overlapping line segments cancel.

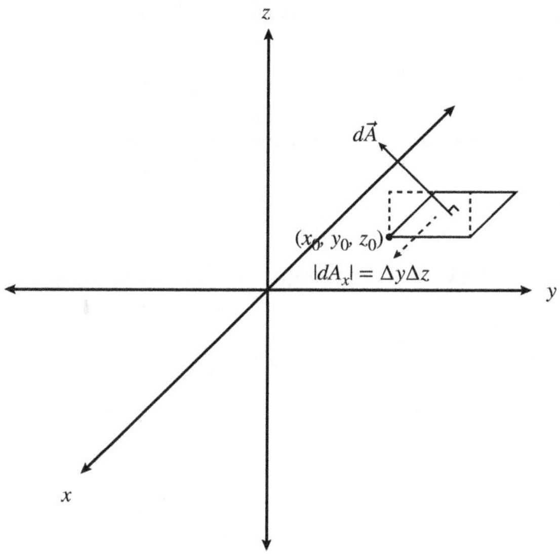

Figure 7.23 Examining the curl contributions of an area element on an arbitrary surface.

7.2.3 Maxwell's Equations in Differential Form

Using our newfound understanding of divergence and curl, we can rewrite Maxwell's equations in differential form. From here we can jump to the mathematics of wave propagation.

Gauss's law is restated here as

$$\oiint_S \vec{E} \cdot d\vec{A} = \frac{1}{\epsilon_0} Q_{\text{encl}}. \tag{7.31}$$

We are going to replace the right side, the Q_{encl}, with a volume integral over the charge density $\rho(x, y, z)$ in the region, and then rewrite the left side expressing the net flux out of the enclosed volume with the volume integral of the divergence of the vector field:

$$\iiint_V \nabla \cdot \vec{E}\, dV = \frac{1}{\epsilon_0} \iiint_V \rho(x, y, z)\, dV. \tag{7.32}$$

In order for this equality to hold over *any* arbitrary volume, we must have that

$$\nabla \cdot \vec{E} = \frac{1}{\epsilon_0} \rho, \tag{7.33}$$

which is the differential form of Gauss's law. Gauss's law for magnetic fields is a bit easier:

$$\nabla \cdot \vec{B} = 0. \tag{7.34}$$

For the Maxwell–Faraday equation, we start in the familiar integral form

$$\oint \vec{E} \cdot d\vec{l} = -\frac{d}{dt} \iint \vec{B} \cdot d\vec{A}. \tag{7.35}$$

We replace the left side using the surface integral of the curl of the vector field:

$$\iint \nabla \times \vec{E} \cdot d\vec{A} = -\frac{\partial}{\partial t} \iint \vec{B} \cdot d\vec{A}. \tag{7.36}$$

We are free to bring the time derivative inside the integral on the right,

$$\iint \nabla \times \vec{E} \cdot d\vec{A} = -\iint \frac{\partial}{\partial t}\vec{B} \cdot d\vec{A}. \tag{7.37}$$

Again, requiring that this equality holds over an arbitrary area of integration gives us that

$$\nabla \times \vec{E} = -\frac{\partial}{\partial t}\vec{B}. \tag{7.38}$$

Finally, for the Ampere–Maxwell equation, we start with

$$\oint \vec{B} \cdot d\vec{l} = \mu_0 I_{\text{encl}} + \mu_0 \epsilon_0 \frac{\partial}{\partial t} \iint \vec{E} \cdot d\vec{A}. \tag{7.39}$$

As we did with Gauss's law and charge, we're going to replace the enclosed current I_{encl} with integrating a current density $\vec{J}(x, y, z)$ integrated over the area inside the loop. This plus using Stokes's theorem again gives us

$$\iint \nabla \times \vec{B} \cdot d\vec{A} = \mu_0 \iint \vec{J} \cdot d\vec{A} + \mu_0 \epsilon_0 \frac{\partial}{\partial t} \iint \vec{E} \cdot d\vec{A}. \tag{7.40}$$

You know the drill:

$$\nabla \times \vec{B} = \mu_0 \left(\vec{J} + \epsilon_0 \frac{\partial}{\partial t}\vec{E} \right). \tag{7.41}$$

7.3 General Wave Solutions of Maxwell's Equations

An astounding feature of physics is the consistent mapping between mathematics and physical meaning. Sometimes, the isomorphism at play can be so astonishing that those with academic backgrounds go too far, and insist that the math *is* the understanding. It is not. There are many ways for a human mind to achieve a useful understanding of Nature. Nevertheless, that the typographical rules of algebra can yield expressions with physical meanings beyond the prior understanding of the scribbler is...remarkable.

Wave solutions of Maxwell's equations are a great example of this dynamic. We did not always think of light as an example of propagating EM waves. But by following a development similar in spirit to what we are going to do here, Maxwell himself found that traveling EM waves were consistent with the equations that would bear his name. Moreover, he calculated that these waves travel at a speed very close to what was then known as the speed of light. He therefore suggested that light is a traveling EM wave. In his own words[6]:

> The agreement of the results seems to show that light and magnetism are affectations of the same substance, and that light is an EM disturbance propagated through the field according to EM laws.

Let's get into it.

Start with Maxwell's equations in differential form, repeated and grouped together here for convenience. We stipulate that we are in free space with no charges and no currents, and simplify accordingly:

$$\begin{aligned} \nabla \cdot \vec{E} &= 0 \\ \nabla \cdot \vec{B} &= 0 \\ \nabla \times \vec{E} &= -\frac{\partial}{\partial t}\vec{B} \\ \nabla \times \vec{B} &= \mu_0 \epsilon_0 \frac{\partial}{\partial t}\vec{E}. \end{aligned} \tag{7.42}$$

Now, the first step in our path to getting to wave equations will be to take the curl of the curl equations in Eq. 7.42:

$$\begin{aligned} \nabla \times \left(\nabla \times \vec{E}\right) &= -\frac{\partial}{\partial t}\left(\nabla \times \vec{B}\right) \\ \nabla \times \left(\nabla \times \vec{B}\right) &= \mu_0 \epsilon_0 \frac{\partial}{\partial t}\left(\nabla \times \vec{E}\right). \end{aligned} \tag{7.43}$$

Wait! Come back! Casually "taking the curl of the curl equations" is not the blind swipe that it appears at first. When we look at Maxwell's equations in differential form, we are nowhere at first because the equations are coupled. E is defined in terms of B, and vice versa. But if we look at, for example,

$$\nabla \times \left(\nabla \times \vec{E}\right) = -\frac{\partial}{\partial t}\left(\nabla \times \vec{B}\right), \tag{7.44}$$

6 Maxwell, A dynamical theory of the electromagnetic field, *Philosophical Transactions of the Royal Society of London,* **155**: pp.459–512. 1865.

we see immense promise. The curl of the B field can be written as a time derivative of the E field! While we may not know quite what to do with $\nabla \times (\nabla \times \vec{E})$, we at least now have an uncoupled equation solely in E and that is victory enough to start. With fresh enthusiasm, now rewrite Eq. 7.43 as

$$\nabla \times \left(\nabla \times \vec{E}\right) = -\mu_0 \epsilon_0 \frac{\partial^2 \vec{E}}{\partial t^2} \qquad (7.45)$$
$$\nabla \times \left(\nabla \times \vec{B}\right) = -\mu_0 \epsilon_0 \frac{\partial^2 \vec{B}}{\partial t^2}.$$

Now about $\nabla \times (\nabla \times \vec{E})$ and $\nabla \times (\nabla \times \vec{B})$, which don't look so friendly, especially after Section 7.2.2. It turns out there is little to be gained by an expansive treatment of the curl of the curl. We instead employ a vector calculus identity which you can look up if you choose:

$$\nabla \times \left(\nabla \times \vec{E}\right) = \nabla\left(\nabla \cdot \vec{E}\right) - \nabla^2 \vec{E} \qquad (7.46)$$
$$\nabla \times \left(\nabla \times \vec{B}\right) = \nabla\left(\nabla \cdot \vec{B}\right) - \nabla^2 \vec{B}.$$

$\nabla(\nabla \cdot \vec{E})$ reads as "the gradient of the divergence of E," which we might get into if $\nabla \cdot \vec{E}$ weren't zero. Another time, then. Using the fact that $\nabla \cdot \vec{B}$ is also zero, we can return to Eq. 7.45 and write

$$\mu_0 \epsilon_0 \frac{\partial^2 \vec{E}}{\partial t^2} - \nabla^2 \vec{E} = 0 \qquad (7.47)$$
$$\mu_0 \epsilon_0 \frac{\partial^2 \vec{B}}{\partial t^2} - \nabla^2 \vec{B} = 0.$$

The last piece we need is ∇^2, the so-called "Laplacian," which is

$$\nabla^2 \vec{E} = \left(\frac{\partial^2}{\partial x^2} + \frac{\partial^2}{\partial y^2} + \frac{\partial^2}{\partial z^2}\right) \vec{E}. \qquad (7.48)$$

One other handy result is that $1/\sqrt{\mu_0 \epsilon_0}$ is the speed of light in a vacuum, or c,[7] which is 3.0×10^8 m/s. We'll use this to clean up what we are going to call our wave equations:

$$\frac{1}{c^2} \frac{\partial^2 \vec{E}}{\partial t^2} - \nabla^2 \vec{E} = 0 \qquad (7.49)$$
$$\frac{1}{c^2} \frac{\partial^2 \vec{B}}{\partial t^2} - \nabla^2 \vec{B} = 0.$$

7 ϵ_0 is the permittivity of free space, 8.85×10^{-12} F/m or $\mathrm{m^{-3} kg^{-1} s^4 A^2}$, while μ_0 is the permeability of free space, $4\pi \times 10^{-7}$ H/m, or $\mathrm{m\,kg\,s^{-2} A^{-2}}$.

Why do we call these "wave equations"? We call them that because they turn out to have very special solutions, which are of the form

$$\vec{E}(\vec{r},t) = \vec{E}_0 \cdot g\left(t - \frac{1}{c}\hat{k}\cdot\vec{r}\right) \tag{7.50}$$
$$\vec{B}(\vec{r},t) = \vec{B}_0 \cdot f\left(t - \frac{1}{c}\hat{k}\cdot\vec{r}\right).$$

We must satisfy ourselves of two things. First, we have to understand why the form $g(t - \frac{1}{c}\hat{k}\cdot\vec{r})$ is rightly described as a traveling wave. Second, we must show that solutions of this form in fact satisfy Eq. 7.49. We'll take these in order.

First, let's figure out how to think about $g(t - \frac{1}{c}\hat{k}\cdot\vec{r})$. Notice that we put no restrictions whatsoever on the form of the function $g(\cdot)$. We require only that the argument of the function be $t - \frac{1}{c}\hat{k}\cdot\vec{r}$. That is to say, we take spatial coordinates x, y, and z (recall that $\vec{r} = x\hat{x} + y\hat{y} + z\hat{z}$), and a time t, and combine them according to $t - \frac{1}{c}\hat{k}\cdot\vec{r}$. This scalar quantity is fed as the input to $g(\cdot)$, which yields another scalar quantity. We multiply this scalar quantity by the constant vector $\vec{E}_0$ $(= E_{0x}\hat{x} + E_{0y}\hat{y} + E_{0z}\hat{z})$, and *that* product is the vector electric field at (x, y, z, t).

We say that this is a traveling wave. For that to make any sense, then as time progresses we must move in space somehow to keep up with the waves. Certainly, if we are "keeping up with" the wave as it travels, then that means that we are moving through space (changing (x, y, z)) such that the argument of $g(\cdot)$ is constant:

$$\left(t - \frac{1}{c}\hat{k}\cdot\vec{r}\right) = C. \tag{7.51}$$

Saying that the argument is a constant is the same thing as saying that the time derivative of the argument is zero:

$$\frac{d}{dt}\left(t - \frac{1}{c}\hat{k}\cdot\vec{r}\right) = 0. \tag{7.52}$$

Let's solve for that.

$$1 - \frac{1}{c}\frac{d}{dt}\left(k_x x + k_y y + k_z z\right) = 0 \tag{7.53}$$
$$1 - \frac{1}{c}\left(k_x v_x + k_y v_y + k_z v_z\right) = 0$$
$$1 - \frac{1}{c}\hat{k}\cdot\vec{v} = 0.$$

We are saying mathematically that if we as observers move with a particular velocity vector, the argument of $g(\cdot)$ remains constant.

Recall that the vector dot product is $\hat{k}\cdot\vec{v} = 1\cdot|v|\cos\theta$, where θ is the angle between the velocity vector and $\hat{k}$. It becomes clear that to keep up with the wave, we must move with speed

$$|v| = \frac{c}{\cos\theta}. \tag{7.54}$$

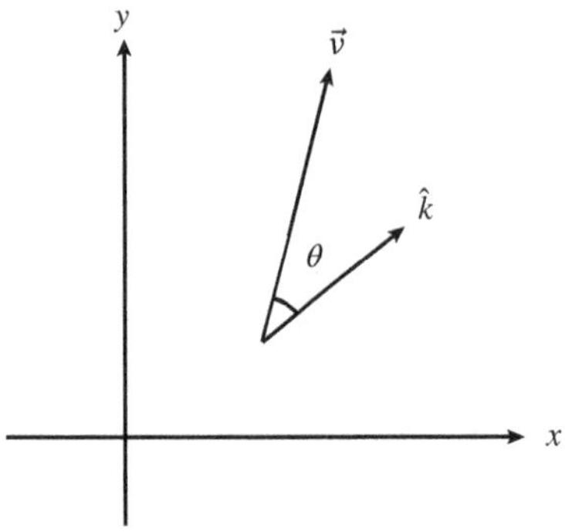

Figure 7.24 Moving in space along $\vec{v}$ to keep $t - \frac{1}{c}\hat{k} \cdot \vec{r}$ constant. It is natural to interpret $\hat{k}$ as the direction of propagation for the wave front.

The speed $|v|$ is smallest when $\vec{v}$ is aligned with $\hat{k}$, with $\theta = 0$ (and $\cos\theta = 1$). At the other extreme, when $\theta = 90°$ and our velocity vector is orthogonal to $\hat{k}$, it is impossible to keep up with what we will now call the wave front. A helpful visualization is given in Figure 7.24.

When you look at the figure, you can see the point that we have been building up to: for a traveling wave $g(t - \frac{1}{c}\hat{k} \cdot \vec{r})$, regardless of the details of $g(\cdot)$, $\hat{k}$ is properly interpreted as *the direction of propagation*. Following along the wave front in the direction of propagation, we must move with speed c, as we expect.

The next thing we promised is that $g(t - \frac{1}{c}\hat{k} \cdot \vec{r})$ is, in fact, a solution to wave equations Eq. 7.49. We'll be aided by writing our expressions with $\hat{k}$ spelled out explicitly:

$$\hat{k} = k_x\hat{x} + k_y\hat{y} + k_z\hat{z}. \tag{7.55}$$

We pick one of the equations from Eq. 7.49, and barrel ahead:

$$\begin{aligned} &\frac{1}{c^2}\frac{\partial^2}{\partial t^2}\left[\vec{E}_0 \cdot g\left(t - \frac{1}{c}(k_x x + k_y y + k_z z)\right)\right] \\ &- \left(\frac{\partial^2}{\partial x^2} + \frac{\partial^2}{\partial y^2} + \frac{\partial^2}{\partial z^2}\right)\vec{E}_0 \cdot g\left(t - \frac{1}{c}(k_x x + k_y y + k_z z)\right) = 0. \end{aligned} \tag{7.56}$$

From here we employ the chain rule from calculus,

$$\frac{\partial}{\partial t}g(f(t)) = g'(f(t)) \cdot \frac{\partial f}{\partial t} \tag{7.57}$$

where $g'(\cdot)$ is the first derivative of g with respect to the whole argument. If we follow our analytical nose starting from Eq. 7.56 and apply the chain rule twice in rapid succession, we arrive at

$$\frac{1}{c^2}\vec{E}_0 g''(\cdot) - \frac{1}{c^2}(k_x^2 + k_y^2 + k_z^2)\vec{E}_0 g''(\cdot) = 0. \tag{7.58}$$

Recall that $\hat{k}$ is a unit vector, so

$$k_x^2 + k_y^2 + k_z^2 = 1. \tag{7.59}$$

We are left with obvious equality

$$\frac{1}{c^2}\vec{E}_0 g''(\cdot) - \frac{1}{c^2}\vec{E}_0 g''(\cdot) = 0, \tag{7.60}$$

and this *means* that traveling waves of the form $g(t - \frac{1}{c}\hat{k} \cdot \vec{r})$ are solutions of the wave equations!

And what, finally, to say about $g(\cdot)$ itself? It brings clarity to first tackle the visualization problem by evaluating $g(\cdot)$ at a fixed time:

$$g\left(t_0 - \frac{1}{c}(k_x x + k_y y + k_z z)\right). \tag{7.61}$$

At this fixed time, t_0, $g(\cdot)$ tells you what the field is for all points in space. It is the spatial shape, in other words. When we release t to progress forward in time as usual, that whole shape moves through space in the $\hat{k}$ direction with speed c. For ease of visualizing, Figure 7.25 illustrates a propagating wave whose only spatial dependence is on x. It turns out that we like this trick...going forward, we are going to reduce the amount of bookkeeping we must do by aligning, say, the x-axis of our coordinate system with the direction of propagation. This allows us to write the field of a plane wave as

$$\vec{E}(\vec{r}, t) = \vec{E}_0 \cdot g\left(t - \frac{1}{c}x\right). \tag{7.62}$$

It bears emphasis that we do this without loss of generality. In your mind, you can overlay $\hat{x}$ now with "the direction of propagation," and the yz plane with "the plane orthogonal to the direction of propagation."

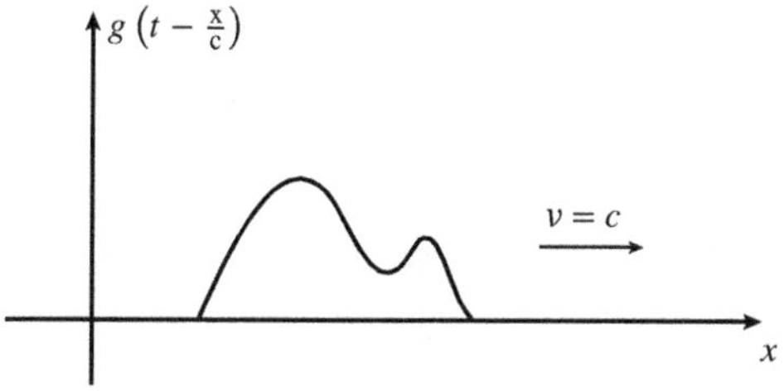

Figure 7.25 A pulse traveling in the $\hat{x}$ direction.

Having determined that traveling waves are consistent with Maxwell's equations, we go back to those equations and mine them for other insights. Start with Gauss's law in a charge-free region:

$$\nabla \cdot \vec{E} = 0 \tag{7.63}$$

$$\frac{\partial E_x}{\partial x} + \frac{\partial E_y}{\partial y} + \frac{\partial E_z}{\partial z} = 0.$$

We know from Eq. 7.62 that $\frac{\partial E_y}{\partial y} = \frac{\partial E_z}{\partial z} = 0$. Applying Gauss's law tells us further that

$$\frac{\partial E_x}{\partial x} = 0. \tag{7.64}$$

That is to say, the x component of the E field does not vary with x, and is constant in space. While allowed, a constant E_x is not of interest to us, focused as we are on dynamic fields that travel. For purposes of analysis, we put a constant E_x to the side as not part of the traveling wave. When we are thinking about the traveling wave, then, $E_x = 0$.

This leads to an important aspect of traveling EM waves, which is that the E field is *orthogonal to the direction of propagation*. And while we have that $\frac{\partial E_y}{\partial y} = \frac{\partial E_z}{\partial z} = 0$, we do *not* have that $\frac{\partial E_y}{\partial x}$ or $\frac{\partial E_z}{\partial x}$ are necessarily zero.

In the interest of keeping our bookkeeping simpler while giving up nothing on generality, we're going to focus on a solution for which $E_z = 0$ an $E_y \neq 0$. (Remember that we can always write our final, general wave as the sum of a wave with an E_z and a wave with an E_y.) What we find as we move forward is that each of Maxwell's equations has something to teach us about this new wave solution. Let's take the next step with

$$\nabla \times \vec{E} = -\frac{\partial \vec{B}}{\partial t}. \tag{7.65}$$

Remember that $\nabla \times \vec{E}$ is itself a vector field, so we have three components to keep track of. Note also that we have chosen $E_z = 0$, and that the form of Eq. 7.62 gives immediate guidance as to certain partial derivatives being zero by construction (that is, *all* derivatives with respect to y or z). With all of that in mind, we devote ourselves to $\nabla \times \vec{E}$:

$$\left(\nabla \times \vec{E}\right)_x = \frac{\partial E_z}{\partial y} - \frac{\partial E_y}{\partial z} = -\frac{\partial B_x}{\partial t} = 0 \tag{7.66}$$

$$\left(\nabla \times \vec{E}\right)_y = \frac{\partial E_x}{\partial z} - \frac{\partial E_z}{\partial x} = -\frac{\partial B_y}{\partial t} = 0$$

$$\left(\nabla \times \vec{E}\right)_z = \frac{\partial E_y}{\partial x} - \frac{\partial E_x}{\partial y} = \frac{\partial E_y}{\partial x} = -\frac{\partial B_z}{\partial t}.$$

Our careful construction apparently only leaves room for a nonstatic B_z. Writing this in isolation, we have

$$\frac{\partial E_y}{\partial x} = -\frac{\partial B_z}{\partial t}. \tag{7.67}$$

So...both E and B are orthogonal to $\hat{k}$, the direction of propagation, *and* they are orthogonal to each other.

Two more of Maxwell's equations to go. When we evaluate $\nabla \cdot \vec{B}$, we do so knowing that B_x and B_y are zero from previous work. This leaves us with

$$\nabla \cdot \vec{B} = \frac{\partial B_z}{\partial z} = 0. \tag{7.68}$$

This removes the possibility of B_z varying with z, and so it is safe to say that the B field everywhere in space, like the E field we constructed, varies only with x and not with y or z.

Finally, let's look at what $\nabla \times \vec{B} = \mu_0 \epsilon_0 \frac{\partial \vec{E}}{\partial t}$ has to teach us. We remember that $c = 1/\sqrt{\mu_0 \epsilon_0}$, and so we can write $\nabla \times \vec{B} = (1/c^2)\frac{\partial}{\partial t}\vec{E}$. From here we must write out all the components

$$\begin{aligned}
\left(\nabla \times \vec{B}\right)_x &= \frac{\partial B_z}{\partial y} - \frac{\partial B_y}{\partial z} = \frac{1}{c^2}\frac{\partial E_x}{\partial t} = 0 \\
\left(\nabla \times \vec{B}\right)_y &= \frac{\partial B_x}{\partial z} - \frac{\partial B_z}{\partial x} = \frac{1}{c^2}\frac{\partial E_y}{\partial t} \\
\left(\nabla \times \vec{B}\right)_z &= \frac{\partial B_y}{\partial x} - \frac{\partial B_x}{\partial y} = \frac{1}{c^2}\frac{\partial E_z}{\partial t} = 0.
\end{aligned} \tag{7.69}$$

We have that $\vec{B}$ has only a z component, that partial derivatives with respect to y and z are zero, and $\vec{E}$ has only a y component. This makes a lot of terms zero out, and we are left with

$$-\frac{\partial B_z}{\partial x} = \frac{1}{c^2}\frac{\partial E_y}{\partial t}. \tag{7.70}$$

At last, we get to the part of "solving" differential equations that is irritating for many students, but there is no getting around it. That is, we look at our equations and guess, or more properly, identify, a solution, and then plug that solution into our equations to see if everything works out.

We start with our decision to look at the case of $\vec{E}$ pointing solely in the $\hat{y}$ direction. From Eq. 7.62, we write

$$E_y(x,t) = E_{y1} g\left(t - \frac{1}{c}x\right), \tag{7.71}$$

where E_{y1} is a constant coefficient. Now, we did not explore before but we will point out now that it is equally valid to have a wave traveling in the opposite direction.

Whenever this happens with linear equations, the discipline is to write the "most general solution" as a linear combination of the two, knowing that you can always set one coefficient to zero:

$$E_y(x,t) = E_{y1} g\left(t - \frac{1}{c}x\right) + E_{y2} f\left(t + \frac{1}{c}x\right). \tag{7.72}$$

How about the B field, now? Here is where the guessing comes in. We suppose that the B field has the same shape as the E field (i.e. the same $g(\cdot)$ and $f(\cdot)$). How will we know if that is a good guess? The answer is that it must be consistent with both Eqs. 7.67 and 7.70. So let's say that

$$B_z(x,t) = B_{z1} g\left(t - \frac{1}{c}x\right) + B_{z2} f\left(t + \frac{1}{c}x\right). \tag{7.73}$$

Let's see now what Eq. 7.70 requires:

$$\begin{aligned} \frac{1}{c}B_{z1}g(\cdot) - \frac{1}{c}B_{z2}f(\cdot) &= \frac{1}{c^2}E_{y1}g(\cdot) + \frac{1}{c^2}E_{y2}f(\cdot) \\ B_{z1}g(\cdot) - B_{z2}f(\cdot) &= \frac{1}{c}E_{y1}g(\cdot) + \frac{1}{c}E_{y2}f(\cdot). \end{aligned} \tag{7.74}$$

We get equality if we have

$$\begin{aligned} B_{z1} &= \frac{1}{c}E_{y1} \\ B_{z2} &= -\frac{1}{c}E_{y2}. \end{aligned} \tag{7.75}$$

To be extra sure, you can check to make sure that with these choices we also satisfy Eq. 7.67.

After all that, at the end of the day, we have two choices to make that completely determine the wave solution of Maxwell's equations. Once we choose the amplitude of the forward-going y component of the E field, E_{y1} and the reverse-going y component of the E field, E_{y2}, all else follows:

$$\begin{aligned} E_y(x,t) &= E_{y1} g\left(t - \frac{1}{c}x\right) + E_{y2} f\left(t + \frac{1}{c}x\right) \\ B_z(x,t) &= \frac{1}{c}E_{y1} g\left(t - \frac{1}{c}x\right) - \frac{1}{c}E_{y2} f\left(t + \frac{1}{c}x\right). \end{aligned} \tag{7.76}$$

We could do the whole exercise again, this time for an E_z and a B_y:

$$\begin{aligned} E_z(x,t) &= E_{z1} g\left(t - \frac{1}{c}x\right) + E_{z2} f\left(t + \frac{1}{c}x\right) \\ B_y(x,t) &= -\frac{1}{c}E_{z1} g\left(t - \frac{1}{c}x\right) + \frac{1}{c}E_{z2} f\left(t + \frac{1}{c}x\right). \end{aligned} \tag{7.77}$$

We arrive at last at the most general solution for a plane wave traveling along the x-axis as

$$\vec{E}(\vec{r},t) = \hat{y}E_y(x,t) + \hat{z}E_z(x,t)$$
$$\vec{B}(\vec{r},t) = \hat{y}B_y(x,t) + \hat{z}B_z(x,t). \tag{7.78}$$

And there you have it, the most general wave solution of Maxwell's equations. It is probably true that it is useful for most people in our field to go through this development at least once. Now you have it, though, and can come back to this section any time as a reference if you get curious. In the next section, we will not hold ourselves to such rigorous generality. It's the sinusoidal traveling wave that will pop up most often, and it is the special case that you should pay most attention to.

7.4 A Wave Solution of Particular Interest

We have now satisfied the conditions of honor, having looked deeply at Maxwell's equations to find that traveling waves are consistent with them. You will find in practical work with radio that one *particular* solution is...pretty much all you need. And that is the traveling sinusoid:

$$E_y = E_{yI}\sin(\omega t - kx) + E_{yQ}\cos(\omega t - kx). \tag{7.79}$$

You have seen a version of this before, in Figures 6.1 and 6.2. We are thinking of transmission only one way, from a dedicated transmitter to a dedicated receiver, and we'll suppose that they are far enough apart that a plane wave is a decent approximation of the EM field (see Figure 7.26).

The thing to take away from here is how we co-opt the wave solutions of Maxwell's equations to build our communications apparatus. You will notice the

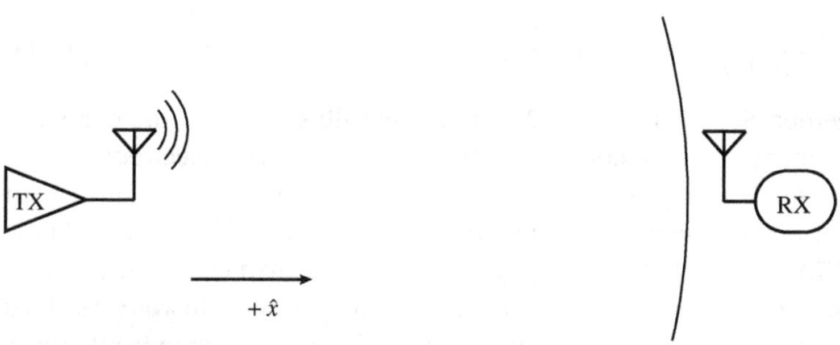

Figure 7.26 When the transmitter and receiver are far enough apart, the radiated waves look like plane waves at the receiver.

form of the argument $(\omega t - kx)$, differs slightly from the agreed upon $(t - x/c)$. This is okay provided that we stick to very specific conditions. First, a helpful rewrite

$$E_y = E_{yI} \sin\left(\omega\left(t - \frac{k}{\omega}x\right)\right) + E_{yQ} \cos\left(\omega\left(t - \frac{k}{\omega}x\right)\right), \tag{7.80}$$

which looks immediately more promising. To make it all work out, we require

$$\frac{k}{\omega} = \frac{1}{c}. \tag{7.81}$$

The parameter k is sometimes called the "wave number," and it is analogous to the angular frequency ω for the time variable. The spatial period is called the "wavelength," is conventionally labeled λ, and is related to the wave number according to

$$\lambda = \frac{2\pi}{k}. \tag{7.82}$$

Remembering that $\omega = 2\pi f$, we can combine Eqs. 7.81 and 7.82 to relate frequency, wavelength, and the speed of light as follows:

$$\lambda = \frac{c}{f}. \tag{7.83}$$

This is one of a small handful of relations that are useful to commit to memory. Sometimes, instead of talking about communications bands in terms of frequencies, practitioners will refer to the relevant wavelength. So, for example, ham radio operators may talk of the "20 m band," which you now see refers to a carrier frequency of

$$f = \frac{3.0 \times 10^8 \text{ m/s}}{20 \text{ m}} = 15 \text{ MHz}. \tag{7.84}$$

Elsewhere, fans of short wavelengths will talk of carrier frequencies in the 60 GHz range as a being in the "millimeter wave" part of the spectrum:

$$f = \frac{3.0 \times 10^8 \text{ m/s}}{60 \times 10^9 \text{ Hz}} = 0.005 \text{ m} = 5 \text{ mm}. \tag{7.85}$$

Taken together, Sections 7.3 and 7.4 are a great illustration of the challenge facing a newcomer to wireless communications. The underlying electrodynamics, which we have made a mighty effort to condense and summarize here, take a fair amount of time and mathematical sophistication to digest. And then after all that, Eqs. 7.79 and 7.83 are all that regularly get used in the day to day. Decide for yourself how much of the whole picture you need to carry around in your head. At minimum, hopefully you will find the foregoing treatment to be a handy reference.

7.5 EM Fields and Energy

In this section, we treat briefly an idea that you see quite often, which is that energy is "stored" in electric and magnetic fields. This turns out to somewhat overstate things, as Purcell comments in his classic text on electricity and magnetism:[8]

> The potential energy U of a system of charges, which is the total work required to assemble the system, can be calculated from the electric field itself simply by assigning an amount of energy $E^2/8\pi\ dv$ to any volume element dv and integrating over all space where there is an electric field.[9]

He goes to describe this as an accounting trick, but a trick that nevertheless has its benefits:

> One may think of this energy as "stored" in the field. The system being conservative, that amount of energy can of course be recovered by allowing the charges to go apart; so it is nice to think of the energy as "being somewhere" meanwhile. Our accounting comes out right if we think of it as stored in space with a density of $E^2/8\pi$, in ergs/cm^3. There is no harm in this, but in fact we have no way of identifying, quite independently of anything else, the energy stored in a particular cubic centimeter of space. Only the total energy is physically measurable, that is, the work required to bring the charge into some configuration, starting from some other configuration...Sometimes a change in viewpoint, even if is at first only a change in bookkeeping, can stimulate new ideas and deeper understanding.

Thinking of energy as stored in fields is particularly useful when we get to energy flow and EM radiation. Let's get acquainted, then, with this form of bookkeeping.

7.5.1 Energy Associated with Static Fields

We're now going to consider the problem of calculating how much energy it takes to assemble a particular charge distribution. We choose this particular charge distribution, you will see, because it makes our calculations clear and compelling.

The setup is that we start with a perfectly spherical, insulating ball. We then suppose that we have a way of transporting charge from some long distance away to the surface of the sphere, whereupon it dutifully (1) spreads itself uniformly

---

8 Purcell. *Electricity and Magnetism, 2e,* USA: McGraw-Hill, Inc., 1985. This quotation is taken from page 32.

9 Purcell uses CGS units in his text. In SI units, the energy density is $\frac{1}{2}\epsilon_0 E^2$.

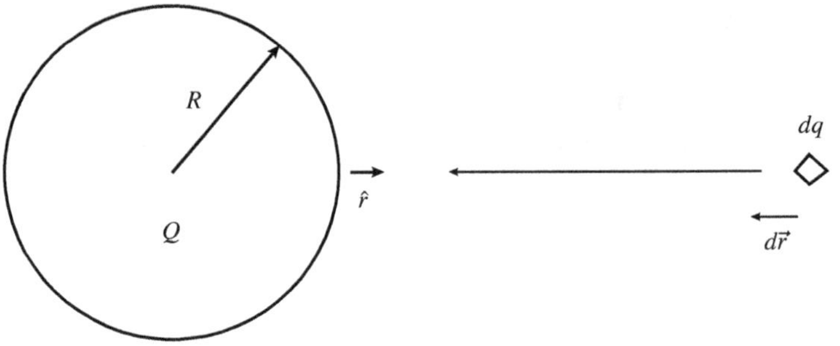

Figure 7.27 Charging a sphere by bringing charge elements *dq* from a long distance away.

over the surface of the sphere, and (2) stays there (see Figure 7.27). Because of the spherical symmetry, the electric field takes the form

$$\vec{E} = \begin{cases} 0 & r < R \\ \frac{1}{4\pi\epsilon_0}\frac{Q_{\text{tot}}}{r^2}\hat{r} & r \geq R. \end{cases} \tag{7.86}$$

Now we ask: how much energy does it take to bring an infinitesimal charge dq to the surface of the sphere from a long distance away, when the sphere already has charge Q? The answer is

$$dU = dq \int_{\infty}^{R} -\frac{1}{4\pi\epsilon_0}\frac{Q}{r^2}\, dr. \tag{7.87}$$

Carrying out the definite integral gives us the increment of work that we did to add the charge dq:

$$\begin{aligned} dU &= -\frac{1}{4\pi\epsilon_0} Q\, dq \left[-\frac{1}{r}\right]_{\infty}^{R} \\ &= \frac{1}{4\pi\epsilon_0}\frac{Q}{R}\, dq. \end{aligned} \tag{7.88}$$

To get the total energy to bring the sphere from uncharged to charge Q_F, we must integrate again:

$$\begin{aligned} U &= \int_0^{Q_F} \frac{1}{4\pi\epsilon_0}\frac{q}{R}\, dq \\ &= \frac{1}{4\pi\epsilon_0}\frac{1}{R}\left[\frac{1}{2}q^2\right]_0^{Q_F} \\ &= \frac{1}{8\pi\epsilon_0}\frac{1}{R}Q_F^2. \end{aligned} \tag{7.89}$$

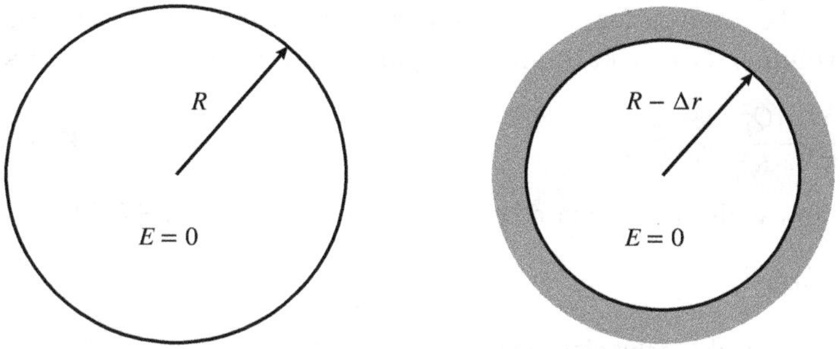

Figure 7.28 Charging a sphere with the same total charge, but on a sphere with radius smaller by Δr. It requires more energy to assemble this tighter charge distribution, and the result is a region "filled" with electric field that before had an E field of zero.

That was our first step. Now, suppose that we repeat the experiment with a new sphere, this time of radius $R - \Delta r$, with $\Delta r << R$. Compared to our first run at this, we now have a new region filled with an electric field that we did not have before (see Figure 7.28). We know that the energy that it took to charge this new, slightly smaller sphere is given by

$$U_{\text{new}} = \frac{1}{8\pi\epsilon_0}\frac{1}{R - \Delta r}Q_{\text{F}}^2. \tag{7.90}$$

The energy difference between the old an new spheres is given by

$$\begin{aligned}\Delta U &= \frac{1}{8\pi\epsilon_0}Q_{\text{F}}^2\left[\frac{1}{R-\Delta r} - \frac{1}{R}\right] \\ &= \frac{1}{8\pi\epsilon_0}Q_{\text{F}}^2\frac{R-(R-\Delta r)}{R(R-\Delta r)} \\ &= \frac{1}{8\pi\epsilon_0}Q_{\text{F}}^2\frac{\Delta r}{R(R-\Delta r)}.\end{aligned} \tag{7.91}$$

For Δr sufficiently small compared to R, we have

$$\Delta U \approx \frac{1}{8\pi\epsilon_0}Q_{\text{F}}^2\frac{\Delta r}{R^2}. \tag{7.92}$$

This next step is where the bookkeeping decision gets made. Equation 7.92 answers the question of fundamental importance, which is the difference in work required to assemble two different charge distributions. In the second setup, we *do* have a region in space that is filled with an electric field that was not filled with an electric field in the first setup. To an excellent approximation, the volume of that space is given by

$$\Delta V = 4\pi R^2 \Delta r. \tag{7.93}$$

We could certainly be forgiven for associating this energy delta with the electric field in the new volume. If we make this bookkeeping choice, the energy density is

$$\frac{\Delta U}{\Delta V} = \frac{1}{32\pi^2\epsilon_0} \frac{Q_F^2}{R^4}. \tag{7.94}$$

Compare this now with the "standard" energy density given for an E field:

$$U_{\text{density}} = \frac{\epsilon_0}{2}|E|^2. \tag{7.95}$$

In our case, the magnitude of the electric field at radius R is

$$|E| = \frac{1}{4\pi\epsilon_0} \frac{Q_F}{R^2}. \tag{7.96}$$

Plugging in gives us the energy density

$$\begin{aligned} U_{\text{density}} &= \frac{\epsilon_0}{2}\left|\frac{1}{4\pi\epsilon_0}\frac{Q_F}{R^2}\right|^2 \\ &= \frac{1}{32\pi^2\epsilon_0}\frac{Q_F^2}{R^4}. \end{aligned} \tag{7.97}$$

Equations 7.94 and 7.97 match, as we hoped.

So there we are. There are any number of structures you can contrive to validate this form of energy accounting. (A parallel plate capacitor is an especially popular choice.) And then, given that what we experience as an E versus B field depends on the frame of reference, you would expect an energy density to be associated with the B field as well. It turns out that the complete accounting for energy in the presence of EM fields is given by

$$U_{\text{density}} = \frac{1}{2}\left(\epsilon_0|E|^2 + \frac{1}{\mu_0}|B|^2\right). \tag{7.98}$$

7.5.2 Poynting Flux: When Field Energy Is On the Move

If we look back to our wave solutions in Section 7.3, we see that once we commit to Eq. 7.98 as an energy density, we also lock into the idea of EM energy traveling at speed c in the direction of $\hat{k}$.

It turns out that a *consistent* definition of energy flux in EM fields is given by the Poynting vector, named after John Poynting:

$$\vec{S} = \frac{1}{\mu_0}\vec{E} \times \vec{B}. \tag{7.99}$$

The units of the Poynting vector are watts per square meter, and is the number of joules per second of energy flowing through 1 m^2 of a surface perpendicular to $\hat{k}$. You can check the general wave solutions for EM waves, by the way, and verify that the vector cross product $\vec{E} \times \vec{B}$ quite helpfully points in the direction of $\hat{k}$.

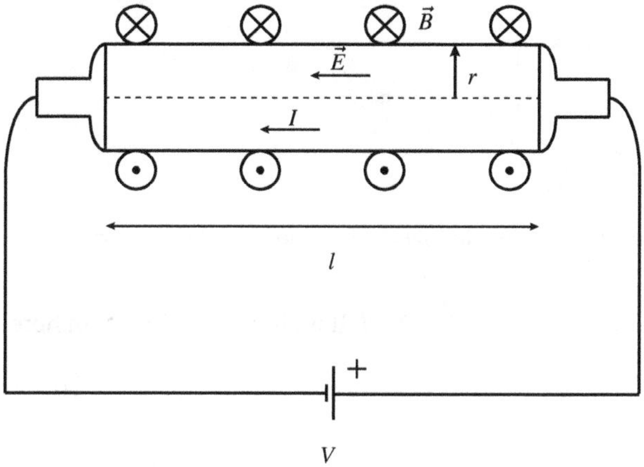

Figure 7.29 Passing a current through a resistor, and sketching the resulting electric and magnetic fields.

7.5.3 Poynting Flux and the Humble Resistor

We'll illustrate Poynting flux with a familiar example that may surprise some readers. The setup this time is a resistor, which we will posit as perfectly cylindrical in the style that was everywhere in Radio Shack stores in the 1980s and 1990s (see Figure 7.29). We can determine a couple of things right off the bat. The electric field is determined by $El = V$, so $E = \frac{V}{l}$. Ohm's law gives us the current, $I = \frac{V}{R}$.

We are interested in the B field and the E field right at the surface of the resistor. We already have the E field; for the B field, we use the result in Eq. 7.15:

$$B = \frac{\mu_0 I}{2\pi r}. \tag{7.100}$$

(Now this is not *quite* right, since when we derived the field then we had the perfect cylindrical symmetry that only an infinitely long wire can provide. We do not have that here, but we shall see that we get away with it by construction because of the way that we set up our E field and our current distribution. Such abuses are quite often very useful when applying EM theory in practical settings, and turn out to be minor.)

We look now at the Poynting vector on the surface of our resistor as shown in Figure 7.30. We can use the Poynting vector to calculate the total energy flow into the body of the resistor using

$$P = |\vec{S}| \cdot A, \tag{7.101}$$

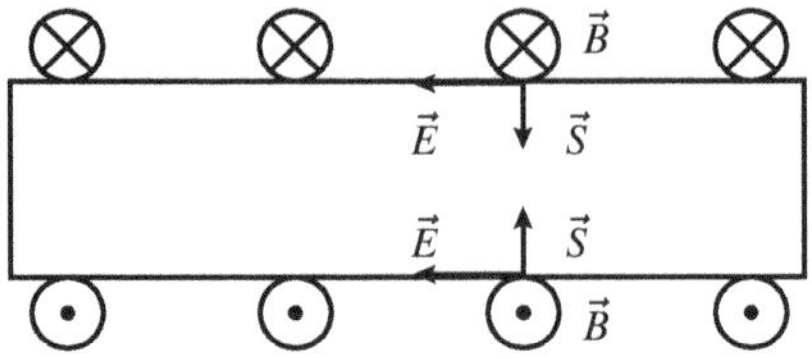

Figure 7.30 The Poynting flux vector points in toward the long axis of the resistor.

where A is the surface area of the resistor, $A = 2\pi rl$. It is simple algebra from here:

$$\begin{aligned} P &= \left|\vec{S}\right| \cdot A \\ &= \frac{1}{\mu_0} \left|\vec{E}\right| \left|\vec{B}\right| \cdot A \\ &= \frac{1}{\mu_0} \frac{V}{l} \cdot \frac{\mu_0 I}{2\pi r} \cdot 2\pi rl \\ &= VI. \end{aligned} \tag{7.102}$$

Exactly as it should be! According to our fields bookkeeping, VI joules per second flow into the body of the resistor. We know from experience that the resistor heats up. It must be true that the resistor temperature rises as high as it needs to so that the heat flow from the resistor to the environment *exactly* balances the Poynting flux flowing into the resistor. As a bonus insight, notice that this is a case of energy flowing even when the fields are *static*...

Interesting, and consistent. Always keep an eye on things like this.

7.6 Accelerating Charges: Why We Can Have Wireless Links

At this point, we have looked deep into the heart of Maxwell's equations and discovered that traveling waves are consistent with our understanding of EM fields. Very good. But how do you launch a wave?

At the end of the day, it turns out that we create EM waves by *accelerating* charges. It is not enough for them to move with a constant velocity: that velocity must vary with time. We illustrate in Figure 7.31. If the charge has instantaneous acceleration $a(t)$ along the $\hat{z}$ axis, the magnitude of the E field will be strongest for $\theta = 90°$, or $\pi/2$ radians. We say that the transverse component of the E field, E_θ, is given by

$$E_\theta(r, t) = \frac{q}{4\pi\epsilon_0 c^2} \left(\frac{1}{r}\right) \sin\theta \cdot a_z \left(t - \frac{r}{c}\right). \tag{7.103}$$

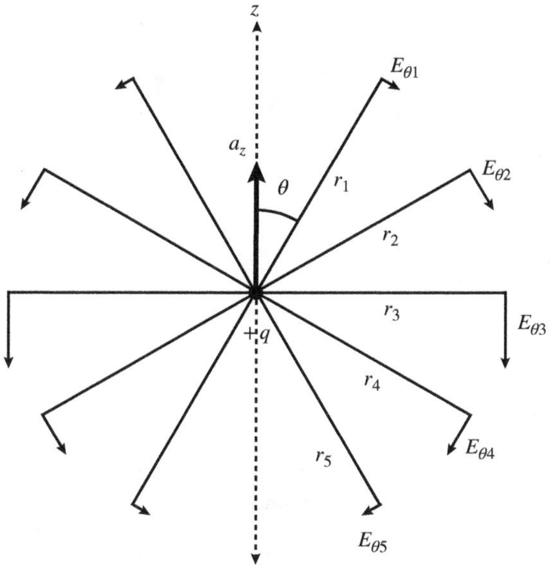

Figure 7.31 The radiating E field due to a charge $+q$ at the origin accelerating along the z-axis. We use spherical coordinates here, so θ should be understood to always be the angle between a given position vector and the positive vertical (z) axis.

The first interesting aspect of this formula is the $1/r$ dependence on field strength. This is a surprise and delight and changes everything. Based on Coulomb's law (Eq. 7.2), we might have expected something that goes as $1/r^2$.

So what happened to the $1/r^2$ field?! Nothing happened, it is still there. It's just that it falls off so much faster than the $1/r$ dependence that you don't have to get very far from the charge in order for the latter to dominate the E field.

Actually, that's only *part* of the reason we ignore the $1/r^2$ field. We'll have much more to say about this in Chapter 8, but consider briefly a wire segment that carries a sinusoidal current. The current itself is caused by moving electrons, of course (see Figure 7.32). The wire is electrically neutral, so in any given wire segment there are as many protons as there are electrons. This means that for every Coulomb term $-\frac{q}{4\pi\epsilon_0}\frac{\hat{r}}{r^2}$ due to an electron, there is also a $+\frac{q}{4\pi\epsilon_0}\frac{\hat{r}}{r^2}$ nearby. Between this near-perfect cancellation and the $1/r^2$ dependence, the $1/r$ radiating fields are all that matter.

The next thing that we see in Eq. 7.103 is the $\sin\theta$. This means that the radiating field is zero along the axis of acceleration, and maximum for $\vec{r}$'s perpendicular to the axis of acceleration. Additionally, we see that, for a given $\vec{r}$, the E field is determined not by the current acceleration of the charged particle, but by the acceleration at a time r/c in the past. This is because a transverse EM wave launched by

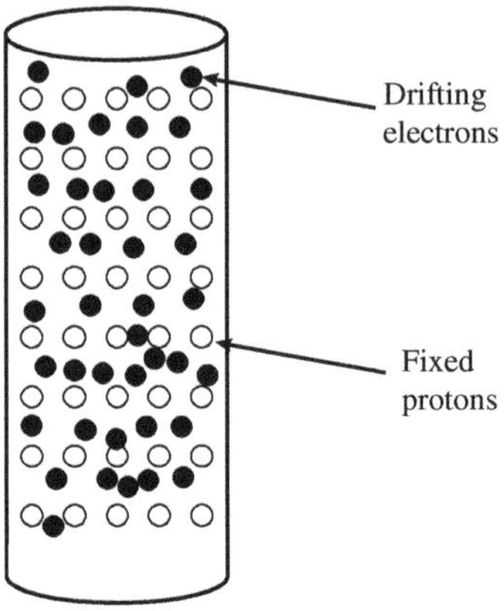

Figure 7.32 A current carrying wire segment. All charged particles are buzzing with an average kinetic energy of $\frac{3}{2}kT$, of course. But the current we measure is due to an average drift velocity of the electrons, generally on the order of centimeters per second along the wire.

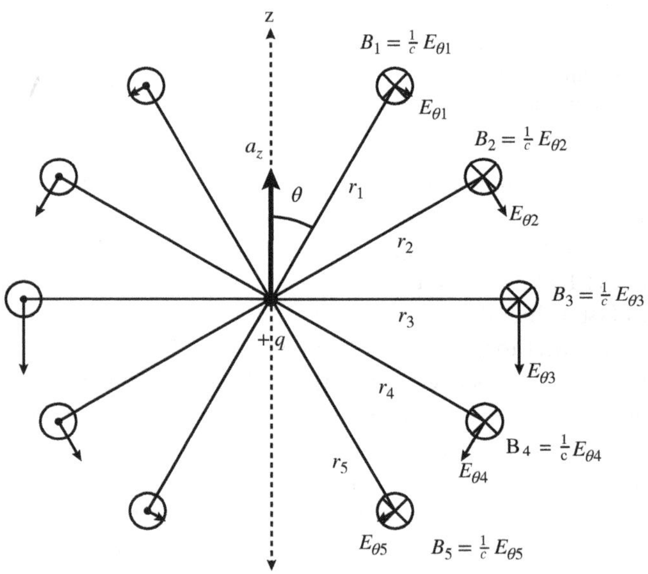

Figure 7.33 The radiating E field due to a charge $+q$ at the origin accelerating along the z-axis, this time showing the accompanying B field.

the acceleration of the particle propagates with speed c. News of its acceleration, therefore, takes r/c seconds to reach us at position $\vec{r}$.

There's more. Thanks to our painstaking work in Section 7.3, we know that if we have a traveling wave E field there must be an accompanying traveling B field.

We know its magnitude to be $|B| = \frac{1}{c}|E_\theta|$, and that its direction is perpendicular to both the E field and the direction of propagation. We can use these facts to flesh out Figure 7.31 a little more, with the result being Figure 7.33.

We close with a famous result that will be useful to us going forward. It is called the Larmor formula, and it gives us the total EM power radiated by an accelerating charge. In SI units (watts), that power is

$$P = \frac{q^2 a^2}{6\pi\epsilon_0 c^3} = \mu_0 \frac{q^2 a^2}{6\pi c}. \tag{7.104}$$

7.7 EM Theory Summary

Theory has been the focus of this chapter, but it is useful and important to come back and make the application explicit. In wireless communications, we exploit the fact that by accelerating charged particles, usually by generating sinusoidal electrical currents, we launch radiating EM waves that decay with distance astonishingly slowly ($1/r$). By modulating the sinusoidal currents, we modulate the acceleration of the charged particles and therefore the magnitudes of the radiating fields. The modulation of these fields is how we encode information. Assuming that we have ways of detecting and demodulating these waves, we have on our hands a means of communicating over vast distances.

And we do mean *vast*. At the time of this writing, the Voyager 1 and Voyager 2 satellites, launched in 1977, are at a distance of 15 billion and 13 billion miles away from Earth, respectively. We continue to receive telemetry from them, even though at those distances EM waves take 23 hours (and 19 hours), traveling at the speed of *light*, to reach us. It is astonishing what radiating fields due to accelerating charges allow us to do.

8

The Least You Need to Know About Antennas

As with pretty much all of the other chapters in this book, our ambitions here are big. We seek to equip the reader with a minimal, yet *powerful*, basis for understanding antennas.

Antennas are the physical structures that launch our modulated electromagnetic waves into space at the output of the transmitter. Other antennas then gather those modulated waves for subsequent decoding in the receiver. The stock in trade of E&M theory up to this point—the perfect sphere, the infinitely long cylinder, the point charge, and, of course, the infinite plane—these abstractions abandon us here. We are stuck instead with finite structures that lack the symmetry that so often simplifies Maxwell's equations. We must make do.

It turns out that the Hertzian dipole, combined with superposition, will get us a long way. Let's get into it.

8.1 The Friis Equation

We're going to kick off our antenna study by examining something of a "mystery" in the telecommunications field. It is not a mystery in the ordinary sense of the word: almost everyone knows the final answer, and there is no question that that final answer is correct. But figuring out the why of things proves to be an odyssey, or at least it did for the author.

Our mystery concerns the famous Friis equation,[1] which governs the received signal power in a wireless link. We show the setup in Figure 8.1. The Friis equation

1 H. T. Friis, "A Note on a Simple Transmission Formula," *Proc. of the I.R.E and Waves and Electrons*, pp. 254–256, May 1946.

Energy Fundamentals of Radio, First Edition. Joel L. Dawson.

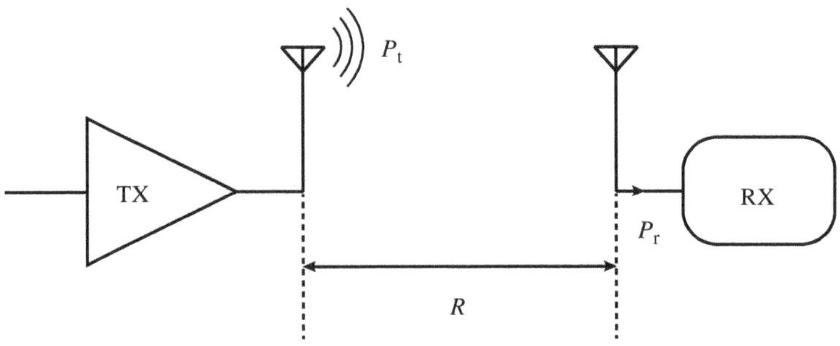

Figure 8.1 The setup for examining the Friis equation.

answers the following question. For a given transmit power P_t, what is the received power P_r? The "modern" form of the answer is given by

$$P_r = P_t \cdot \frac{G_t G_r \lambda^2}{(4\pi R)^2}. \tag{8.1}$$

So where is the mystery? Let's go over the multiplicative terms one by one, and see if we can reconcile with our conceptual understanding:

- P_t: Absolutely makes sense. The more power we transmit, all else being equal, the more power we should receive.
- G_t and G_r: The default, ideal antenna is an isotropic radiator (i.e. radiates/receives equally in all directions). While useful for analysis and building understanding, when we get to real-world calculations we must modify our model to account for our antennas transmitting/receiving preferentially in certain directions. The convention is that for a given P_t, the *additional* power that we get in the preferred direction compared to that from an isotropic radiator is captured by antenna gain G. Note that when we use the Friis equation, we assume that we aligned the receive and transmit antennas so that their preferred directions face each other. Note also that this additional power in the preferred direction must be balanced, perfectly, by reduced power in other, non-preferred directions.
- $\frac{1}{R^2}$: This also makes sense after a little bit of thought. We found in Chapter 7 that the radiating E fields fall as $1/r$; the radiating B fields fall as $1/r$: therefore, the radiating Poynting flux must go as $1/r^2$. We also note with satisfaction that the λ^2 in the numerator, having units of length squared, pairs nicely with the $1/r^2$ to make the units of Eq. 8.1 consistent.
- λ^2: Wait...what?! Aside from the note above about units, this is weird. The basic question: Does Nature preferentially propagate EM waves at longer wavelengths? There was no hint of this in Chapter 7!

For that last bullet, it is worthwhile to write the Friis equation in a slightly different form. Recalling that $\lambda = c/f$, where c is the speed of light and f is the frequency of the propagating wave, we have for Eq. 8.1 instead

$$P_r = P_t \cdot \frac{G_t G_r}{(4\pi R)^2} \cdot \frac{c^2}{f^2}. \tag{8.2}$$

The received power goes down as $1/f^2$! This is the infamous "path loss" that network planners refer to when they are making link budgets. This also explains why, in the cellular world, people talk about low bands (600–700 MHz) being "good for broad area coverage" compared to higher bands, like 6 GHz. The Friis equation makes this blindingly obvious: going from 600 MHz to 6 GHz apparently costs us a factor of 100(!) in received power, all other things being equal.

Something seems off. Just based on the ordinary meaning of the words "path loss," you might think of some sort of attenuation based on frequency-dependent absorption in the atmosphere. But if that were the case, you would expect to see some parameters in the Friis equation related to, say, the composition of gases in the path of propagation, *and* expect these parameters to vanish in the vacuum of space.

But you do not see such parameters.

What you have instead is a blanket assertion that the received power falls off according to an inverse square law in frequency. Play around with numbers to see how dramatic this is. Even going from 1 to 3 GHz earns you a path loss penalty of almost an order of magnitude. An inverse square law is dramatic and felt keenly by every network planner whose job it is to ensure coverage over large geographical areas.

Yikes. It bears repeating that nothing we saw in Sections 7.3 or 7.4 prepared us for this.

It turns out that our big mistake is taking the words "path loss" so seriously: it has us looking in the wrong place. The origin of this confounding inverse square law actually lies in Nature's fundamental mechanism for launching radiating fields, which is the acceleration of charged particles. When we remember that our waves originate from charges whose displacements oscillate sinusoidally, what we find is that higher frequencies of oscillation *will* require greater instantaneous accelerations of the charged particles. For now, the thing to keep your eye on is that higher accelerations have energy implications, as exemplified by the Larmor formula in Eq. 7.104. It appears that Nature does not treat all frequencies equally after all!

We will get into the details shortly. Meanwhile, it is interesting to note that Harald Friis himself was not confused on this point in the slightest. Right at the beginning of his paper, where he describes the power received by a small dipole as depending on the radiation resistance, the λ^2 springs to life with no ambiguity concerning its origins. It is the subsequent labeling of this dependence as path loss that causes so much confusion.

8.2 Focus on SNR Rather Than Received Power

A departure for this text from most other treatments is that we will not focus on transferring power to the receiver. We will see that both approaches get the same answer. But in the author's experience, focus on power transfer leads to subtle but impactful misunderstandings in those who design the circuits of the radio receive chain. The essence of the confusion is this: the low noise amplifier at the front end of the receiver is ultimately responsive to the voltage swing across the gate-source or base-emitter junction of the device, and is indifferent to how many joules per second it takes to maintain that swing.

Let's start with an example problem to illustrate the difference in mindset. Consider the circuit of Figure 8.2, with signal voltage source v_S, source impedance R_S, and load impedance R_L. For this circuit we derive an expression for the SNR at the output, and then calculate the SNR for the cases of $R_L = 0\ \Omega$, $R_L = R_A$, $R_L = \infty\ \Omega$. Start with an expression for the mean squared noise voltage:

$$\overline{v_n^2} = \left(\frac{R_L}{R_A + R_L}\right)^2 \cdot 4kT_A R_A B + \left(\frac{R_A}{R_A + R_L}\right)^2 \cdot 4kT_L R_L B. \tag{8.3}$$

The mean squared signal voltage is meanwhile

$$\overline{v_{sig}^2} = \left(\frac{R_L}{R_A + R_L}\right)^2 \overline{v_S^2}. \tag{8.4}$$

Doing all of the algebra carefully ultimately gives us the SNR

$$\text{SNR} = \frac{R_L \overline{v_S^2}}{4kB(R_L R_A T_A + R_A^2 T_L)}. \tag{8.5}$$

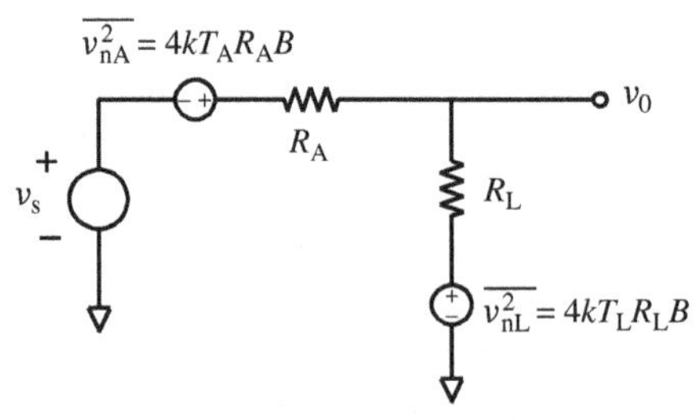

Figure 8.2 Calculating SNR with a source and two resistors. This is a model of the very front end of a radio receiver.

Now we can look at our illustrative cases. $R_L = 0$ is easy: the SNR is zero, as it should be if we short the output to ground. $R_L = \infty\ \Omega$ comes quickly as well. We set $T_A = T_L = T$ for thermal equilibrium and write

$$\mathrm{SNR} = \frac{\overline{v_s^2}}{4kTBR_A}. \tag{8.6}$$

And finally, we look at the case for $R_L = R_A$, matching the load to the source and therefore, maximizing power transfer as we discussed in Chapter 3. If you are expecting to maximize SNR in this righteous sounding case, hold on to your hat:

$$\mathrm{SNR} = \frac{1}{2} \cdot \frac{\overline{v_s^2}}{4kTBR_A}. \tag{8.7}$$

Surprise! The SNR in the matched load case is half that of the case where we do as big a mismatch as possible.

If you are surprised, it is likely because you have some experience with real-world RF circuits and know how religious we are about matching source and load impedances. And you are *correct*: we normally do match impedances for many practical reasons, including making sure that the SNR is robustly high under a wide variety of conditions. Let's explore a little bit.

Most of the time, the system that we are modeling in Figure 8.2 is actually a receive antenna driving the input impedance of a low noise amplifier. See Figure 8.3. The resistor R_A is associated with the antenna, and is what we call a "radiation resistance." It actually has an f^2 dependence, which we will get to in

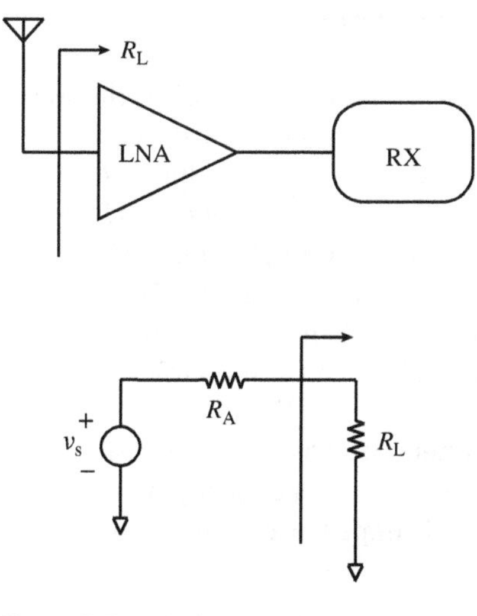

Figure 8.3 A little more detail on the model of a receiver front end.

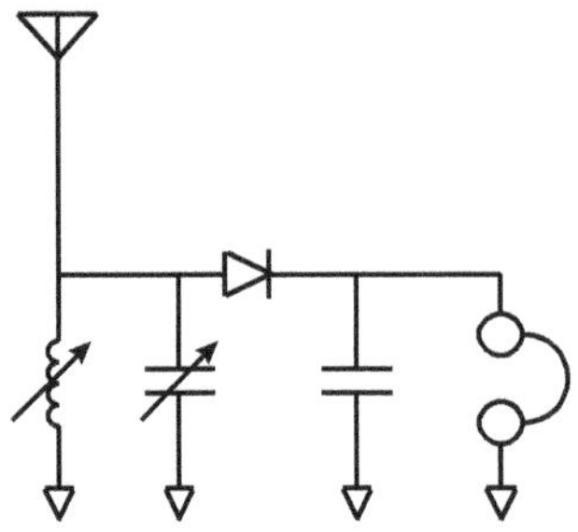

Figure 8.4 The classic crystal radio.

great detail shortly. But the question remains, given the importance of SNR in communications links: why don't we make R_L as close to infinity as we can?

Part of the answer is that sometimes, we *do* do this. Many of us electronics hobbyists who came of age in the 1970s and 1980s got our introduction to wireless communications through the venerable crystal radio. A diagram is shown in Figure 8.4. At its heart is a parallel *LC* circuit, which of course is an open circuit at resonance (i.e. the station we are tuning for). No conjugate match here!

The crystal radio is ideal for the AM broadcast band in the US, which runs from 540 to 1600 kHz. These carrier frequencies correspond to wavelengths of 556 and 188 m, respectively; roughly six American football fields on the low end, and just shy of two football fields on the high end. We would say that in this frequency range we are comfortably in the "lumped element" region of how electronic components behave. That is to say, the resistor in the schematic is a resistor in real life, a capacitor is a capacitor, and, most critically, *a wire is a wire with the voltage uniform all along its surface.*

But in common cellular bands the situation is different. At 1 GHz, the wavelength is 30 cm in free space and 15 cm in a common dielectric called FR4. At these frequencies, we are down in the region where "transmission line effects" are quite strong. It turns out that our receiver front end is much better modeled by Figure 8.5. The component between the antenna and the amplifier input is known as a transmission line. The most important thing about this transmission line is that because the wavelength of our signal is shorter than, or on the order of, the length of the transmission line, the voltage along the conducting surfaces actually varies along the length of the line.

There is no need to get into a detailed treatment of transmission line structures here. What is important is that if the radiation resistance and the input impedance of the LNA are not matched to the "characteristic impedance" of the transmission

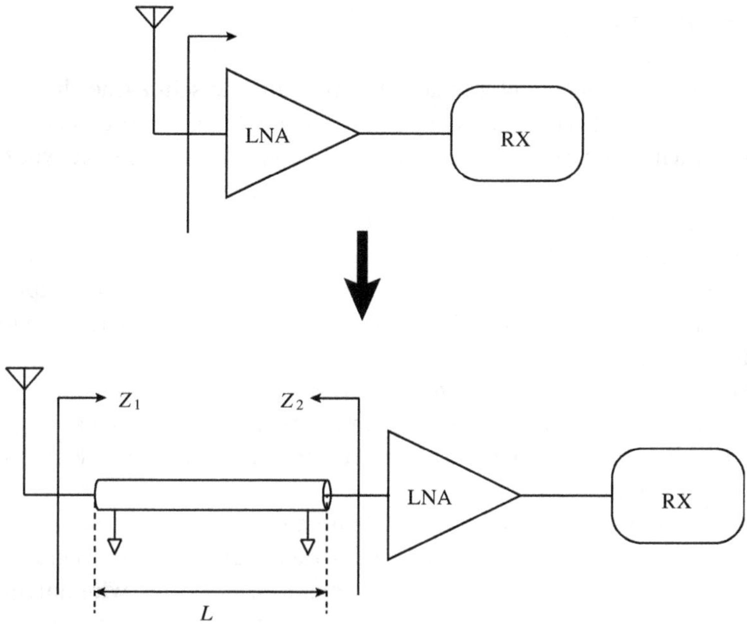

Figure 8.5 The receiver front end, with a transmission line of length L between the antenna and the input to the LNA.

line, then Z_1 and Z_2, the impedances seen looking into either end of the line will vary quite strongly as a function of L and frequency f. This is a huge inconvenience in the real world, perhaps more than it sounds at first.[2]

We avoid all this by imposing a strict design discipline. We choose a characteristic impedance for our transmission lines (50 ohms is nearly universal, but sometimes people use 75 ohms), and require that *all* our circuit blocks present input/output impedances that match this choice. If we are strict, we find that we can connect our blocks together using transmission lines of any convenient length and process signals with widely varying spectral content without difficulty. It is for this practical reason, rather than maximum power transfer, that we are fanatical about matching impedances in RF systems.

2 Imagine, instead of buying coaxial cables of standard lengths, you had to always trim your cables to *exact* lengths. And then having to do it all over again as soon as you change to a different radio station!

8.3 The Hertzian Dipole

Antenna theory is a big topic, and a major challenge for the self-learner lies in determining where to start. There are many truly excellent books on antenna theory. They are all, without exception, a heavy lift when you are trying to get your bearings.

The thing to keep sight of is the fundamental mechanism for launching EM waves, which is accelerating charged particles. A physicist might start with a cartoon like Figure 8.6 For the electrical engineer, of course, this is unsatisfactory because we do not routinely accelerate an isolated charge through space. What we do instead is move current through wires, as shown in Figure 8.7. Now, if $I(t)$ is a constant current, we have electrons moving with a constant drift velocity and therefore no radiation. If, however, we have a time varying $I(t)$, then that average drift velocity is changing. And changing drift velocity is exactly the acceleration we need to launch radiating waves.

A good starting point for analysis is to decide on a simple primitive structure with which, by scaling and replicating, we can construct larger antenna structures through the magic of superposition. For many antennas, especially wire antennas, a *fantastic* primitive is what is known as the Hertzian dipole, shown in Figure 8.8. We need not worry, for the moment, about where the current comes from and where it goes to. The answer for now is "it goes to the adjacent Hertzian dipole." And if the adjacent dipole does not have the exact same current, charge accumulates. That is allowed!

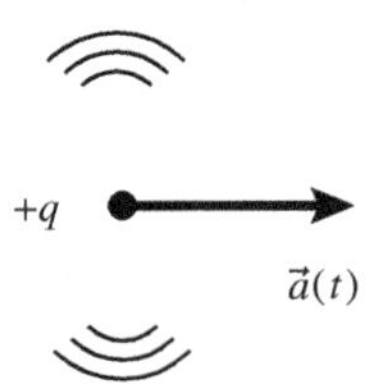

Figure 8.6 Isolated charge, accelerating in space.

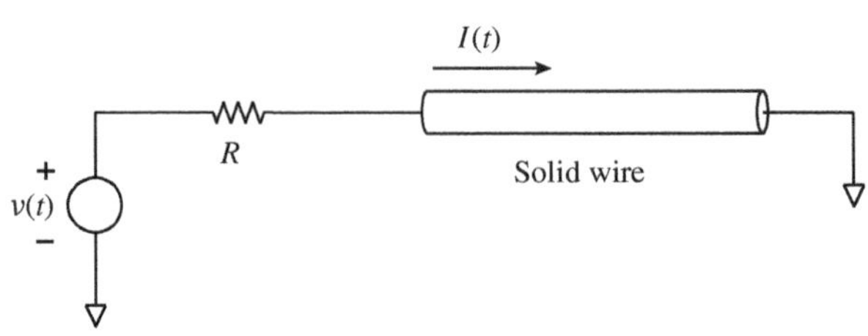

Figure 8.7 A simple radiator using a varying current $I(t)$.

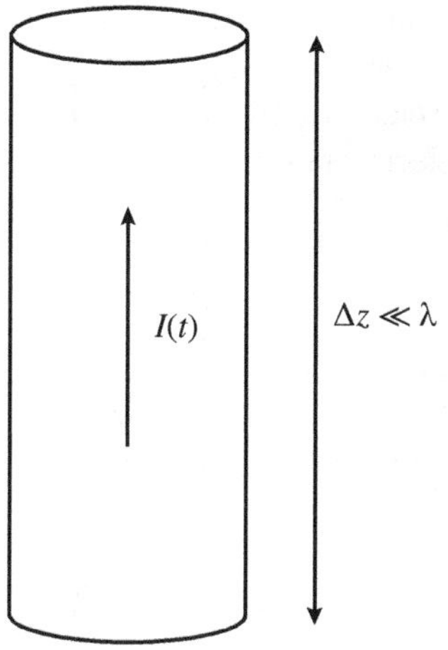

Figure 8.8 A Hertzian dipole.

The overall game for many types of antennas is that we know, even if only by empirical means, the current distribution on the structure. When we have that, we break the structure into Δz size chunks; compute the radiation pattern due to each chunk; and then superpose our way to the overall radiation pattern. Let's take a deep look, then, at the Hertzian dipole.

8.3.1 The Hertzian Dipole as It Is Usually Analyzed

We will not spend a lot of time on this approach simply because, for the author, it is a less intuitive way to obtain the detailed E and B fields. Nevertheless, the start of the derivation is an opportunity to introduce the magnetic vector potential $\vec{A}$, which we never got to in Chapter 7, but which you will see a fair amount of if your curiosity leads you to a deeper study of antennas.

The traditional approach starts with something called the magnetic vector potential, notated $\vec{A}$, whose relation to the B field is given by

$$\vec{B} = \nabla \times \vec{A}. \tag{8.8}$$

What we love about the vector potential is its relatively straightforward relation to a current element:

$$d\vec{A} = \frac{\mu_0 I}{4\pi r} d\vec{l}, \tag{8.9}$$

which for a current element situated on the z axis as we posited in Figure 8.8 becomes

$$d\vec{A} = \frac{\mu_0 I}{4\pi r} d\vec{z}. \tag{8.10}$$

Contrast this starting point with that of the Biot–Savart law of Eq. 7.4. It's just cleaner and easier.

It also helps to have on hand another tool called the electric potential, which we touched on but did not dwell on in Chapter 7. The electric potential is related to charge distribution $\rho(x, y, z)$ according to

$$\phi = \frac{1}{4\pi\epsilon_0} \iiint_V \frac{\rho(x, y, z)}{r} dV. \tag{8.11}$$

In the case of our wire segment, we have this on hand because charge will be accumulating at the ends of our wire segment. The grand conclusion of this treatment is to derive the E and B fields through relations

$$\vec{E} = -\nabla\phi - \frac{\partial\vec{A}}{\partial t} \tag{8.12}$$

and Eq. 8.8.

There are important nuances, such as being careful to use "retarded" potentials to account for the speed of electromagnetic wave propagation when you get far from the current element. For clarity of approach, it is hard to do better than John Kraus's treatment in his antennas text.[3]

8.3.2 The Hertzian Dipole as Accelerating Charges

Very good. Now it turns out that another way to analyze this problem is to focus on the fact that a time-varying current is in fact composed of a bunch of accelerating charged particles. Recall from Chapter 7 that the E field (far field) of a single accelerating charge along the z axis is given by Eq. 7.103, repeated here:

$$E_\theta(r, t) = \frac{q}{4\pi\epsilon_0 c^2}\left(\frac{1}{r}\right)\sin\theta \cdot a_z\left(t - \frac{r}{c}\right). \tag{8.13}$$

(We did not dive into details at the time, but this turns out to be a simplification of a super-sophisticated expression, quoted here from the Feynman lectures:[4]

$$\vec{E}(\vec{r}\,) = -\frac{q}{4\pi\epsilon_0}\left[\frac{\hat{r}'}{r'^2} + \frac{r'}{c}\frac{d}{dt}\left(\frac{\hat{r}'}{r^2}\right) + \frac{1}{c^2}\frac{d^2}{dt^2}\hat{r}'\right]. \tag{8.14}$$

In this expression, r' is the distance from the charge to the point P (whose position vector is $\vec{r}$) at the "retarded time," which is to say, at the time when the field that we feel at point P "left" the charge. We adjust, in other words, for the fact that changes in the E field due to the motion of our point charge can travel no faster than the speed of light. The vector $\hat{r}$ is just the unit vector of the same. We include

3 J. D. Kraus, *Antennas* USA: McGraw-Hill, Inc., 1950. The derivation of the fields for a Hertzian dipole begins on p. 127.

4 Ibid. This expression comes from Volume 1, p. 28–32.

this here only by way of emphasizing that Eq. 8.13 did not arise in some stealthy way from the magnetic vector potential development of Section 8.3.1, yet is entirely consistent. The author finds this altogether incredible.)

We also wrote down the Larmor formula, which gives us the power radiated by a single accelerating particle, repeated here from Eq. 7.104:

$$P = \frac{q^2 a^2}{6\pi\epsilon_0 c^3} = \mu_0 \frac{q^2 a^2}{6\pi c}. \tag{8.15}$$

Taken together, Eqs. 8.13 and 8.15 give us quite a lot to work with when it comes to analyzing the Hertzian dipole. We lack only an organized way to relate current flow to the movement of a bunch of individual charges.

That wait ends here. Figure 8.9 is a redrawing of Figure 7.32, but relabeled with dimensions that we need for analysis. We know from Chapter 4 to expect that all the individual charged particles are shaking and buzzing with average kinetic energy $\frac{3}{2}kT$. For our present purposes, though, we are completely uninterested in this shaking. The reason is that the contribution to the radiating fields due to any single, say, electron, will tend to be canceled out, on average, by another nearby electron accelerating in the opposite direction. When it comes to radiating fields, we don't think about thermal agitation much.

What we are *very* interested in is the macroscopic currents that we deliberately induce in the wire. For a conductor with n free electrons per unit volume (cubic meters, if the current is to be in Amperes), the average macroscopic current can be expressed in terms of the average drift velocity of the electrons according to

$$I = neAv_{\text{drift}}, \tag{8.16}$$

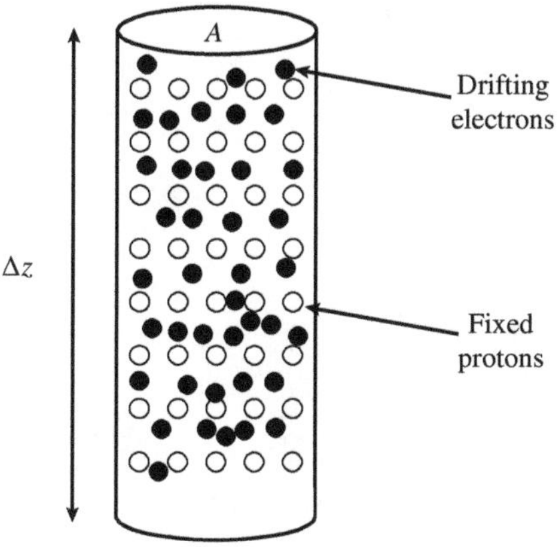

Figure 8.9 A current carrying wire segment, which we analyze as a collection of accelerating charges.

where e is the charge in coulombs per electron (1.602×10^{-19} C), and A is the cross-sectional area in square meters.

To keep our bearings, it is helpful to run with some typical numbers. Copper at room temperature has a free electron density of 8.5×10^{28} electrons per cubic meter. For a wire of cross-sectional area equal to one square millimeter, and a current of 1 A, we calculate an average drift velocity of

$$v_{\text{drift}} = \frac{I}{neA} = \frac{1\text{ A}}{(8.5 \times 10^{28}\text{ m}^{-3})(1.602 \times 10^{-19}\text{ C})(10^{-6}\text{ m}^2)} = 0.0735\text{ mm/ sec}. \tag{8.17}$$

A snail's pace![5]

From here we may write the average drift acceleration a_{drift} in terms of the element current as follows:

$$a_{\text{drift}} = \frac{d}{dt} v_{\text{drift}} = \frac{1}{neA} \cdot \frac{dI}{dt}. \tag{8.18}$$

The final piece to know is that we choose Δz to be both small compared to the wavelength of the EM wave, and small compared to $|\vec{r}|$, the distance from the current element at which we calculate the fields. See Figure 8.10. An important part of our approximation, and a big reason that we insist on making Δz small, is that we'll now lump all the accelerating charge into a single point charge of quantity

$$Q = neA\Delta z \tag{8.19}$$

which accelerates according to Eq. 8.18.

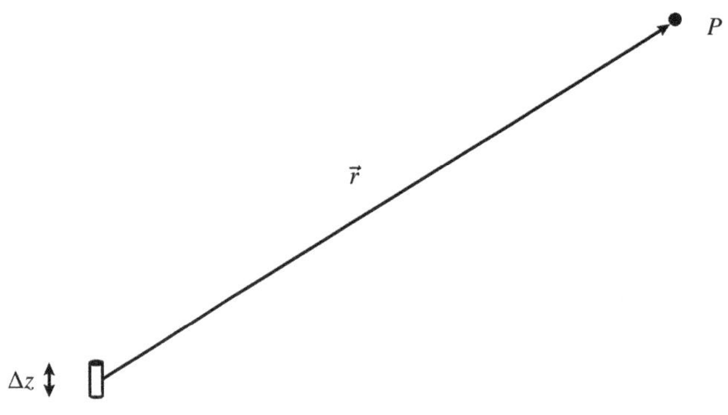

Figure 8.10 We are interested in the fields far from our current segment.

5 For reference, the pace of an actual snail is on the order of one full millimeter per second.

Now things get easy and fun. We first write out the fields and deal also with the total power radiated from this current element. From here, generalizing to bigger wire structures, such as a half-wave dipole or quarter-wave monopole, becomes a three-step process: (1) determine, by some means (often empirically), the current distribution along the structure; (2) discretize the current distribution into a bunch of suitably small Hertzian dipoles such as those we analyze here; and (3) use superposition to combine the effects of these dipoles for any point in space at which you want to know the fields.

Let us suppose that the current drive is sinusoidal, according to

$$I = I_0 \sin \omega t. \tag{8.20}$$

Again, from Eq. 8.18, we have that the acceleration of our charged particle Q is

$$a_{\text{drift}} = \frac{I_0}{neA} \cdot \omega \cos \omega t. \tag{8.21}$$

(Note that it is easy to slip, by deeply ingrained habit, into the convention of defining the current as though the charge carriers were positively charged instead of being electrons. This is all okay and harmless for the most part. Just keep in your mind, when and if it matters, that the actual electrons will be flowing in the opposite direction of the stated current.)

Equation 8.13 gives us the electric field due to our accelerating charge. We just have to plug in for q and the drift acceleration:

$$\begin{aligned} E_\theta(r,t) &= \frac{neA\Delta z}{4\pi\epsilon_0 c^2}\left(\frac{1}{r}\right)\sin\theta \cdot \frac{I_0}{neA} \cdot \omega \cos \omega\left(t - \frac{r}{c}\right) \\ &= \frac{I_0 \Delta z}{4\pi\epsilon_0 c^2}\left(\frac{1}{r}\right)\sin\theta \cdot \omega \cos\omega\left(t - \frac{r}{c}\right). \end{aligned} \tag{8.22}$$

Referring to Figure 7.31, the corresponding B field at any given θ is

$$\begin{aligned} |B| &= \frac{1}{c} \cdot \frac{I_0 \Delta z}{4\pi\epsilon_0 c^2}\left(\frac{1}{r}\right)\sin\theta \cdot \omega\cos\omega\left(t - \frac{r}{c}\right) \\ &= \frac{I_0 \Delta z \mu_0}{4\pi c}\left(\frac{1}{r}\right)\sin\theta \cdot \omega \cos\omega\left(t - \frac{r}{c}\right). \end{aligned} \tag{8.23}$$

And here we notice, for the first time, something to keep your eye on. For a given current amplitude I_0, the strength of the electric and magnetic fields scales *up* with frequency, and therefore so does the Poynting flux. This will be important when we come back to treating radiated power for a Hertzian dipole. Meanwhile, reflect on the fact that the average heat dissipated in a resistor carrying current $I_0 \sin \omega t$ depends *not at all* on frequency.

Now we could integrate the Poynting flux over a surface surrounding the dipole to figure out the radiated power, but fortunately the Larmor formula gives us exactly that in a neatly packaged expression. We have for the instantaneous radiated power

$$P(t) = \frac{Q^2 a^2}{6\pi\epsilon_0 c^3} \tag{8.24}$$
$$= \frac{(neA\Delta z)^2}{6\pi\epsilon_0 c^3} \cdot \left(\frac{I_0}{neA} \cdot \omega \cos \omega t\right)^2$$
$$= \frac{(I_0 \Delta z)^2}{6\pi\epsilon_0 c^3} \cdot \omega^2 \cos^2 \omega t.$$

For the average radiated power, we make use of the math fact that the time average value of $\cos^2 \omega t$ is $1/2$:

$$\langle P(t)\rangle = \frac{1}{2}\frac{(I_0 \Delta z)^2}{6\pi\epsilon_0 c^3} \cdot \omega^2 \tag{8.25}$$
$$= \frac{1}{2} I_0^2 \frac{(\Delta z)^2}{6\pi\epsilon_0 c^3} \cdot \omega^2.$$

We are on the doorstep of a major result for Hertzian dipoles. From the standpoint of the terminals of the dipole, we inject a current and have energy dissipate, never to be recovered. If all we know is terminal currents and voltages, we are unable to distinguish between an antenna and a resistor between these terminals. What resistance would dissipate the equivalent amount of energy in heat that we dissipate here in integrated Poynting flux?

First, recall that the power dissipated in a sinusoidally driven resistor is

$$\langle P_{\text{diss}}\rangle = \frac{1}{2} I_0^2 R. \tag{8.26}$$

Looking back at Eq. 8.25, we identify a "radiation resistance" as

$$R_{\text{rad}} = \frac{(\Delta z)^2}{6\pi\epsilon_0 c^3} \cdot \omega^2 \tag{8.27}$$
$$= \frac{2\pi(\Delta z)^2}{3\epsilon_0 c^3} \cdot f^2.$$

And here, with some explanation still needed, is what we were really looking for at the end of Section 8.1, which is a major characteristic of the antenna that scales as f^2. We'll do a few more manipulations to massage this expression into the form that is most common in the literature. Start with $f = \frac{c}{\lambda}$:

$$R_{\text{rad}} = \frac{2\pi(\Delta z)^2}{3\epsilon_0 c^3}\frac{c^2}{\lambda^2} \tag{8.28}$$
$$= \left(\frac{\Delta z}{\lambda}\right)^2 \frac{2\pi}{3\epsilon_0 c}.$$

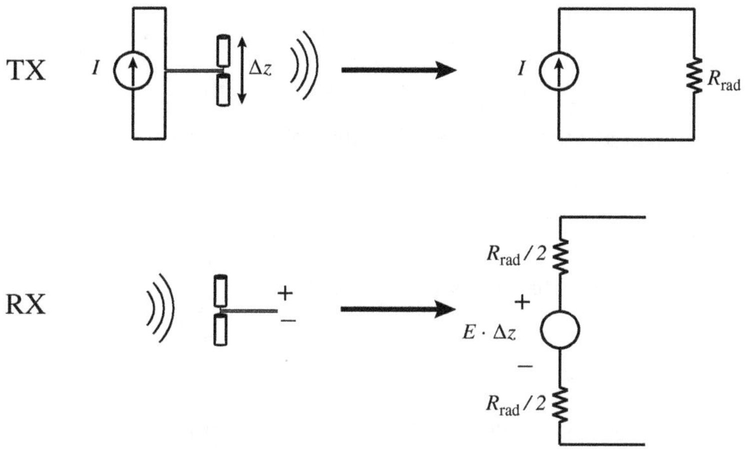

Figure 8.11 The Hertzian dipole as a transmitter and a receiver.

We have that $\epsilon_0 = \frac{1}{\mu_0 c^2}$, so

$$R_{rad} = \left(\frac{\Delta z}{\lambda}\right)^2 \frac{2\pi\mu_0 c}{3}. \tag{8.29}$$

As a last step, we plug in for constants $\mu_0 = 4\pi \times 10^{-7}$ H/m and $c = 3.0 \times 10^8$ m/s. This gives us at last the standard result for the radiation resistance of a Hertzian dipole:

$$R_{rad} = 80\pi^2\left(\frac{\Delta z}{\lambda}\right)^2. \tag{8.30}$$

Now we work so hard to come up with a radiation resistance because a circuit model of the antenna winds up yielding a great deal of understanding. Figure 8.11 shows how our Hertzian dipole looks as both a transmitter and a receiver. The picture on the transmit side will not surprise you, but you may have to think a bit to be comfortable with the picture of the receive antenna, a voltage source with a source resistance R_{rad}. Do take the time to get comfortable, because it turns out that this radiation resistance in the receive antenna is the *whole game* when it comes to understanding path loss. We'll tie it all up neatly after a brief detour to show that the radiated field strengths are solely a function of drive power, and independent of frequency.

8.3.3 Field Strengths Are Solely a Function of Transmit Power

To show this, we start with the radiated electric field from Eq. 8.22

$$E_\theta(r,t) = \frac{I_0 \Delta z}{4\pi\epsilon_0 c^2}\left(\frac{1}{r}\right) \sin\theta \cdot \omega \cos\omega\left(t - \frac{r}{c}\right), \tag{8.31}$$

but we must tie I_0 to radiated power via

$$\frac{1}{2}|I_0|^2 R_{\text{rad}} = P_{\text{t}}. \tag{8.32}$$

This expression allows us to write I_0 in terms of the transmit power P_{t}

$$|I_0| = \sqrt{\frac{2P_{\text{t}}}{R_{\text{rad}}}}, \tag{8.33}$$

which we will then modify using Eq. 8.30

$$|I_0| = \frac{\lambda}{\pi \Delta z} \cdot \sqrt{\frac{P_{\text{t}}}{40}}. \tag{8.34}$$

Recalling that λ and ω are related through $\omega = 2\pi f = \frac{2\pi c}{\lambda}$, careful algebra gives us

$$\begin{aligned} |I_0| &= \frac{1}{\pi \Delta z} \cdot \frac{2\pi c}{\omega} \cdot \sqrt{\frac{P_{\text{t}}}{40}} \\ &= \frac{c}{\omega \Delta z} \cdot \sqrt{\frac{P_{\text{t}}}{10}}. \end{aligned} \tag{8.35}$$

Our study from Chapter 7 leads us to believe that the radiated field strength should be a function of transmitted power, regardless of frequency. We get the result we are looking for when we substitute Eq. 8.35 into Eq. 8.22:

$$\begin{aligned} E_\theta(r,t) &= \frac{I_0 \Delta z}{4\pi \epsilon_0 c^2}\left(\frac{1}{r}\right)\sin\theta \cdot \omega \cos\omega\left(t - \frac{r}{c}\right) \\ &= \frac{c}{\omega \Delta z} \cdot \sqrt{\frac{P_{\text{t}}}{10}} \cdot \frac{\Delta z}{4\pi \epsilon_0 c^2}\left(\frac{1}{r}\right)\sin\theta \cdot \omega \cos\omega\left(t - \frac{r}{c}\right) \\ &= \sqrt{\frac{P_{\text{t}}}{10}} \cdot \left[\frac{\Delta z}{4\pi \epsilon_0 c}\left(\frac{1}{r}\right)\sin\theta \cdot \cos\omega\left(t - \frac{r}{c}\right)\right]. \end{aligned} \tag{8.36}$$

No path loss here.

8.4 A Friis-like Equation for the SNR in a Wireless Link

Now is a good time to remind ourselves of our objective with this chapter. Certainly, we do want to know a little bit about antennas, and so it makes sense to look at the Hertzian dipole as a useful primitive and building block.

But in this book, we mostly care about antenna theory as a means to an end. And that end is, we want to know how many bits per second of information we can get through our wireless link, and we want to do so with a great awareness of the energy cost per bit. We have ample reason now to believe that our choice of carrier frequency has a great effect on our end goal. Let's see how this plays out.

Our receive system, for purposes of analysis, looks like Figure 8.12. In the figure, R_L is the impedance seen looking into the input of the LNA. If you are quick with your circuit analysis, you can write the mean-squared noise voltage at the input of the LNA as

$$\overline{v_n^2} = \left(\frac{R_{\text{rad}}}{R_{\text{rad}} + R_L}\right)^2 \cdot 4kTR_L B + \left(\frac{R_L}{R_{\text{rad}} + R_L}\right)^2 \cdot 4kT_N R_{\text{rad}} B. \tag{8.37}$$

Meanwhile, the mean squared signal voltage at the input of the LNA is

$$\overline{v_{\text{sig}}^2} = \left(\frac{R_L}{R_{\text{rad}} + R_L}\right)^2 \overline{v_A^2}. \tag{8.38}$$

And within a few algebraic steps, we arrive at the SNR

$$\text{SNR} = \frac{R_L^2 \overline{v_A^2}}{4kTR_L R_{\text{rad}}^2 B + 4kT_N R_{\text{rad}} R_L^2 B}. \tag{8.39}$$

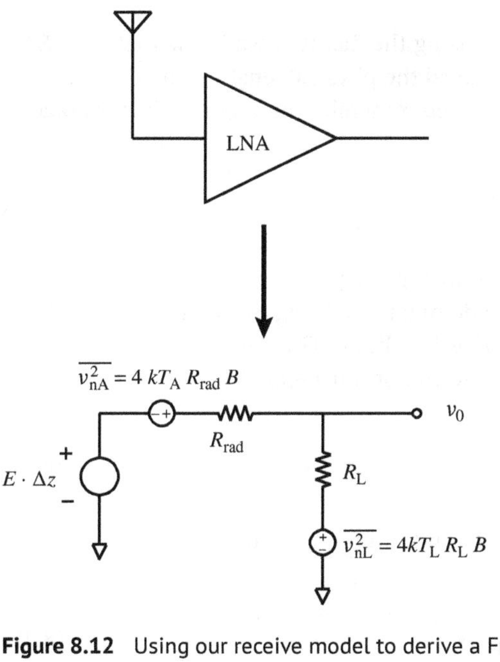

Figure 8.12 Using our receive model to derive a Friis-like equation.

It brings some clarity to rearrange terms slightly:

$$\mathrm{SNR} = \frac{1}{4kTB} \cdot \frac{1}{R_{\mathrm{rad}}} \cdot \frac{R_{\mathrm{L}}/R_{\mathrm{rad}}}{1 + \frac{T_{\mathrm{N}}}{T}\frac{R_{\mathrm{L}}}{R_{\mathrm{rad}}}} \cdot \overline{v_{\mathrm{A}}^2}. \tag{8.40}$$

Let's follow convention, now, and set $T = T_{\mathrm{N}}$ and match impedances so that $R_{\mathrm{L}} = R_{\mathrm{rad}}$. This gives us the SNR

$$\mathrm{SNR} = \frac{1}{8kTB} \cdot \frac{1}{R_{\mathrm{rad}}} \cdot \overline{v_{\mathrm{A}}^2}. \tag{8.41}$$

We can use our expression here for the radiation resistance, $R_{\mathrm{rad}} = 80\pi^2\left(\frac{\Delta z}{\lambda}\right)^2$. Substituting in, we have

$$\mathrm{SNR} = \frac{1}{8kTB}\frac{1}{80\pi^2}\frac{\lambda^2}{(\Delta z)^2}\overline{v_{\mathrm{A}}^2}. \tag{8.42}$$

And we arrive, albeit not in final form, at a result that we have been expecting all along: the SNR in our link goes as $\lambda^2 = c^2/f^2$, as the Friis equation says it must. But what you should take away from this development is where the λ^2 dependence comes from, which is the f^2 dependence of the source impedance of the receive antenna.

But let's go all the way to the end, using the fact that we know that $v_{\mathrm{A}} = E\Delta z$. Using Eq. 8.36 for E, being careful to label the physical length element associated with the transmitting antenna as Δz_{t}, *and* remembering that the time average of $\cos^2\omega t$ is 1/2, we get

$$\overline{v_{\mathrm{A}}^2} = \frac{1}{2}\frac{P_{\mathrm{t}}}{10} \cdot \left[\frac{\Delta z_{\mathrm{t}}}{4\pi\epsilon_0 c}\left(\frac{1}{r}\right)\sin\theta\right]^2 (\Delta z)^2. \tag{8.43}$$

Let us suppose that we give ourselves the highest possible SNR by setting $\theta = 90°$, so $\sin\theta = 1$. And then from here we do one more thing, which is step back from equalities and speak instead of proportionalities. This lets us drop a bunch of numerical constants and focus on the important dependencies. Our next step becomes

$$\overline{v_{\mathrm{A}}^2} \propto \frac{P_{\mathrm{t}}}{r^2} \cdot (\Delta z)^2. \tag{8.44}$$

Substituting in for $\overline{v_{\mathrm{A}}^2}$ in Eq. 8.42, and dropping constants with abandon, gives us

$$\mathrm{SNR} \propto \frac{1}{kTB} \cdot \frac{P_{\mathrm{t}}}{r^2}\lambda^2. \tag{8.45}$$

And at the end, we will make one more important modification. Obviously, our effective P_{t} is enhanced if we were to magically add antenna gains somehow in our system, and make them more directive. Assuming that we have aligned the directionality of the transmitter and receiver, it adds insight to our derived proportionality if we modify according to

$$\mathrm{SNR} \propto \frac{1}{kTB} \cdot \frac{P_t G_t G_r}{r^2} \lambda^2. \tag{8.46}$$

We arrive at the finish line at last. If we compare Eq. 8.46 to the original Friis equation,

$$P_r = P_t \cdot \frac{G_t G_r \lambda^2}{(4\pi r)^2}, \tag{8.47}$$

we see that they agree! They are not identical, of course, because SNR and received power are not the same things, and moreover, proportionalities are not the same as equalities. But from the standpoint of calculating channel capacity, you see that we capture the essential dependencies either way.

8.5 A First Look at Friis and the Shannon Limit

We're now getting to a major conceptual milestone in this book. Not for the last time, we will examine the connection between the energy considerations elucidated by the Friis equation and the information theoretic considerations elucidated by the Shannon limit. Recall that for data transmitted over a thermal noise floor, we have, according to Shannon

$$C = B \cdot \log_2 \left(\frac{P_{sig}}{kTB} + 1 \right). \tag{8.48}$$

To the extent that the *power consumption* of our system is proportional to the transmit power,[6] and assuming an SNR that is large compared to 1, we can say that

$$P_{sys} \propto kTB \cdot 2^{C/B}. \tag{8.49}$$

Based on the Section 8.4, now, we can replace $\frac{P_{sig}}{kTB}$ in Equation 8.48 with the proportionality

$$\begin{aligned} \mathrm{SNR} &\propto \frac{1}{kTB} \cdot \frac{P_t G_t G_r}{r^2} \lambda^2 \\ &\propto \frac{1}{kTB} \cdot \frac{P_t G_t G_r}{r^2} \frac{1}{f^2}, \end{aligned} \tag{8.50}$$

6 This is a more complicated topic than it might seem like it should be at first. We will discuss extensively in Chapter 11.

and write an *extremely* useful scaling rule for scaling power consumption in our wireless links:

$$P_{\text{sys}} \propto kTB \cdot f^{2} \cdot 2^{C/B} \cdot \frac{1}{G_{\text{t}}G_{\text{r}}} \cdot r^{2}. \tag{8.51}$$

It is always helpful in understanding expressions like this to practice making declarative statements that adhere to the rule under examination. To wit:

- All else being equal, if I double the distance between my transmitter and receiver I quadruple the power I need to keep the same bit rate.
- All else being equal, I can drop my system power in proportion to how much I raise my antenna gain.

And a big one:

- All else being equal, if I double my carrier frequency f, I quadruple my system power. Just in case I missed that...if I triple my carrier frequency, my system power goes up by nearly an order of magnitude (9×).

You will hear that last one spoken often by engineers who plan cellular network deployments. They will talk of the "900 MHz band being good for broad geographical coverage," whereas bands located at 3.5 or 5 GHz are "better for dense urban settings." This f^2 dependence on the link SNR budget is far from an academic concern. It is felt very strongly in the real world.

Coming to terms with Eq. 8.51 is time extremely well spent. When you have absorbed it, you have taken a major step forward in your understanding of radio links.

8.6 A Few Loose Ends

Even by the standards of this text, we are leaving out a *lot* about antennas. What hurts more with antennas versus, say, linear system theory, is that it is not so straightforward to extend one's knowledge armed only with skill in algebra, calculus, and differential equations. You're not helpless, exactly, but more so than for many of the topics in this book, the study of antennas benefits from experience, pragmatism, and, above all, knowing the answer ahead of time!

To help keep your bearings with exotic structures, cling to the fact that accelerating charges are what launch radiating fields, and also that superposition holds. No matter what antenna structure presents itself, be it a horn antenna, a parabolic dish, a loop, or a patch, you could do worse for a start than determining (or asking) what the current distribution is on the structure. Then in your mind's eye, break that continuous current distribution up into a bunch of Hertzian dipoles...and see what you can visualize. If you can identify, even roughly, what directions ought to be strong or weak as a radiator, you have cause to pat yourself on the back and declare a small victory. From there, learn from experience and/or a good electromagnetic simulator, paying careful attention to the aspects of the radiation pattern that surprise you. Do your best to reconcile the real radiation pattern with what you imagined in your head. Over time, you will deepen your understanding (and you may just catch errors in your simulations along the way).

Another big hole in our treatment here relative to the literature is that we have said nothing about the "effective area" of the antenna. True antenna people love effective area, but for the author it caused untold confusion at first. The reason can be found immediately in Friis' original formulation of his famous equation, where he writes, accurately, that

$$P_{\mathrm{r}} = \frac{A_{\mathrm{r}} A_{\mathrm{t}}}{r^2 \lambda^2} P_{\mathrm{t}}. \tag{8.52}$$

In this expression, P_{r} and P_{t} are the received and transmitted power, and A_{r} and A_{t} are the effective areas of the antennas.

At first glance, this expression seems to indicate that the received power goes as $1/\lambda^2$, not as λ^2 as we know it does. It is only after delving further that one finds that the "effective" area of our antennas each goes as λ^2, so the numerator in our equation actually goes as λ^4. Together with our λ^2 denominator, things do in fact work out. But for the author, the wavelength dependence that *looks* upside down causes a painful intuitive wrong footing at first. If this is not you, proceed according to your preference.

The major strength of effective area as a concept becomes clear when you want to calculate received signal strength for a wide variety of antennas. In our treatment of the Hertzian dipole, recall that in Eq. 8.43 we had to get down to the very specific geometry of the dipole and the electric field. But all antenna structures in common use have effective area numbers that you can look up, and if you have that in hand, then the power available at the output terminals of the receive antenna is simply

$$P_{\mathrm{r}} = S \cdot A_{\mathrm{eff}}, \tag{8.53}$$

where S is the Poynting flux in Watts per square meter at the receive antenna. This result will come in handy in Chapter 12, when we take all that we have learned and apply it to the energy analysis of complete wireless links.

9

A Note on Linear Algebra

With this chapter, we continue with the familiar challenge of dealing with a big, rich topic in a short span. Our approach remains constant, which is that we start with applications that we need to understand and then work backward, prioritizing the topics that are needed. For radio, the most important applications are the DFT (or FFT), least squares, and multi-antenna systems. We had a preview of our analytical needs with the discussion of the DFT in Chapter 2. Singular value decomposition is central to the multi-antenna techniques of Chapter 10, and it will turn out that least squares techniques figure prominently in understanding digital predistortion in Chapter 11.

9.1 What Problem Are We Solving with a Matrix?

Most engineers run across matrices at some point in their professional trajectory. If you are a control theorist or practitioner, you live with this stuff and can recite more matrix fun facts in a minute than most other engineers ever learn. And if you subjected yourself to a formal linear algebra course in college, you may have vague memories of how matrices helped to solve systems of equations; the importance of something called the "determinant," and how it had something to do with inversion; special vectors called eigenvectors; etc.

After many years of study and application in various contexts, the author suggests that the broadest application, and the most useful foundational understanding, of matrices, is to see them as a means of *mechanizing* linear transformations. What do we mean by this? We are very familiar with functions operating on scalars, to the tune of $f(x) = y$. And if f is a linear function, we have that $f(u + v) = f(u) + f(v)$. Well, a matrix is the equivalent of a linear function, only it takes

Energy Fundamentals of Radio, First Edition. Joel L. Dawson.

a vector as an input and produces as its output another vector. If we have a matrix A, and vectors $\vec{u}$ and $\vec{v}$, then it is true that

$$A\left(c\vec{u} + d\vec{v}\right) = cA\vec{u} + dA\vec{v}, \tag{9.1}$$

where c and d are scalars. In this formulation, $A\vec{u}$ and $A\vec{v}$ are also vectors. So just as $f(\cdot)$ takes scalar x as an input and produces scalar $f(x)$ as an output, the matrix A takes vector $\vec{u}$ as an input and produces vector $A\vec{u}$ as an output.

Before we get into the mechanics of how a matrix realizes a linear transformation, it will help to fix in your mind just what a linear transformation is in the first place. Suppose that we had a matrix in two dimensions that affected a rotation counterclockwise of 45°. Figure 9.1 shows what the matrix would do on vectors that lie purely on $\hat{x}$ or $\hat{y}$. Now suppose that we started with the 2D vector $\left(\frac{\sqrt{2}}{2}, \frac{\sqrt{2}}{2}\right)$. This is, of course, a sum of the scaled basis vectors $\hat{x}$ and $\hat{y}$ that we just considered:

$$\left(\frac{\sqrt{2}}{2}, \frac{\sqrt{2}}{2}\right) = \frac{\sqrt{2}}{2}\hat{x} + \frac{\sqrt{2}}{2}\hat{y}. \tag{9.2}$$

The logic of our linear transformation says that we should be able to rotate each component by 45°, and then the sum of these rotated components tells us the result of rotating the original $\left(\frac{\sqrt{2}}{2}, \frac{\sqrt{2}}{2}\right)$. See Figure 9.2. By now you should start to appreciate the magic of the rotation matrix. It is a linear operator, and this linearity

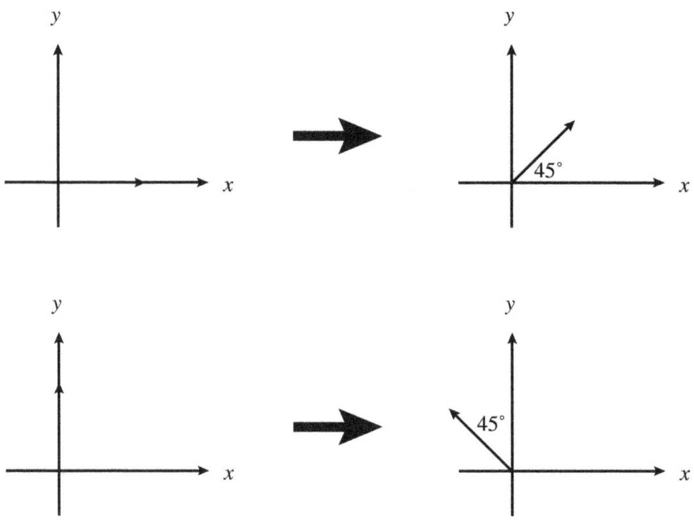

Figure 9.1 A rotation as a first example of a linear transformation.

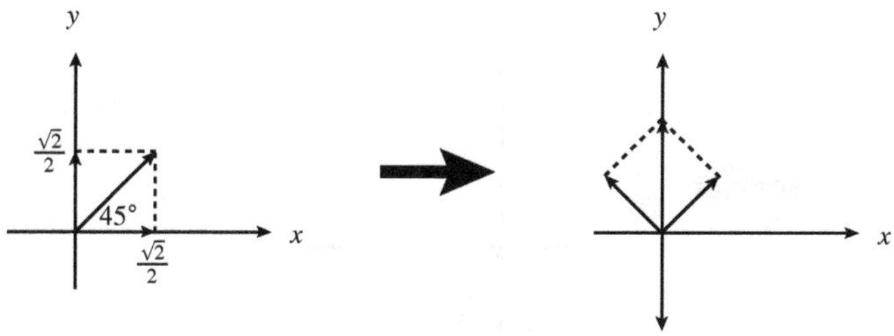

Figure 9.2 Rotating an arbitrary vector by 45°.

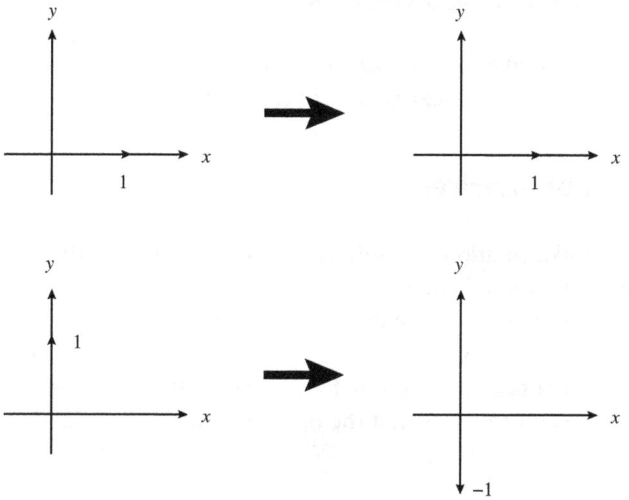

Figure 9.3 Reflecting across the x axis.

makes it very easy for us to rotate *any* input vector, regardless of its initial angle and length. We can easily break the input vector into its components. We know how the operator treats each component. Therefore, we know how the rotation matrix treats any input vector. Kind of a neat trick, and worth appreciating before you get into the choreography of matrix multiplication. Also as you ponder this, consider that the Fourier transform is itself a linear transformation. It may therefore be less surprising than it was before that this transform finds expression as FFT and IFFT matrices.

One more example to help cement the concept. What if we were interested in a transformation that *reflects* an input vector across, say, the x axis? Figure 9.3 shows how such a transformation would affect vectors $1 \cdot \hat{x} + 0 \cdot \hat{y}$ and $0 \cdot \hat{x} + 1 \cdot \hat{y}$. If we can do this operation with a matrix, it is necessarily a linear operator and we can easily visualize its effect on any general input vector. See Figure 9.4. Fix in your mind this "why" of matrices, which is that they are a way for us to mechanize a

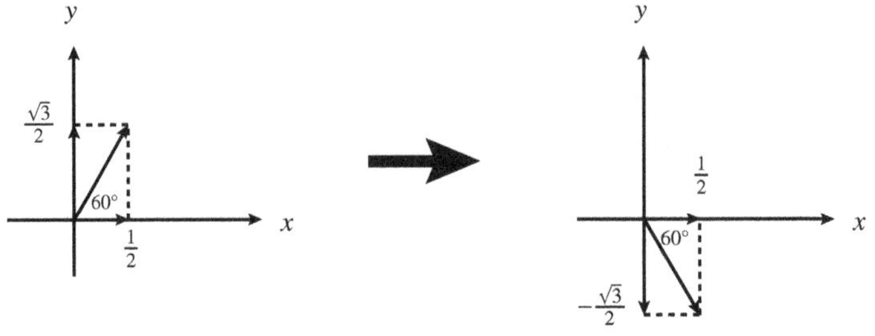

Figure 9.4 Reflecting an arbitrary vector across the *x* axis.

linear transformation. A matrix takes a vector as an input, produces a vector as an output, and strictly binds itself to the linearity code of conduct.

9.2 The Mechanics of Matrices

Having provided examples like rotation and reflection and visualizing things in two-space, let's move one step beyond and default to three-space, or vectors with three elements. Let us further default to the convention of nominally using column vectors, or vectors whose elements are stacked vertically on the page, rather than row vectors, whose elements are placed side by side. We will see shortly that operating on a 3D column vector requires that the operating matrix have no less than three columns. Behold the matrix A:

$$A = \begin{bmatrix} a_{11} & a_{12} & a_{13} \\ a_{21} & a_{22} & a_{23} \\ a_{31} & a_{32} & a_{33} \end{bmatrix}. \tag{9.3}$$

(We did not have to choose to make this a "square" matrix, with the same number of rows and columns, but we are at the beginning and why not?) Note that a row is a horizontal series of coefficients, so in the matrix A row 1 is $[a_{11}\ a_{12}\ a_{13}]$, row 2 is $[a_{21}\ a_{22}\ a_{23}]$, etc. Meanwhile, a column in the matrix is a vertical slice of coefficients, so column 1 is given by

$$\begin{bmatrix} a_{11} \\ a_{21} \\ a_{31} \end{bmatrix}. \tag{9.4}$$

Our subscript convention is "row-column," so a_{ij} refers to the coefficient of the ith row and j^{th} column.[1]

Very good. We move on to what you really want to know, which is how this vector operator actually operates on a vector in three-space. The setup is as follows:

$$\begin{bmatrix} a_{11} & a_{12} & a_{13} \\ a_{21} & a_{22} & a_{23} \\ a_{31} & a_{32} & a_{33} \end{bmatrix} \begin{bmatrix} x_1 \\ x_2 \\ x_3 \end{bmatrix} = ? \tag{9.5}$$

If we were not to write out all the details, we would instead write

$$A\vec{x} = ? \tag{9.6}$$

There are two useful ways to proceed, both of which involve breaking the matrix up into its constituent vectors, either row vectors or column vectors. The first is to look at the matrix as three row vectors stacked on top of one another. With this approach, we get

$$\begin{bmatrix} a_{11} & a_{12} & a_{13} \\ a_{21} & a_{22} & a_{23} \\ a_{31} & a_{32} & a_{33} \end{bmatrix} \begin{bmatrix} x_1 \\ x_2 \\ x_3 \end{bmatrix} = \begin{bmatrix} \vec{a}_{\text{row1}} \cdot \vec{x} \\ \vec{a}_{\text{row2}} \cdot \vec{x} \\ \vec{a}_{\text{row3}} \cdot \vec{x} \end{bmatrix}. \tag{9.7}$$

We mean here the dot product in the ordinary sense, which results in a scalar every time it is applied. So the net result of applying a 3×3 matrix to a column vector in three-space is another column vector in three-space

$$\begin{bmatrix} \vec{a}_{\text{row1}} \cdot \vec{x} \\ \vec{a}_{\text{row2}} \cdot \vec{x} \\ \vec{a}_{\text{row3}} \cdot \vec{x} \end{bmatrix} = \begin{bmatrix} a_{11}x_1 + a_{12}x_2 + a_{13}x_3 \\ a_{21}x_1 + a_{22}x_2 + a_{23}x_3 \\ a_{31}x_1 + a_{32}x_2 + a_{33}x_3 \end{bmatrix}. \tag{9.8}$$

A second way of viewing this operation is to see the matrix A as a collection of column vectors. This route to the same answer looks as follows:

$$\begin{bmatrix} a_{11} & a_{12} & a_{13} \\ a_{21} & a_{22} & a_{23} \\ a_{31} & a_{32} & a_{33} \end{bmatrix} \begin{bmatrix} x_1 \\ x_2 \\ x_3 \end{bmatrix} = \begin{bmatrix} a_{11} \\ a_{21} \\ a_{31} \end{bmatrix} x_1 + \begin{bmatrix} a_{12} \\ a_{22} \\ a_{32} \end{bmatrix} x_2 + \begin{bmatrix} a_{13} \\ a_{23} \\ a_{33} \end{bmatrix} x_3. \tag{9.9}$$

---

1 If you have trouble remembering this convention, then (1) you have plenty of company, and (2) bookmark this page for easy reference until you do remember it. It's a little thing, but failure to keep this convention straight greatly magnifies the difficulty of following any mathematical discussion of matrices.

Comparing Eqs. 9.9 and 9.16, we should get the same answer, and we do:

$$\begin{bmatrix} a_{11} \\ a_{21} \\ a_{31} \end{bmatrix} x_1 + \begin{bmatrix} a_{12} \\ a_{22} \\ a_{32} \end{bmatrix} x_2 + \begin{bmatrix} a_{13} \\ a_{23} \\ a_{33} \end{bmatrix} x_3 = \begin{bmatrix} a_{11}x_1 + a_{12}x_2 + a_{13}x_3 \\ a_{21}x_1 + a_{22}x_2 + a_{23}x_3 \\ a_{31}x_1 + a_{32}x_2 + a_{33}x_3 \end{bmatrix}. \tag{9.10}$$

Having shown two different ways to do the matrix operation on a vector, we must resolve the question of which method one uses at a given time. Given that they both give the same answer, as they must, the resolution to this question is no more complicated than: "whichever one makes the most sense to you."

An example helps. For a matrix multiplication that realizes a DFT, such as in Eq. 6.33, the author would tend in his mind toward the first method, where we see the matrix as a vertical stack of row vectors. This is because thinking this way maps so strongly onto our DFT "analysis" equation, which you'll recall from Eq. 2.104 was

$$C_k = \frac{1}{N} \sum_{n=0}^{N-1} x[n] \cdot e^{-j\left(\frac{2\pi}{N} \cdot k\right)n}. \tag{9.11}$$

Similarly, when doing an IDFT, as in Eq. 6.35, it is more natural for the author to think of the matrix multiplication in terms of summing column vectors, referring to Eq. 2.104:

$$x[n] = \sum_{k=0}^{N-1} C_k e^{j\left(\frac{2\pi}{N} \cdot k\right)n}. \tag{9.12}$$

The only real rule here is not to be too rigid about it. Allow yourself to go toward whatever method you feel strongly drawn to, and if neither seems preferable pick one at random and forge ahead. You can always revise your methods as you learn more.

Thankfully, there are easier examples to ponder if the DFT proves to be a heavy lift at first. We will come back to this; let's return to the rotation matrix concept that we discussed in Section 9.1. Back then we formulated the problem in two-space, not three-space, but pay attention to how we formulate things now.

For the rotation matrix, we suppose that we wish to rotate a vector in the xy plane counterclockwise by angle θ. The right matrix turns out to be given by

$$A_{\text{rot}} = \begin{bmatrix} \cos\theta & -\sin\theta & 0 \\ \sin\theta & \cos\theta & 0 \\ 0 & 0 & 1 \end{bmatrix}. \tag{9.13}$$

If we apply this rotation matrix to vector in three-space $\vec{r} = (x, y, z)$, we have

$$\begin{bmatrix} \cos\theta & -\sin\theta & 0 \\ \sin\theta & \cos\theta & 0 \\ 0 & 0 & 1 \end{bmatrix} \begin{bmatrix} x \\ y \\ z \end{bmatrix} = \begin{bmatrix} \cos\theta \\ \sin\theta \\ 0 \end{bmatrix} x + \begin{bmatrix} -\sin\theta \\ \cos\theta \\ 0 \end{bmatrix} y + \begin{bmatrix} 0 \\ 0 \\ 1 \end{bmatrix} z. \tag{9.14}$$

Notice that breaking down the computation using the rotation matrix as a collection of column vectors is very natural here. We see that the x component of $\vec{r}$ will be moved to $(\cos\theta, \sin\theta, 0)$; the y component will be moved to $(-\sin\theta, \cos\theta, 0)$; and the z component stays put. If you want to compare our operations here to the 45° rotation of Section 9.1, recall that $\cos(45°) = \sin(45°) = \sqrt{2}/2$.

For the reflection matrix of Section 9.1, the correct matrix is

$$A_{\text{refl}} = \begin{bmatrix} 1 & 0 & 0 \\ 0 & -1 & 0 \\ 0 & 0 & 1 \end{bmatrix}. \tag{9.15}$$

There is much to be gained from pencil-and-paper experimentation with the simple rotation and reflection matrices shown here. Another key thing for the novice to navigate is the balance between the mechanics of matrix calculations and the conceptual exercise of treating matrices as linear transformations. Time and application will gradually get you comfortable with this balance.

9.3 A Few Facts About Matrices

Sometimes it will be true that $A\vec{x} = 0$. Spelling it out:

$$A\vec{x} = \begin{bmatrix} a_{11} & a_{12} & a_{13} \\ a_{21} & a_{22} & a_{23} \\ a_{31} & a_{32} & a_{33} \end{bmatrix} \begin{bmatrix} x_1 \\ x_2 \\ x_3 \end{bmatrix} = \begin{bmatrix} \vec{a}_{\text{row1}} \cdot \vec{x} \\ \vec{a}_{\text{row2}} \cdot \vec{x} \\ \vec{a}_{\text{row3}} \cdot \vec{x} \end{bmatrix} = \begin{bmatrix} 0 \\ 0 \\ 0 \end{bmatrix}. \tag{9.16}$$

What does this mean if we carve out the trivial case of $\vec{x} = (0, 0, 0)$, and the equally trivial case of the matrix having all zero elements? In words, we say that "$\vec{x}$ is orthogonal to every row vector of A." This, in turn, means that the three row vectors of A do not completely span the space of all possible 3-vectors. If this were not true, at least one of the dot products would be nonzero. We say that this particular $\vec{x}$, along with all scalings $c\vec{x}$, sit "outside the row space" of A.

Now let's take this same $\vec{x}$, and go to the view of A as a collection of column vectors. We have

$$\begin{bmatrix} a_{11} & a_{12} & a_{13} \\ a_{21} & a_{22} & a_{23} \\ a_{31} & a_{32} & a_{33} \end{bmatrix} \begin{bmatrix} x_1 \\ x_2 \\ x_3 \end{bmatrix} = \begin{bmatrix} a_{11} \\ a_{21} \\ a_{31} \end{bmatrix} x_1 + \begin{bmatrix} a_{12} \\ a_{22} \\ a_{32} \end{bmatrix} x_2 + \begin{bmatrix} a_{13} \\ a_{23} \\ a_{33} \end{bmatrix} x_3 = \begin{bmatrix} 0 \\ 0 \\ 0 \end{bmatrix}. \tag{9.17}$$

Again in words, we say that "the columns of A are not linearly independent." If they *were* linearly independent, then there is no combination of x_1, x_2, and x_3 for which the above equality holds.

It is vitally important to know not only whether an $\vec{x}$ exists for which $A\vec{x} = 0$, but also how many *linearly independent* $\vec{x}$'s exist. We then use this tally to speak of the "rank" of matrix A. Some matrix rank facts:

1. If a matrix has m rows and n columns, the maximum rank r is the lesser of m and n.
2. The rank of a matrix is the number of linearly independent rows, which is always equal to the number of linearly independent columns.
3. If there is no $\vec{x}$ for which $A\vec{x} = 0$, then the rows of A "fully span" the space of n-vectors, where n is the number of columns. Put another way, we can get any n-vector we like using an appropriately weighted sum of the rows of A.
4. Again if there is no $\vec{x}$ for which $A\vec{x} = 0$, we conclude that the columns of A are linearly independent.

There is a great deal to get into about matrices from this jumping off point. For more, see Strang's book on linear algebra.[2] We stick here to the bare minimum to understand the application at hand. As always, the reader is encouraged to explore further as curiosity dictates.

9.4 Orthonormal Basis Vectors

We first encountered the concept of orthonormal basis vectors in Section 2.3.4 when we discussed the DFT. Suppose that we have vectors v_i in m-space. Suppose further that we have exactly m of them, and that we scale them so that they are not only *orthogonal*, but *orthonormal*. That is

$$\vec{v}_i \cdot \vec{v}_j = \begin{cases} 0 & i \neq j \\ 1 & i = j \end{cases}. \tag{9.18}$$

2 G. Strang, *Introduction to Linear Algebra*. Wellesley, MA: Wellesley-Cambridge Press, 1993.

(Notes: We reserve the term "orthonormal" for when the vectors are "normalized" such that when $i = j$, the dot product $\vec{v}_i \cdot \vec{v}_i$ is 1 as opposed to some other nonzero constant; when the vectors v_i are complex, the right expression of orthonormality is

$$\vec{v}_i \cdot (\vec{v}_j)^* = \begin{cases} 0 & i \neq j \\ 1 & i = j \end{cases}. \tag{9.19}$$

Finally, this choice of normalization means that the $\vec{v}_i$ are unit vectors, so $|\vec{v}_i| = 1$ for all i.) If these conditions are satisfied, you can produce any m-vector with some linear combination of the v_is. Accordingly, we say that the v_is are an *orthonormal basis* for the space of m-vectors. Determining the exact linear combination of v_is to synthesize a given x is as easy as

$$\begin{bmatrix} \leftarrow & v_1^* & \rightarrow \\ \leftarrow & v_2^* & \rightarrow \\ \leftarrow & \vdots & \rightarrow \\ \leftarrow & v_n^* & \rightarrow \end{bmatrix} \begin{bmatrix} \uparrow \\ x \\ \downarrow \end{bmatrix} = \begin{bmatrix} C_1 \\ \vdots \\ C_n \end{bmatrix}. \tag{9.20}$$

If $\vec{x}$ is a sum of the $\vec{v}_i$s, then Eq. 9.20 is a perfectly elegant way of extracting how much of each $\vec{v}_i$ is in $\vec{x}$. If this seems light on details for you, carefully compare Eqs. 9.20 and 6.32, which are exactly the same conceptually but in different contexts.

With respect to orthogonality and orthonormality, there is at least one more concept worth getting comfortable with. We have emphasized here a complete set of m vectors of dimension m that are all mutually orthogonal, and because there are m of them, they completely span the space. Now suppose that out of our set of m vectors, we *removed* one or more vectors from the collection? We would say then that the remaining vectors no longer span the space of m-vectors. (It may seem a little silly and obvious to put this into words, but there is clarity to be had here.) Indeed, if, say, $\vec{v}_1$ and $\vec{v}_2$ are removed from the collection, then any vector $c_1\vec{v}_1 + c_2\vec{v}_2$ will lie outside the reach of the space spanned by $(\vec{v}_3, \cdots, \vec{v}_m)$. This is because, by construction, $c_1\vec{v}_1 + c_2\vec{v}_2$ will be orthogonal to any of the $(\vec{v}_3, \cdots, \vec{v}_m)$ individually, and therefore to any linear combination of the same.

Get used to this idea of the span of a vector space, by which we mean the space of vectors that can be synthesized by a linear combination of our member, or basis, vectors. Note finally we do not need our basis vectors to be orthogonal in order to span a given space, only that they be linearly independent. However, when they *are* orthogonal, our mathematics is neat and tidy and allows us to do things like Eq. 9.20.

9.5 Matrix Inversion

As a last step before we get to the singular value decomposition, it is useful to have some understanding of what it means to invert a matrix. The very simple idea is that the inverse of a matrix undoes the linear transformation that the matrix does. If you follow this, it will not surprise you that in equations we would have

$$A^{-1}A\vec{x} = \vec{x}. \tag{9.21}$$

A^{-1} is the inverse of matrix A, and when you cascade them it's as though you never did anything at all to $\vec{x}$.

Another way of appreciating matrix inversion is to consider the matrix $A^{-1}A$ by itself. When you do this, what you find is that

$$A^{-1}A = I = \begin{bmatrix} 1 & 0 & 0 & 0 & \cdots \\ 0 & 1 & 0 & \cdots & \cdots \\ 0 & 0 & 1 & \cdots & \cdots \\ 0 & \vdots & \vdots & \ddots & \vdots \\ 0 & \cdots & \cdots & \cdots & 1 \end{bmatrix}. \tag{9.22}$$

Here I is a special matrix called the identity matrix, and it always has a 1 along the diagonal and zeros everywhere else. (Check that this is the right form of an identity matrix with your new understanding of the mechanics of matrices from Section 9.2).

Now we have not reviewed how to multiply one matrix by another, so it's time to remedy that. For the author, it has always been easiest to remember how to do matrix multiplication AB by mentally organizing A as a stack of row vectors and B as a side-by-side collection of column vectors. To wit:

$$AB = \begin{bmatrix} \leftarrow & a_1 & \rightarrow \\ \leftarrow & a_2 & \rightarrow \\ \leftarrow & a_3 & \rightarrow \\ & \vdots & \\ \leftarrow & a_m & \rightarrow \end{bmatrix} \begin{bmatrix} \uparrow & \uparrow & & & \uparrow \\ & & & & \\ b_1 & b_2 & \cdots & \cdots & b_n \\ & & & & \\ \downarrow & \downarrow & & & \downarrow \end{bmatrix}. \tag{9.23}$$

Set up this way, the elements of the matrix AB are given by

$$AB = \begin{bmatrix} \vec{a}_1 \cdot \vec{b}_1 & \vec{a}_1 \cdot \vec{b}_2 & \cdots & \cdots & \vec{a}_1 \cdot \vec{b}_n \\ \vec{a}_2 \cdot \vec{b}_1 & \vec{a}_2 \cdot \vec{b}_2 & \cdots & \cdots & \vec{a}_2 \cdot \vec{b}_n \\ \vdots & & \vdots & & \vdots \\ \vdots & & \vdots & & \vdots \\ \vec{a}_m \cdot \vec{b}_1 & \vec{a}_n \cdot \vec{b}_2 & \cdots & \cdots & \vec{a}_m \cdot \vec{b}_n \end{bmatrix}. \tag{9.24}$$

So now, how should we think about matrix inversion? It can be very grounding to try this out on a matrix for which we can logically deduce the inverse. Recall that the matrix for doing a rotation about the z axis was given in Eq. 9.13:

$$A_{\text{rot}} = \begin{bmatrix} \cos\theta & -\sin\theta & 0 \\ \sin\theta & \cos\theta & 0 \\ 0 & 0 & 1 \end{bmatrix}. \tag{9.25}$$

Well, the inverse of a rotation matrix ought to just be...a rotation matrix that rotates by the same amount but in the opposite direction. Remembering that $\cos(-\theta) = \cos\theta$ and $\sin(-\theta) = -\sin\theta$, we have every right to expect that the inverse matrix is given by

$$A_{\text{rot}}^{-1} = \begin{bmatrix} \cos\theta & \sin\theta & 0 \\ -\sin\theta & \cos\theta & 0 \\ 0 & 0 & 1 \end{bmatrix}. \tag{9.26}$$

Let's try it:

$$A_{\text{rot}}^{-1} A_{\text{rot}} = \begin{bmatrix} \cos^2\theta + \sin^2\theta & -\cos\theta\sin\theta + \sin\theta\cos\theta & 0 \\ -\sin\theta\cos\theta + \cos\theta\sin\theta & \sin^2\theta + \cos^2\theta & 0 \\ 0 & 0 & 1 \end{bmatrix}. \tag{9.27}$$

Recalling that $\cos^2\theta + \sin^2\theta = 1$, we have in fact that

$$A_{\text{rot}}^{-1}A_{\text{rot}} = \begin{bmatrix} 1 & 0 & 0 \\ 0 & 1 & 0 \\ 0 & 0 & 1 \end{bmatrix} = I. \tag{9.28}$$

Excellent.

Note that in general, matrices are *not* this easy to invert, and readers are referred again to Strang[3] or to their favorite linear algebra reference. But the logic of matrix inversion described here is important and worth internalizing. Also, note that matrices are not restricted to having the same number of rows and columns (being "square"). Looking specifically at matrix multiplication in Eq. 9.23, you can see that two matrices AB can be multiplied provided that A has as many columns as B has rows.

9.6 The Singular Value Decomposition

Linear algebra is a big subject, and now we come to the singular value decomposition (SVD) of matrix A while bypassing many of the customary preliminaries. Indeed, during the author's studies his introductory course never made it to the SVD. That said, SVD is so useful and important that for a long time the author struggled to keep up in engineering settings long after his schooling was done. For this reason, in this text we skip over the usual discussions of things like eigenvalues and eigenvectors, solutions to sets of linear equations, determinants, and the like, to get to this important and useful topic. See what you think.

As with matrix inversion, we will not concern ourselves with the actual computation of the SVD. Computation of the SVD *is* important. But if one has to choose, and in a volume of this size one *does* have to choose, better to focus on understanding the SVD and leave it to the reader to delve into the details of computation.

Off we go, then. It turns out that all matrices can be written as the following decomposition:

$$A = U\Sigma V^{\mathrm{T}}. \tag{9.29}$$

The "T" superscript on the matrix V means "transpose," and to do a transpose you basically take every element v_{ij} and for the transpose matrix move it to new position v_{ji}. (Tall, skinny matrices become short and fat, and vice versa.) But that's not the important part of putting the transpose label here. Often with matrices by convention we default to seeing them as a collection of column vectors. The "T" in

3 Ibid.

this case is an admonition that the conceptually important vectors this time will be in the *rows*. To clarify, let's write Eq. 9.29 in expanded form:

$$A = U\Sigma V^{\mathrm{T}} = \begin{bmatrix} \uparrow & \uparrow & & & \uparrow \\ & & & & \\ \vec{u}_1 & \vec{u}_2 & \cdots & \cdots & \vec{u}_{\mathrm{R}} \\ & & & & \\ \downarrow & \downarrow & & & \downarrow \end{bmatrix} \begin{bmatrix} \sigma_1 & 0 & 0 & 0 & \cdots \\ 0 & \sigma_2 & 0 & \cdots & \cdots \\ 0 & 0 & \sigma_3 & \cdots & \cdots \\ 0 & \vdots & \vdots & \ddots & \vdots \\ 0 & \cdots & \cdots & \cdots & \sigma_{\mathrm{R}} \end{bmatrix} \begin{bmatrix} \leftarrow & \vec{v}_1 & \rightarrow \\ \leftarrow & \vec{v}_2 & \rightarrow \\ \leftarrow & \vdots & \rightarrow \\ \leftarrow & \vec{v}_{\mathrm{R}} & \rightarrow \end{bmatrix}. \tag{9.30}$$

At the end of a properly executed SVD, there are R singular values, which are the σs, and R is the rank of the matrix as we discussed in Section 9.3. The vectors $\vec{v}_i$ are chosen to be an orthonormal basis that spans the space of row vectors in the original matrix A.[4] Similarly, the vectors $\vec{u}_i$ are chosen to be an orthonormal basis that spans the vector space of the columns of A. The σ_i are the singular values themselves, whose meaning we will show.

Now the *best* way to understand SVD is to see it in action: posit an input vector $\vec{x}$, and follow the decomposed A as it operates on that vector step by step.

Step 1:

$$U\Sigma V^{\mathrm{T}}\vec{x} = \begin{bmatrix} \uparrow & \uparrow & & & \uparrow \\ & & & & \\ \vec{u}_1 & \vec{u}_2 & \cdots & \cdots & \vec{u}_{\mathrm{R}} \\ & & & & \\ \downarrow & \downarrow & & & \downarrow \end{bmatrix} \begin{bmatrix} \sigma_1 & 0 & 0 & 0 & \cdots \\ 0 & \sigma_2 & 0 & \cdots & \cdots \\ 0 & 0 & \sigma_3 & \cdots & \cdots \\ 0 & \vdots & \vdots & \ddots & \vdots \\ 0 & \cdots & \cdots & \cdots & \sigma_{\mathrm{R}} \end{bmatrix} \begin{bmatrix} \vec{v}_1 \cdot \vec{x} \\ \vec{v}_2 \cdot \vec{x} \\ \vdots \\ \vec{v}_{\mathrm{R}} \cdot \vec{x} \end{bmatrix}. \tag{9.31}$$

Step 1, then, is that we decompose $\vec{x}$ into its $\vec{v}_i$ components. Remember, if we ask, "How much of the (unit vector) $\vec{v}_i$ is in $\vec{x}$?", the answer is $\vec{v}_i \cdot \vec{x}$. If the $\vec{v}_i$ happen to span the space of $\vec{x}$ vectors, then the fact that the $\vec{v}_i$ form an orthonormal basis means that we could turn around and write with authority:

$$\vec{x} = (\vec{v}_1 \cdot \vec{x})\,\vec{v}_1 + (\vec{v}_2 \cdot \vec{x})\,\vec{v}_2 + \cdots + (\vec{v}_{\mathrm{R}} \cdot \vec{x})\,\vec{v}_{\mathrm{R}}. \tag{9.32}$$

On the other hand, if the $\vec{v}_i$s don't span the space of $\vec{x}$ vectors, and we get unlucky and have $\vec{x}$ orthogonal to all the $\vec{v}_i$s, then our very first step collapses into a zero

4 If SVD is new to you, it is helpful at this point to deliberately resist focusing on *how* we derive the v_is; suffice it to say, there is a recipe to follow. Focus instead on the fact that the v_is span the row space of A. If a vector $\vec{x}$ lies outside of the row space of A, then $\vec{v}_i \cdot \vec{x} = 0$ for all i.

vector. In the linear algebra vernacular, we say that $\vec{x}$ "falls into the nullspace" of the matrix A.

In Step 2, we apply the matrix of singular values. Being a diagonal matrix, this is straightforward:

$$U\Sigma V^{\mathrm{T}}\vec{x} = \begin{bmatrix} \uparrow & \uparrow & & & \uparrow \\ \vec{u}_1 & \vec{u}_2 & \cdots & \cdots & \vec{u}_{\mathrm{R}} \\ \downarrow & \downarrow & & & \downarrow \end{bmatrix} \begin{bmatrix} \sigma_1\vec{v}_1 \cdot \vec{x} \\ \sigma_2\vec{v}_2 \cdot \vec{x} \\ \vdots \\ \sigma_{\mathrm{R}}\vec{v}_{\mathrm{R}} \cdot \vec{x} \end{bmatrix}. \tag{9.33}$$

Interesting: we have decomposed $\vec{x}$ into its separate $\vec{v}_i$ components, and then scaled each of those components by the corresponding singular value.

We finish things off dramatically in Step 3 as follows:

$$U\Sigma V^{\mathrm{T}}\vec{x} = \begin{bmatrix} \uparrow \\ \vec{u}_1 \\ \downarrow \end{bmatrix} \sigma_1\vec{v}_1 \cdot \vec{x} + \begin{bmatrix} \uparrow \\ \vec{u}_2 \\ \downarrow \end{bmatrix} \sigma_2\vec{v}_2 \cdot \vec{x} + \cdots + \begin{bmatrix} \uparrow \\ \vec{u}_{\mathrm{R}} \\ \downarrow \end{bmatrix} \sigma_{\mathrm{R}}\vec{v}_{\mathrm{R}} \cdot \vec{x}. \tag{9.34}$$

Remember now that any matrix can be broken down this way, as a singular value decomposition. The value of the SVD is that it *lays bare the inner workings of the matrix as a linear transformation.* In other words, we may say that the matrix A "grabs the $\vec{v}_1$ component of the input vector $\vec{x}$, scales it by σ_1, and outputs it as $\vec{u}_1$ scaled by $\sigma_1\vec{v}_1 \cdot \vec{x}$; it repeats this procedure for all the singular values; and at last the output of the linear transformation is the sum of the scaled $\vec{u}_i$s."

Once you get this understanding of SVD in your head, you will find that in some very special cases you can write down the SVD of a matrix by inspection. Suppose, for example, that we return to the rotation matrix of Section 9.2, and throw in for good measure a scaling of the length by a factor of 2. See Figure 9.5. Applying the words that we use to describe how SVD breaks a matrix down into an SVD, convince yourself that our rotate and scale can be represented by

$$U\Sigma V^{\mathrm{T}} = \begin{bmatrix} \cos\theta & -\sin\theta & 0 \\ \sin\theta & \cos\theta & 0 \\ 0 & 0 & 1 \end{bmatrix} \begin{bmatrix} \sqrt{2} & 0 & 0 \\ 0 & \sqrt{2} & 0 \\ 0 & 0 & \sqrt{2} \end{bmatrix} \begin{bmatrix} 1 & 0 & 0 \\ 0 & 1 & 0 \\ 0 & 0 & 1 \end{bmatrix}. \tag{9.35}$$

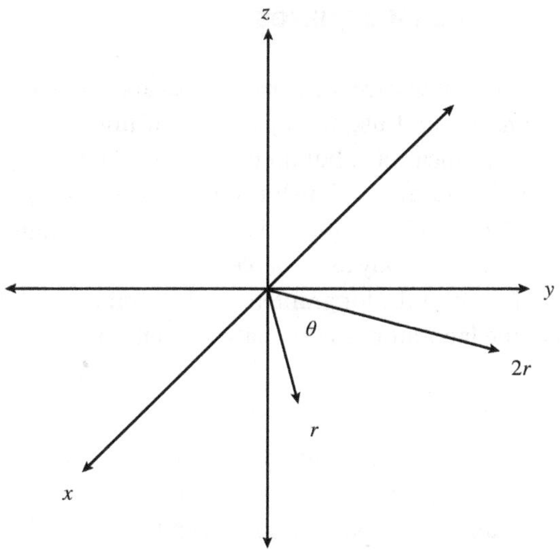

Figure 9.5 Rotating an arbitrary vector by θ, and also scaling it by 2×.

Now with both Σ and V^T diagonal, it hardly seems worth the trouble. We could just write the far simpler

$$A = \sqrt{2}\begin{bmatrix} \cos\theta & -\sin\theta & 0 \\ \sin\theta & \cos\theta & 0 \\ 0 & 0 & 1 \end{bmatrix}. \tag{9.36}$$

But remember that we did this for illustration purposes. In the normal course of business, U and V^T will *look* more complicated. Nevertheless, the conceptual foundation of those more complicated cases is well illustrated here.

Finally, note that it is customary to write an SVD with the σ_is in descending order, with σ_1 being the largest and σ_R being the smallest. Furthermore, when dealing in real life with linear transformation matrix A, sometimes there is value in *approximating* A, saving memory and computation cycles in exchange for a negligible loss in accuracy. SVD gives us a rather organized way of obtaining an approximation. Rather than use the original A, we use an approximate matrix A' whose SVD consists of the largest k magnitude singular values $(\sigma_1, \ldots, \sigma_k)$, and only the corresponding $\vec{v}_i$s and $\vec{u}_i$s.

We will find ready application of SVD in our study of multiple antenna systems in Chapter 10. For now, the hope is that rather than regard SVD as an immensely complicated and esoteric technique, the reader can now follow mathematical treatments based on SVD with considerable insight.

9.7 Least Squares

It is worth discussing least squares approximations here because it is an extremely common and useful application of linear algebra. It has many uses in wireless communications, but an especially prominent one is approximating a nonlinear input–output relationship for purposes of digital predistortion. The best way to understand least squares is by hurrying through the theory and getting to the application as quickly as possible.[5]

To start, take for granted for the moment that we can formulate many problems in the laboratory as the matrix equation

$$A\vec{x} = \vec{b}, \tag{9.37}$$

where we know $\vec{b}$ and need to solve for $\vec{x}$. Also, despite any assumptions you may have made when reading Section 9.5, sometimes, many times, the inverse matrix A^{-1} does not exist, and so we can't just casually write

$$\vec{x} = A^{-1}\vec{b}, \tag{9.38}$$

much as we would like to. When we are unable to get a solution to $A\vec{x} = \vec{b}$, we are now equipped by our study of SVD to say that $\vec{b}$, or some portion thereof, sits outside the space spanned by the $\vec{u}_i$s. This is equivalent to saying that at least some portion of $\vec{b}$ sits outside the column space of A.

Fine: we cannot always solve for an $\vec{x}$ that satisfies $A\vec{x} = \vec{b}$. Must we throw up our hands in despair? Is there an organized way to get "close"?

Well, suppose that we do our best with $A\vec{x}$, and the result is that we don't get $\vec{b}$ exactly but we get $\vec{b}$ plus and additive error $\vec{e}$:

$$A\vec{x} = \vec{b} + \vec{e} \tag{9.39}$$

$$\left(A\vec{x} - \vec{b}\right) = \vec{e}.$$

It turns out that we minimize the squared error $|\vec{e}|^2$ (hence the term "least squares") if we insist that $\vec{e}$ be orthogonal to the columns of A. In other words, we require

$$A^{\mathrm{T}}\vec{e} = 0. \tag{9.40}$$

We look to make $\vec{e}$ orthogonal to every *row* of A^{T} which is the same as requiring $\vec{e}$ to be orthogonal to every *column* of our original matrix A. Following our algebraic nose, we get to

$$A^{\mathrm{T}}\left(A\vec{x} - \vec{b}\right) = 0. \tag{9.41}$$

---

5 With apologies and all due respect to our theorist friends. The author reminds these readers that he is an engineer.

And now we look for the $\vec{x}$ that satisfies *this* equation. We arrive at a solution in just a few steps:

$$\begin{aligned} A^{\mathrm{T}}\left(A\vec{x}-\vec{b}\right) &= 0. \\ A^{\mathrm{T}}A\vec{x}-A^{\mathrm{T}}\vec{b} &= 0 \\ A^{\mathrm{T}}A\vec{x} &= A^{\mathrm{T}}\vec{b} \\ \vec{x} &= \left(A^{\mathrm{T}}A\right)^{-1}A^{\mathrm{T}}\vec{b}. \end{aligned} \tag{9.42}$$

From the standpoint of the least squares problem, we often call $\left(A^{\mathrm{T}}A\right)^{-1}A^{\mathrm{T}}$ the "pseudoinverse" of A, and denote it A^{+}. Many numerical solvers will compute this for you directly.

That is the theory. The application turns out to be immensely clarifying, as it so often is. Suppose that we have a sequence of inputs to a system, and we catalog this succession of inputs as vector $\vec{x}$. We then feed this sequence of inputs into a mystery box or electronic circuit, and that mystery box gives us a corresponding vector of outputs $\vec{y}=f(\vec{x})$ (by which we mean the ith element of $\vec{y}$, output y_i, corresponds to $f(x_i)$). See Figure 9.6. We do not know exactly what is in the box, but let us suppose that we have our reasons to believe that $f(\cdot)$ is a polynomial. Our goal is to determine, as well as we can, the form of $f(\cdot)$ through our input–output measurements.

So we do our experiment and collect our input–output data, and our plot looks something like Figure 9.7. The measurement noise is annoying but a fact of life in the laboratory. The experienced eye will see that this mystery box is behaving a lot like a third-order polynomial along the lines of[6]

$$f(x) \approx C_0 + C_1x + C_2x^2 + C_3x^3. \tag{9.43}$$

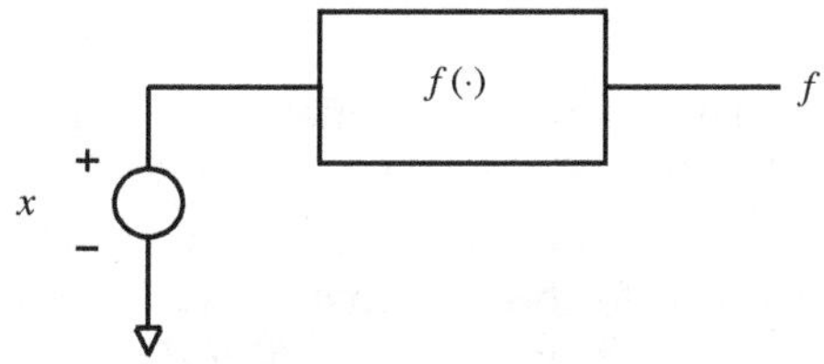

Figure 9.6 An experimental system, whose input–output function $f(x)$ we will estimate using least squares.

6 A *really* experienced eye might lead us to assert that C_0 is effectively zero, since $f(x)$ seems to be trying mightily to pass through the origin. But fortunately, with least squares you don't have to commit ahead of time: if C_0 turns out to be orders of magnitude smaller than the other C_is, then you can rerun the pseudoinverse fitting only for C_1, C_2, and C_3.

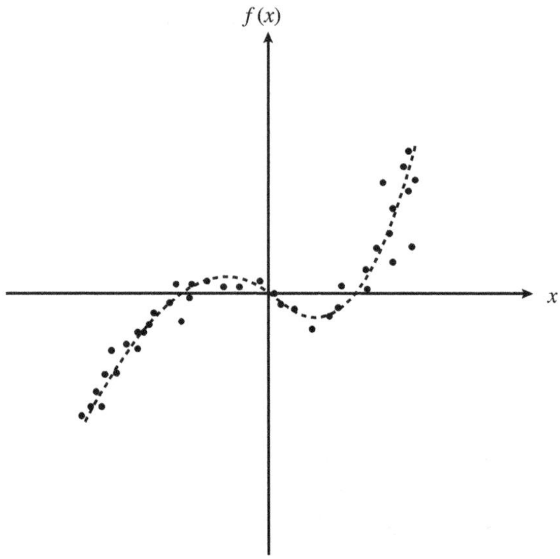

Figure 9.7 Input–output data from our experimental system.

Our problem now is to estimate "good" values for the C_is. It turns out that a good setup for this problem is to form a matrix using the values of x, x^2, and x^3 for each data point, and formulate as follows:

$$\begin{bmatrix} 1 & x_1 & x_1^2 & x_1^3 \\ 1 & x_2 & x_2^2 & x_2^3 \\ 1 & x_3 & x_3^2 & x_3^3 \\ \vdots & \vdots & \vdots & \vdots \\ 1 & x_n & x_n^2 & x_n^3 \end{bmatrix} \begin{bmatrix} C_0 \\ C_1 \\ C_2 \\ C_3 \end{bmatrix} = \begin{bmatrix} f(x_1) \\ f(x_2) \\ f(x_3) \\ \vdots \\ f(x_n) \end{bmatrix}. \tag{9.44}$$

The matrix here is the A matrix of our formulation up to this point, and in practice it will be a bunch of numbers. For example, if our second data point is $x = 2$, then the second row of our A matrix will be $[1\ 2\ 4\ 8]$. The vector of C_is is what we are solving for, because that determines for us the polynomial that we committed to in Eq. 9.43, and the vectors of $f(x_i)$s are the output measurements that correspond to the x_is. Note that a valid way of assessing this situation is that we have "n equations and four unknowns"; assuming n is (much) greater than four, we say this is an overdetermined problem and the matrix A cannot be inverted.[7] It follows

7 We have not discussed it here, but matrices are often used to solve systems of equations. When there are an equal number of equations and unknowns, it is often true that a unique set of

that the rows of A are not linearly independent, and that we can have at most four independent columns in this example.

At this point, happily, the hard work is done. You now have your A matrix, and your measured data. The next step is simply sitting down with a computation engine that can do a pseudoinverse, and let fly:

$$\begin{bmatrix} C_0 \\ C_1 \\ C_2 \\ C_3 \end{bmatrix} = \left(A^{\mathrm{T}}A\right)^{-1}A^{\mathrm{T}}\begin{bmatrix} f(x_1) \\ f(x_2) \\ f(x_3) \\ \vdots \\ f(x_n) \end{bmatrix}. \tag{9.45}$$

You should try this trick. It may catch you off guard how satisfying it is. The good news is that you needn't wait for a messy measurement problem in the lab to give it a go. Choose for f some common nonlinear function that is easily differentiable, and use that function to generate your own table of input and output values. Finally, commit to a polynomial expansion with a few terms, maybe 3–5, and unleash the pseudoinverse with your favorite numerical tool. When the dust settles, compare to what you get when you use the Taylor expansion

$$f(x_0 + \Delta x) \approx f(x_0) + f'(x_0)\frac{\Delta x}{1!} + f''(x_0)\frac{(\Delta x)^2}{2!} + f'''(x_0)\frac{(\Delta x)^3}{3!} + \cdots. \tag{9.46}$$

Don't be confused by the x_0, most times it will be easiest to use the Maclaurin expansion, for which $x_0 = 0$. Other tips: don't use too many terms at first, three is good to start and no more than five; don't get too aggressive with the Δx, sticking to values that are on the order of unity, maybe an order of magnitude higher or lower but no more at first (otherwise, third, fourth, and fifth powers in the polynomial expansion make for dramatic numerical ranges for the calculations). Once you get it to work for you a couple of times, you can explore more and learn.

And that's it, that's least squares! It is of immense practical use, and many readers will have heard it described or discussed over much of their professional lives (when someone speaks of "numerically fitting the data," least squares is often what they mean). This section is a good one to bookmark to come back to easily, even if you don't have immediate use for this technique in your work.

---

unknowns (in our case, the C_is) makes the equality happen. In that case, the matrix A has an exact, unique inverse.

9.8 The FFT and the IFFT

Let's close out this note on linear algebra by returning to Section 2.3.4 where we first appealed to the grammar of matrices, at that time in the context of the discrete Fourier transform, or DFT. In the real world we often compute the DFT using matrices, and use an algorithm called the fast Fourier transform, or FFT.

Knowing what you know now, compare the inverse DFT from Eq. 2.104

$$x[n] = \sum_{k=0}^{N-1} C_k e^{j\left(\frac{2\pi}{N}\cdot k\right)n} \tag{9.47}$$

to the matrix multiplication of Eq. 2.96

$$\begin{bmatrix} 1 & 1 & \cdots & 1 \\ 1 & e^{j\left(\frac{2\pi}{N}\cdot 1\right)\cdot 1} & \cdots & e^{j\left(\frac{2\pi}{N}\cdot(N-1)\right)\cdot 1} \\ \vdots & \vdots & \vdots & \vdots \\ 1 & e^{j\left(\frac{2\pi}{N}\cdot 1\right)\cdot(N-1)} & \cdots & e^{j\left(\frac{2\pi}{N}\cdot(N-1)\right)(N-1)} \end{bmatrix} \begin{bmatrix} C_0 \\ C_1 \\ \vdots \\ C_{N-1} \end{bmatrix} = \begin{bmatrix} x[0] \\ x[1] \\ \vdots \\ x[N-1] \end{bmatrix}. \tag{9.48}$$

Compare carefully, and you see that they are the same, as they must be.

Similarly, we can look at the expression for the C_ks from Eq. 2.110

$$C_k = \frac{1}{N}\sum_{n=0}^{N-1} x[n]\cdot e^{-j\left(\frac{2\pi}{N}\cdot k\right)n}, \tag{9.49}$$

and write an equivalent matrix expression

$$\begin{bmatrix} C_0 \\ C_1 \\ \vdots \\ C_{N-1} \end{bmatrix} = \frac{1}{N}\begin{bmatrix} 1 & 1 & \cdots & 1 \\ 1 & e^{-j\left(\frac{2\pi}{N}\cdot 1\right)\cdot 1} & \cdots & e^{-j\left(\frac{2\pi}{N}\cdot(N-1)\right)\cdot 1} \\ \vdots & \vdots & \vdots & \vdots \\ 1 & e^{-j\left(\frac{2\pi}{N}\cdot 1\right)\cdot(N-1)} & \cdots & e^{-j\left(\frac{2\pi}{N}\cdot(N-1)\right)(N-1)} \end{bmatrix} \begin{bmatrix} x[0] \\ x[1] \\ \vdots \\ x[N-1] \end{bmatrix}. \tag{9.50}$$

If you have ever wondered why matrix multiplication and FFTs are always mentioned together, now you know!

We close this section by briefly examining the DFT through the lens of SVD as a means of better understanding SVD. While this may sound like compounding difficulty with difficulty, it turns out to be remarkably simple. With the DFT matrix, for example, we already know that the rows form a complete orthonormal basis based on our Fourier analysis.

It turns out that we can write the SVD for the DFT matrix by inspection:

$$U\Sigma V^{\mathrm{T}} = I \times I \times \frac{1}{N}\begin{bmatrix} 1 & 1 & \cdots & 1 \\ 1 & e^{-j\left(\frac{2\pi}{N}\cdot 1\right)\cdot 1} & \cdots & e^{-j\left(\frac{2\pi}{N}\cdot(N-1)\right)\cdot 1} \\ \vdots & \vdots & \vdots & \vdots \\ 1 & e^{-j\left(\frac{2\pi}{N}\cdot 1\right)\cdot(N-1)} & \cdots & e^{-j\left(\frac{2\pi}{N}\cdot(N-1)\right)(N-1)} \end{bmatrix}, \tag{9.51}$$

where I as usual is the identity matrix. To convince yourself that we have done the SVD properly, trace through the steps of SVD as outlined in Section 9.6.

10

Multiple Antennas in Wireless Links

Up to this point, all of our receiver and transmitter cartoons have been shown with each unit employing a single antenna. You can do things this way, and many times, for a variety of reasons, that is preferred. But especially in wireless links with mobile users, it has become common to use multiple antennas on each transmitter and receiver. As always with high-performance links, our primary concern is maximizing data throughput (bits per second) while minimizing resources (energy and spectrum). As you ponder the techniques described in this chapter, try to tie each technique to this prime directive.

When it comes to multiple antenna techniques, what can be painfully confusing is that all the tricks for exploiting multiple antennas get mashed together and presented at once. This causes no end of confusion to the novice trying to piece together a coherent picture...and has been known to confuse quite a few experts, as well.

The good news is that the basics of multiple antenna techniques do not require a long discussion. Rather, they benefit most from a careful, deliberate separation of techniques. In this chapter we focus on three of the most common. In no particular order:

1. Spatial diversity to combat signal fading in a wireless channel.
2. Phased array techniques (also known as beamforming).
3. Spatial multiplexing using multiple antennas at both the transmitter and the receiver. These are called multiple-input, multiple-output (MIMO) systems, and the singular value decomposition of Chapter 9 turns out to be critical to understanding them. We focus in this book on single-user MIMO.

The usual notes apply here about coverage of the topics not being comprehensive, but extremely representative. Here, though, an added caution is in order: the excitement over single-user MIMO techniques at the time of writing is wildly out

Energy Fundamentals of Radio, First Edition. Joel L. Dawson.

of proportion to the success of the technique in the field. So much so, in fact, that there was a compelling case for leaving it out of a book that aspires to the use of the word "fundamentals" in its title. For at least the last 15 years, however, it has been impossible to go anywhere in the wireless community without the topic of MIMO dominating the discussion. It is important for a reader at this time, then, to at least know what the fuss is about. Specifically with single-user MIMO, but also with MIMO techniques generally, use what you learn in this section and your own further inquiry to ask the right questions. This will help to keep you from being misled by ill-founded hype. It will also equip you to take advantage of a new insight, if applicable, and act with boldness accordingly.

10.1 Spatial Diversity to Combat Fading

It is a fact of life for the unruly wireless channel that there are often multiple paths for an electromagnetic wave to get from a transmitting antenna to a receiving antenna. Consider an indoor link, depicted in Figure 10.1. The figure shows two paths, but there certainly could be others, reflecting off of walls, floors, and other objects. Because the total propagation path lengths p_1 and p_2 differ, we have the addition of two sinusoids at the receive antenna that may not be in phase:

$$r(t) = A(t) \sin \omega (t - p_1/c) + \alpha A(t) \sin \omega (t - p_2/c) . \tag{10.1}$$

We are uninterested in the absolute phase of our carriers and only pay attention to phase differences between carriers arriving via different paths. Let $\delta t = \frac{p_2 - p_1}{c}$, and we are free to choose as expressions for our analysis

$$r(t) = A(t) \sin \omega t + \alpha A(t) \sin \omega (t - \delta t) . \tag{10.2}$$

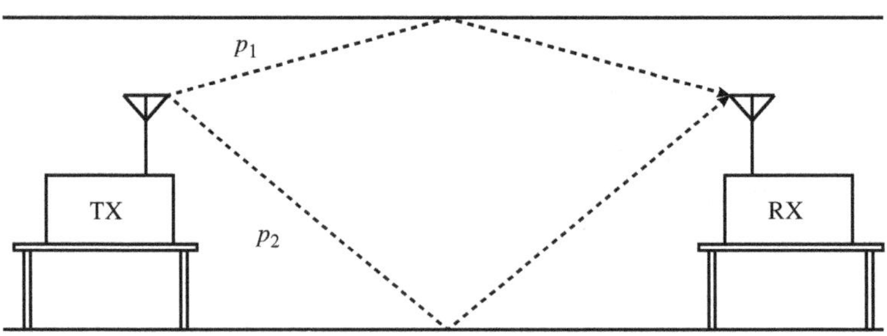

Figure 10.1 Transmission from transmitter to receiver along multiple paths.

We forge ahead with trigonometric identities:

$$\begin{aligned} r(t) &= A(t)\sin\omega t + \alpha A(t)[\sin\omega t\cos\omega\delta t - \cos\omega t\sin\omega\delta t] \\ &= A(t)[1+\alpha\cos\omega\delta t]\sin\omega t - \alpha A(t)\cos\omega t\sin\omega\delta t. \end{aligned} \tag{10.3}$$

This may be a little more complicated than you were expecting: you might be looking for a total and dramatic cancellation if the waves are completely out of phase. That is close to the main takeaway, and in fact if the path lengths differ by an odd number of half wavelengths, you do get maximal cancellation (which becomes total cancellation if $\alpha = 1$). Mathematically, maximal cancellation requires

$$|p_1 - p_2| = \left(k + \frac{1}{2}\right)\lambda. \tag{10.4}$$

for integer k. Now we have that $|\delta t| = \frac{|p_1 - p_2|}{c}$, and so it follows that $|\delta t| = \frac{\left(k+\frac{1}{2}\right)\lambda}{c}$. But $\lambda = \frac{c}{\nu} = \frac{2\pi c}{\omega}$, which allows us to write

$$|\delta t| = \frac{2\pi\left(k+\frac{1}{2}\right)}{\omega}. \tag{10.5}$$

Substituting in to Eq. 10.3 gives

$$r(t) = A(t)\left[1 + \alpha\cos 2\pi\left(k+\frac{1}{2}\right)\right]\sin\omega t - \alpha A(t)\left[\sin 2\pi\left(k+\frac{1}{2}\right)\right]\cos\omega t. \tag{10.6}$$

Getting here allows us to write clearly what happens if the path lengths differ by an odd number of half wavelengths. Recalling that $\sin 2\pi\left(k+\frac{1}{2}\right) = 0$ and that $\cos 2\pi\left(k+\frac{1}{2}\right) = -1$, we simplify at last to

$$r(t) = A(t)[1-\alpha]\sin\omega t. \tag{10.7}$$

This maximal cancellation circumstance is the condition under which we describe our link as "being in a deep fade." It is not at all uncommon for $|\alpha|$ to hover uncomfortably close to unity, so much so that network planners for mobile devices have learned to accept deep fades as an everyday reality.

That said, you can imagine the impossibility of trying to predict (and fully characterize) all possible paths between a transmitter and a receiver, particularly in a mobile use case. You would need perfect knowledge of all surfaces in your environment, not just their angle and position but also their reflection characteristics, and how they are situated relative to the radiation characteristics of your antennas. Faced with this overwhelming difficulty, practitioners do a very sensible thing. They put their analytical and empirical heads together, and settle on a *probabilistic* model of when a deep fade might occur.

Figure 10.2 When the receiver (or the transmitter) is moving, it is possible for the received signal to fluctuate from being very strong to being in a deep fade. Multiple antennas, spaced sufficiently far apart, give us some protection from fades.

It is important to let go of the need to know the details of the probabilistic model and focus on what we might do given that it is, for practical purposes, probabilistic. Suppose that we give ourselves a second antenna on our receiver, as shown in Figure 10.2. Let us suppose for the moment that the probability of a deep fade at a given antenna is P. Let us suppose further that the distance D is large enough that the fade probabilities for antenna 1 are completely independent of the probabilities for antenna 2. If you accept these assumptions, then the probability of a fade simultaneously at both antenna locations is[1]

$$P_{\text{simul}} = P^2. \tag{10.8}$$

Equivalently, the probability that we will *not* have a fade has improved to

$$P_{\text{no-fade}} = 1 - P^2. \tag{10.9}$$

And if we are willing to bear the weight and the complexity of M antennas, we can play this game for as long as we like:

$$P_{\text{no-fade}} = 1 - P^M. \tag{10.10}$$

In practice, the absolute simplest way to act on this idea is to monitor the received signal strength on each antenna at all times. And then at all times, you select only the antenna with the highest signal strength as your actual receive antenna. You can get fancier than this, but this simple implementation is enough to anchor your understanding when it comes to one particular way to make use of multiple receive antennas.[2] Experienced practitioners would say that we used

---

1 A. W. Drake, *Fundamentals of Applied Probability Theory.* New York, NY: McGraw-Hill, Inc., 1967.

2 Q. Chaudhari, *5G Physical Layer: An Easy Guide to Key Technologies.* Melbourne, Australia, 2021.

"multiple receive antennas to give us spatial diversity, which gives us some immunity from signal fades in the mobile environment."

This is one trick.

10.2 Beamforming Basics

If you have ever pondered the topic of multiple antennas on your own, chances are that a phased array is what you had in mind. This topic, and many others concerning antennas, is beautifully treated in Balanis' authoritative book on antenna theory.[3] We'll formalize this topic for now by setting up a linear array of equally spaced, isotropic radiators as shown in Figure 10.3. For our purposes, we are nominally interested in transmitting "broadside," or at a 90° angle to the array. In this text we will adopt the convention that broadside corresponds to $\phi = 0°$, as shown in the figure.[4]

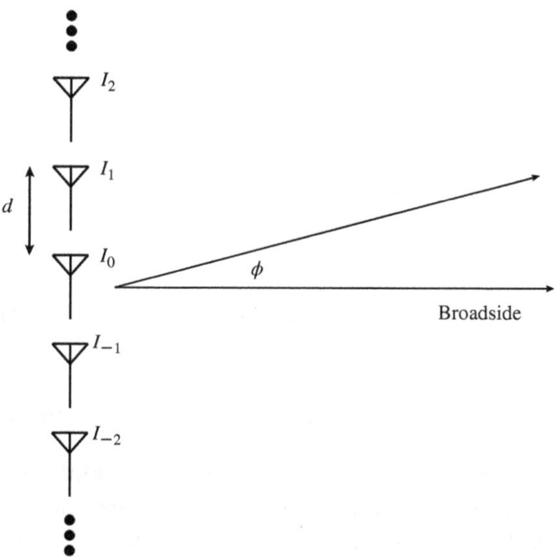

Figure 10.3 A linear antenna array.

3 C. A. Balanis, *Antenna Theory: Analysis and Design*. Hoboken, NJ: John Wiley & Sons, Inc., 2016.

4 As far as the author can tell, this is a slightly unusual choice, as it is more common to have broadside correspond to $\phi = 90°$. This is a purely personal preference. For the author it just seems like broadside, straight-ahead transmission ought to correspond to the angular variable being zero. What this means for the reader is that when comparing the treatment here to treatments elsewhere, you may need to adjust to seeing $\cos\phi$ where you expected $\sin\phi$ and vice versa.

Next, we are going to come up with an analytical stand-in for the electric field of an isotropic radiator. Recall that for a group of charges, we had from Eq. 8.22

$$E_\theta(r,t) = \frac{I_0 \Delta z}{4\pi\epsilon_0 c^2}\left(\frac{1}{r}\right)\sin\theta \cdot \omega\cos\omega\left(t-\frac{r}{c}\right). \tag{10.11}$$

It bears emphasizing that the "isotropic radiator" is an abstraction for which we have no real-world implementation. That is okay, as in our analyses it is often helpful and prudent to start with simple abstractions and add complexity only as we get closer to implementation. In this case, we'll take a step toward our basic radiating element by eliminating the sin θ and replacing it with unity:

$$E(r,t) = \frac{I_0 \Delta z}{4\pi\epsilon_0 c^2}\left(\frac{1}{r}\right)\cdot \omega\cos\omega\left(t-\frac{r}{c}\right). \tag{10.12}$$

From here we will go further and basically eliminate dependencies that are not critical in determining the shape of the radiation pattern or beam. In much of the literature on beamforming, you will see antenna elements represented as the end of the following progression

$$\begin{aligned} E &= I_0\frac{1}{r}\cos\omega\left(t-\frac{r}{c}\right) \\ &= I_0\frac{1}{r}\cos\left(\omega t - 2\pi f\frac{r}{c}\right) \\ &= I_0\frac{1}{r}\cos\left(\omega t - 2\pi\frac{r}{\lambda}\right) \\ &= I_0\frac{1}{r}\cos(\omega t - kr), \end{aligned} \tag{10.13}$$

which we then cap off (for reasons of the algebra to come) by noting we can write this as the *real part* of the complex expression

$$E = I_0\frac{e^{-jkr}}{r}e^{j\omega t}. \tag{10.14}$$

We've gotten a little loose with our symbols here: I_0 cannot, strictly speaking, be a current, because the units do not work out. But it is true that the radiated electric field is proportional to the current drive. It is best to send our inner units auditors on a temporary vacation while we seek insight through our derivations.

A couple of preliminaries remain. The overall problem is that we seek the variation in E as we vary the angle ϕ. When $\phi = 0$, we have the situation depicted in Figure 10.4. Now we make a useful approximation, based on the assumption that

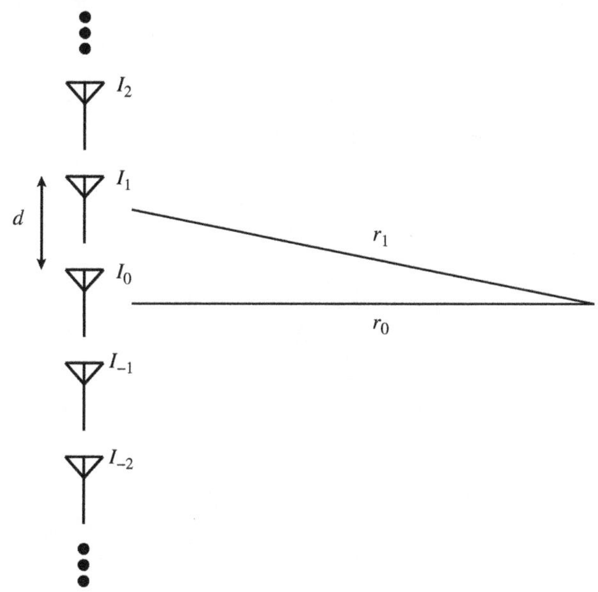

Figure 10.4 Geometry of the linear antenna array, part 1.

r_0 and r_1 are much longer distances than the inter-element distance d. Using the Pythagorean theorem, we have

$$\begin{aligned} r_1^2 &= r_0^2 + d^2 \\ r_1 &= \sqrt{r_0^2 + d^2} \\ &= r_0\sqrt{1 + \frac{d^2}{r_0^2}} \\ &\approx r_0\left(1 + \frac{1}{2}\frac{d^2}{r_0^2}\right). \end{aligned} \tag{10.15}$$

The takeaway here is actually that to first order in $\frac{d}{r_0}$, $r_1 \approx r_0$. We will go with this to the point that for points directly broadside the antenna array, we will regard all radiating elements *equally far away*. If you are new to approximating, this may offend you. But the analytical penalty for not making this approximation is severe and yet gets you nothing at the end of the day.

By way of contrast, consider what happens when we move off of broadside to $\phi \neq 0$ as depicted in Figure 10.5. Once we depart from broadside, we see that the difference in path lengths is a healthy, first-order-in-d, $d \sin \phi$. The geometry that we show in Figure 10.5 allows us to write a very useful, very close approximation

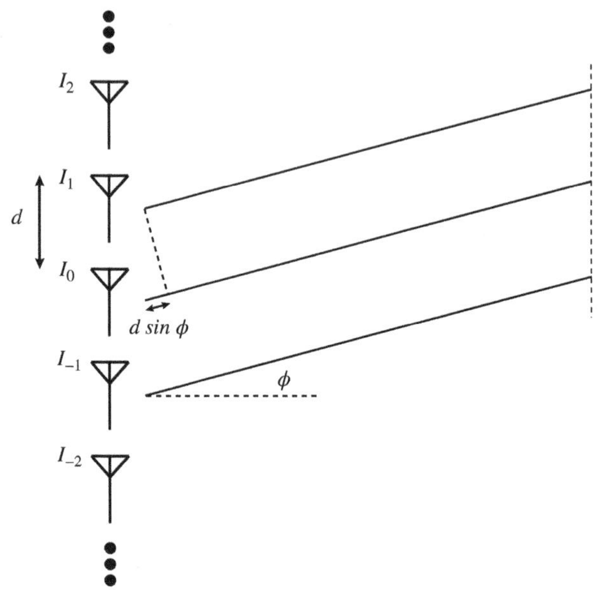

Figure 10.5 Geometry of the linear antenna array, part 2.

to the field strength E at a point (r, ϕ) in the so-called "far-field" region of the array. It is as follows:

$$
\begin{aligned}
E_{\text{tot}} = \Bigg[\ldots I_{-2} \frac{e^{-jk(r-2d\sin\phi)}}{r-2d\sin\phi} + I_{-1} \frac{e^{-jk(r-d\sin\phi)}}{r-d\sin\phi} + I_0 \frac{e^{-jkr}}{r} \\
+ I_1 \frac{e^{-jk(r+d\sin\phi)}}{r+d\sin\phi} + I_2 \frac{e^{-jk(r+2d\sin\phi)}}{r+2d\sin\phi} \cdots \Bigg] e^{j\omega t}.
\end{aligned}
\tag{10.16}
$$

Now a little bit of cleanup, partially in the form of approximations that streamline our equations with no meaningful loss in accuracy. First, we know that the time dependence is $e^{j\omega t}$ so we don't need to carry that around with us. We can just keep it in mind and add it back in at the end when we need it. Second, we continue to concern ourselves with the case where we are far from the array compared to distance d; from the standpoint of magnitude scaling each of these additive terms, there is little practical difference between $r - nd\sin\phi$ and r. These give us for the E field

$$
\begin{aligned}
E_{\text{tot}} = \frac{e^{-jkr}}{r} \big[\ldots I_{-2} e^{j(2kd\sin\phi)} + I_{-1} e^{j(kd\sin\phi)} + I_0 + I_1 e^{-j(kd\sin\phi)} \\
+ I_2 e^{-j(2kd\sin\phi)} \ldots \big].
\end{aligned}
\tag{10.17}
$$

The next helpful thing is to define a new variable.[5] This is almost entirely to emphasize that formally, Eq. 10.17 is identical to the discrete time Fourier transform (DFT), which in turn can be computed using the famous FFT algorithm.

So what is this new variable? We define

$$\psi = kd \sin \phi. \tag{10.18}$$

Then rewrite Eq. 10.17 as

$$E_{\text{tot}} = \frac{e^{-jkr}}{r}\left[\ldots I_{-2}e^{(j2\psi)} + I_{-1}e^{(j\psi)} + I_0 + I_1e^{(-j\psi)} + I_2e^{(-2j\psi)}\ldots\right]. \tag{10.19}$$

We are particularly interested in the quantity in brackets, which we call the *array factor* (AF) and write according to

$$\text{AF} = \sum_{n=-\frac{(N-1)}{2}}^{\frac{(N-1)}{2}} I_n e^{-jn\psi}. \tag{10.20}$$

The new variable ψ is *supposed* to clarify things, and ultimately it does, of course, but in the beginning sometimes things get confusing. So let's help ourselves out by stating loudly that the array factor AF is a continuous function of ψ, and make it explicit in our notation according to

$$\text{AF}(\psi) = \sum_{n=-\frac{(N-1)}{2}}^{\frac{(N-1)}{2}} I_n e^{-jn\psi}, \tag{10.21}$$

remembering always that $\psi = kd \sin \phi$.

Finally, let's make one other change that may seem aesthetic but will actually trip us up later if we don't fix it now. Notice as far back as Eq. 10.16 we set up our mathematics in a way that was especially friendly to arrays with an odd number of elements: you have one element in the geometrical center (corresponding to I_0), and then an equal number of elements on both sides. This choice is actually driven by aesthetics and a desire for balance in our geometrical setup, but is analytically awkward when we have an even number of elements (which is actually

5 Sometimes new students get confused and/or frustrated by authors who just casually define new variables. This is a good time to remind the reader that in the usual course of a complicated derivation, such definitions are almost never made when working things out the first time through. It is only later, when we have had time to digest and make sense of a messy derivation, that we go *back* and define variables from the beginning that make the answer at the end nice and neat.

quite common) and, it turns out, unnecessary. So we rewrite our AF in the form that we will use exclusively going forward:

$$\mathrm{AF}(\psi) = \sum_{n=0}^{N-1} I_n e^{-jn\psi}, \tag{10.22}$$

The whole idea of this expression is that if we want to know the amplitude of the E field for any ϕ, we just choose the ϕ that we are interested in, set $\psi = kd \sin \phi$, and then compute the sum in Eq. 10.22.

To get a feel for this, look at a couple of simple cases. Start with uniform excitation of the isotropic radiators. To within a multiplicative constant, the AF(ψ) is given by

$$\mathrm{AF}(\psi) = \sum_{n=0}^{N-1} e^{-jn\psi}. \tag{10.23}$$

This is nothing more than N consecutive terms of a geometric sequence with common ratio $e^{-j\psi}$, for which we have a handy closed-form expression:

$$\mathrm{AF}(\psi) = \frac{\left(1 - e^{-jN\psi}\right)}{1 - e^{-j\psi}}. \tag{10.24}$$

There are sin functions hidden in here for those who have dealt with enough of these:

$$\begin{aligned}\mathrm{AF}(\psi) &= \frac{e^{-j\frac{N}{2}\psi}\left(e^{j\frac{N}{2}\psi} - e^{-j\frac{N}{2}\psi}\right)}{e^{-j\frac{\psi}{2}}\left(e^{j\frac{\psi}{2}} - e^{-j\frac{\psi}{2}}\right)} \\ &= \frac{e^{-j\frac{(N-1)\psi}{2}}\left(2j \sin \frac{N}{2}\psi\right)}{\left(2j \sin \frac{\psi}{2}\right)} \\ &= e^{-j\frac{(N-1)\psi}{2}} \cdot \frac{\sin \frac{N}{2}\psi}{\sin \frac{\psi}{2}}.\end{aligned} \tag{10.25}$$

We are interested primarily in the magnitude of the array factor, and so we will drop the leading phase term according to

$$\mathrm{AF}(\psi) = \frac{\sin \frac{N}{2}\psi}{\sin \frac{\psi}{2}}. \tag{10.26}$$

This function has a maximum value at $\psi = 0$, where it equals N, and it is customary to normalize the array factor so that it has a maximum value of unity:

$$\mathrm{AF}(\psi) = \frac{1}{N}\frac{\sin\frac{N}{2}\psi}{\sin\frac{\psi}{2}}. \tag{10.27}$$

(Also, as a reminder,

$$\mathrm{AF}(\phi) = \frac{1}{N}\frac{\sin\frac{N}{2}(kd\sin\phi)}{\sin\frac{kd\sin\phi}{2}}. \tag{10.28}$$

You won't often see the array factor written this way, as a function of ϕ, but until you get used to what ψ really means, it may be helpful now and then to revert to this underlying form.) We'll call this maximum, which again occurs at $\psi = \phi = 0$, the "main lobe."

We will also have nulls, or values of ψ for which the array factor is zero. That is, whenever $\frac{N}{2}\psi = m\pi$, for m being a nonzero integer, we have a null. So nulls occur for

$$\psi = \frac{2\pi m}{N}. \tag{10.29}$$

Relating things to ϕ, we have

$$\begin{aligned} kd\sin\phi &= \frac{2\pi m}{N} \\ \sin\phi &= \frac{2\pi m}{kdN}. \end{aligned} \tag{10.30}$$

As you work this out for yourself, keep your head up and remember that we are concerned with launching EM energy in roughly the broadside direction, so $-\frac{\pi}{2} \leq \phi \leq \frac{\pi}{2}$.

We are big on equations in this text, but sometimes there is just no substitute for sitting down with a good numerical tool and pounding out some use cases. Take a look at Figures 10.6 and 10.7.

We'll close by noting that $d = \lambda/2$ is popular choice because of the deep nulls at $\phi = \pm\frac{\pi}{2}$ when the number of elements N is even.

10.2.1 Calculating the Beam Width

One of the important performance metrics for an antenna array is the beam width, commonly defined as the angle (ϕ) range over which the normalized array factor magnitude squared $|\mathrm{AF}|^2$ is 1/2 or greater. If you happen to know where your receiver is then there are great energy benefits to pointing your radiated energy right at the receiver rather than spreading it isotropically through space.[6] This is a

6 As we'll note in Chapter 12, you want to deploy antenna directivity thoughtfully, and not overdo it. A major benefit of wireless communications is not needing to know exactly where your receiver is!

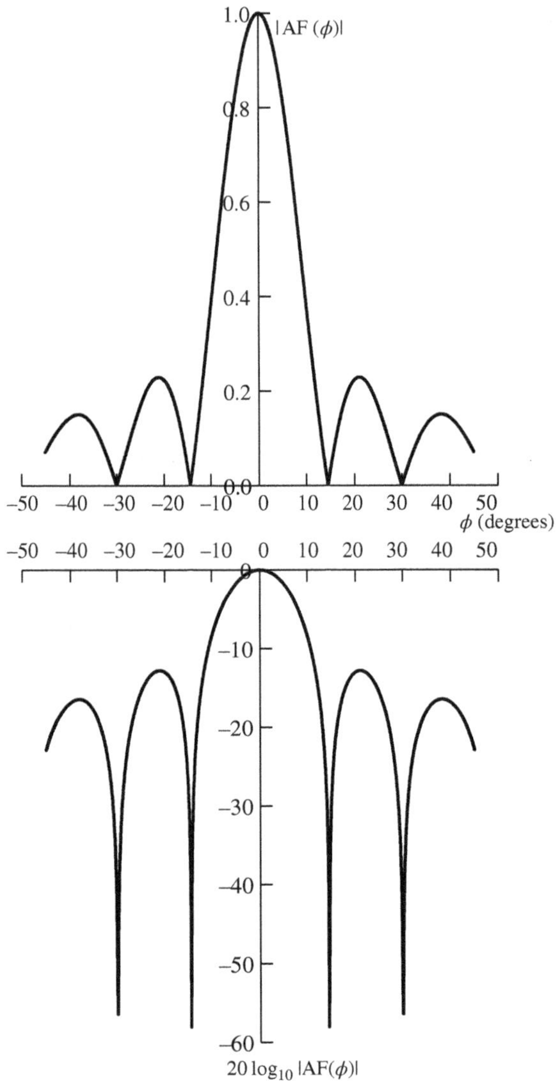

Figure 10.6 Area factor as a function of ϕ for $N = 4$, $d = \lambda/2$.

fixation among practitioners who deploy "fixed wireless access" links, where the transmitter and receiver are fixed, and where, importantly, we would normally deploy an optical fiber if the up-front cost were not prohibitive.[7]

7 Cost over the lifetime of the link is, of course, another matter.

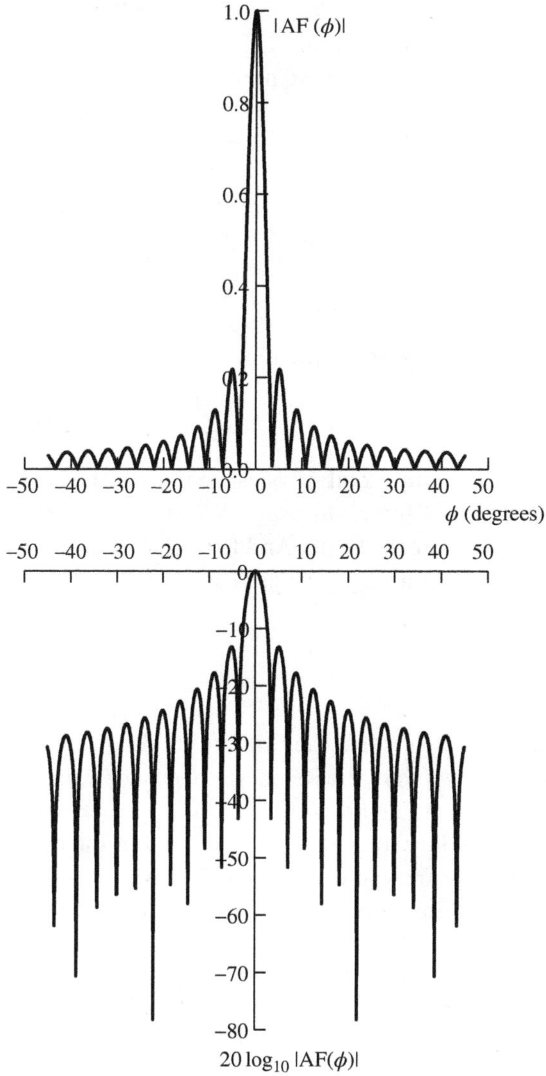

Figure 10.7 Area factor as a function of ϕ for $N = 32$, $d = \lambda/2$.

To get the half power beam width, we must calculate or measure the ψ for which

$$|\mathrm{AF}(\psi)|^2 = \frac{1}{N^2}\frac{\sin^2\left(\frac{N}{2}\psi\right)}{\sin^2\left(\frac{\psi}{2}\right)} = \frac{1}{2}. \tag{10.31}$$

Solving this equation requires numerical means or looking at graphs. We can see a little bit more with pencil and paper if we focus on the angular distance between nulls. Looking back at Eq. 10.30, we recall that we have nulls whenever

$$\sin\phi = \frac{2\pi m}{kdN}. \tag{10.32}$$

The nulls on either side of the main lobe correspond to $m = \pm 1$. The total angular width of this null, taking advantage of symmetry about $\phi = 0$, is given by

$$\phi_{\text{width}} = 2 \times \sin^{-1}\left(\frac{2\pi}{kdN}\right). \tag{10.33}$$

Remembering that $k = 2\pi/\lambda$, a useful form of this equation is

$$\phi_{\text{width}} = 2 \times \sin^{-1}\left(\frac{1}{d/\lambda \cdot N}\right). \tag{10.34}$$

What we see is that for a given λ, it is the total physical length of the array $((N-1)\cdot d)$ that determines the width of the main beam. It is an inverse relationship: the longer the array, the narrower the beam. And for a given number of elements N and desired angular beam width ϕ_{width}, we get a smaller overall form factor if we choose a smaller wavelength λ.[8]

10.2.2 Steering the Beam

One thing that we love about antenna arrays is that they allow us to get directivity. The main lobe lies in the direction that we want the RF energy to go, and we don't waste that energy by sending it to where we do *not* want it to go.

But what we love *even more* is the ability to change the direction of the beam without having to put our hands on the array and mechanically turn it. How do we do this? Well, recall the definition of the array factor (AF) before we decided to drive all the elements uniformly:

$$\text{AF}(\psi) = \frac{1}{N}\sum_{n=0}^{N-1} I_n e^{-jn\psi}. \tag{10.35}$$

It turns out that our control of the beam direction lies in the selection of the I_ns. Through the powerful aid of digital signal processing, we can make a choice other than to have the main lobe of the radiation beam point in the direction broadside to the antenna array (which corresponds to having all of the I_ns equal).

---

8 This basic trade-off motivates enthusiasm for mm-wave carrier frequencies ($f > 30$ GHz). By now, however, readers of this text know to look out for the disastrous "path loss" associated with higher carrier frequencies as illustrated in Eq. 8.50.

So how do we choose the I_ns? This is an interesting and practical special case of radiation pattern synthesis, whose general case we will examine in Section 10.2.3. For steering the beam, the right framework is to posit

$$I_n = Ie^{-jn\beta}, \tag{10.36}$$

where β is a constant. That is to say, we introduce a drive that is uniform across the array in magnitude, but we also introduce a phase factor that increments by β radians as we step sequentially from one element to the next. That is the basic idea, the setup. Now let's examine the details. Start with

$$\begin{aligned} \mathrm{AF}(\psi) &= \frac{1}{N}\sum_{n=0}^{N-1} I_n e^{-jn\psi} \\ &= \frac{1}{N}\sum_{n=0}^{N-1} Ie^{-jn\beta}e^{-jn\psi} \\ &= \frac{1}{N}\sum_{n=0}^{N-1} Ie^{-j(\beta+\psi)n}. \end{aligned} \tag{10.37}$$

If we follow our algebraic nose, we wind up with a new version of Eq. 10.26 as follows:

$$\mathrm{AF} = \frac{1}{N}\frac{\sin\frac{N}{2}(\psi+\beta)}{\sin\frac{1}{2}(\psi+\beta)}. \tag{10.38}$$

You can see that the center of the main lobe has been shifted from $\psi = kd \sin\phi = 0$ to the value of ϕ that satisfies

$$kd\sin\phi_{\text{center}} + \beta = 0. \tag{10.39}$$

Probably not what you were hoping for. On purely aesthetic grounds, you likely hoped for something like the simple substitution $\phi \to \phi + \beta$. It's okay. If you posit ahead of time that you want to steer the beam off of broadside by an amount in radians small compared to unity, you can approximate $\sin\phi \approx \phi$ and so now the center of the beam as a function of β is given approximately by

$$\phi_{\text{center}} \approx -\frac{\beta}{kd}. \tag{10.40}$$

Anyway, the important things to remember are that (1) there are multiple methods for imposing the inter-element phase shift β, including electronic means (digital signal processing) and mechanical means; (2) with this capability in hand, the main lobe of the radiation pattern is easily steered without having to physically turn the antenna array.

Figure 10.8 shows a radiation pattern with $\beta = \pi/6$.

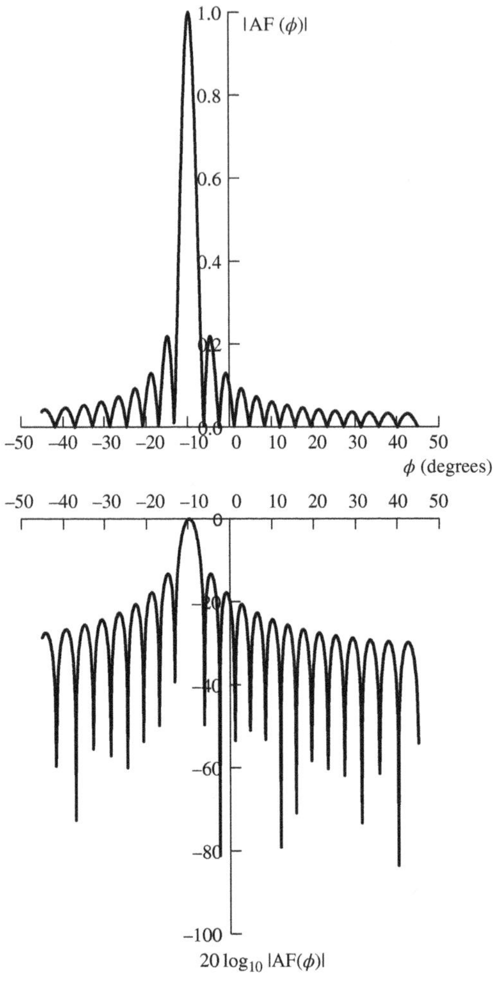

Figure 10.8 Area factor as a function of ϕ for $N = 32$, $d = \lambda/2$, $\beta = +\pi/6$.

10.2.3 Synthesizing a Radiation Pattern

Steering the beam, as we mentioned in Section 10.2.2, is the simplest example of synthesizing a radiation pattern. In this section we will get more general.

Approaching the general radiation pattern synthesis problem, we might start by supposing that we have a "large" number of antenna elements.[9] We then formulate the problem by specifying an AF(ψ) that we want to get as close as possible to.

9 "Large" in practice, tends to be 32 or 64, with dreams of expanding to 128. Notice that we tend to stick to powers of 2 when it comes to the number of elements. This allows us later to repurpose our FFT engines for solving the radiation pattern synthesis problem.

A reasonable way to think about this problem is to see that in the I_ns, we have a total of N variables, or degrees of freedom. If we are used to dealing with systems of equations, we might suppose that if we have N equalities to satisfy, and make all manner of friendly assumptions, we can come up with a unique solution for the I_ns. For the N equalities, we specify N values of ψ for which we must nail the value of AF(ψ) exactly according to our specification.

We start again with our basic definition of the array factor

$$\mathrm{AF}(\psi) = \frac{1}{N}\sum_{n=0}^{N-1} I_n e^{-jn\psi}. \tag{10.41}$$

For our specification, let us take N evenly spaced samples of ψ according to

$$\psi = 2\pi\frac{l}{N}. \tag{10.42}$$

with $l \in \{0, \ldots, N-1\}$. Then we write N equations, one for each value of l:

$$\mathrm{AF}\left(2\pi\frac{l}{N}\right) = \frac{1}{N}\sum_{n=0}^{N-1} I_n e^{-jn\left(2\pi\frac{l}{N}\right)}. \tag{10.43}$$

Depending on your signal processing background, this last equation may look familiar. It is identical to the mathematics of the DFT, which you can see if you compare it with Eq. 2.104. And if the mathematics of a problem are identical to that of the DFT, then it can be computed using the algorithms of the FFT. Huge excitement ensues.[10]

The way that we set this problem up is that we decide ahead of time what values we want for AF at each value of l. We then have N equations in N unknowns, which we write in matrix form according to

$$\begin{bmatrix} 1 & 1 & \cdots & 1 \\ 1 & e^{j\left(\frac{2\pi}{N}\cdot 1\right)\cdot 1} & \cdots & e^{j\left(\frac{2\pi}{N}\cdot(N-1)\right)\cdot 1} \\ \vdots & \vdots & \vdots & \vdots \\ 1 & e^{j\left(\frac{2\pi}{N}\cdot 1\right)\cdot(N-1)} & \cdots & e^{j\left(\frac{2\pi}{N}\cdot(N-1)\right)(N-1)} \end{bmatrix} \begin{bmatrix} I_0 \\ I_1 \\ \vdots \\ I_{N-1} \end{bmatrix} = \begin{bmatrix} \mathrm{AF}\left(2\pi\frac{0}{N}\right) \\ \mathrm{AF}\left(2\pi\frac{1}{N}\right) \\ \vdots \\ \mathrm{AF}\left(2\pi\frac{N-1}{N}\right) \end{bmatrix}. \tag{10.44}$$

The final step is to invert the "DFT" matrix, apply it to the vector of AF(·)s, and out pop your desired excitations I_ns, which are just amplitude scalings and phase shifts of the modulated carriers that are driving the antenna elements.

10 Remember that for the FFT algorithm, it is important to have frequency and time domain samples in groups of powers of 2. If you have ever wondered why multiple antenna arrays at the time of writing are either 2, 4, 8, 16, 32, 64, 128, or 256...this is your answer!

Now you have at least an outline of how to think about using multiple antennas to engineer custom radiation patterns. As a special bonus, the weightings I_i that-give a desired radiation pattern can also be used to create an equivalent directional receive pattern in an array of receive antennas.

Using separate antennas to provide immunity to fading was the first trick. Controlling the excitation of individual antenna elements to provide generalized directionality is the second trick. One trick to go.

10.2.4 On Getting Overly Excited About Beamforming

Beamforming is an extremely useful and practical technique. Sometimes, however, exuberant practitioners get a cartoon in their heads of a huge number of finely focused beams laser focused on a large number of individual users each with their own trajectory through space...

The mathematics of the DFT can help guide your intuition and give you an idea of what is possible. Recall that from Section 2.2.3, for a given sample period T, we were roughly restricted to representing sinusoids whose period was $2T$ or higher. In the case of radiation pattern synthesis, the I_ns play the role of time domain samples, and the values AF $\left(2\pi\frac{l}{N}\right)$ play the role of frequency domain samples. If we say that we have 64 antenna elements, and we allow that we are shaping the radiation pattern over a 180° span, we may (again, roughly speaking) say that we have an "angular domain" sample period of 2.8°, which means that *at the very best* we could have control over feature sizes in our radiation pattern no smaller than 5.6°. Throw in the fact that for ordinary signal processing we insist on high degrees of oversampling (2×, 4×, and higher), and you can get some idea now of what to expect in terms of the specificity of shape *of a single beam*. Add, further, a desire to have multiple beams, and suddenly 64 degrees of freedom doesn't seem like so many after all.

The other thing to pay attention to is the magnitude of the side lobes. In the examples of Section 10.2, notice that the highest side lobes are only 10–15 dB down from the main lobe. From the standpoint of creating unintended interferers, that is not *that* far down. This is only to temper any image you may have of beamforming as creating hyper focused beams that hit the center of their main lobe and vanish elsewhere. Reality in the field is messier than that.

10.3 Spatial Multiplexing for Single User MIMO

Last on this non-exhaustive list of fun things to do with multiple antennas is a technique called spatial multiplexing. To get a feel for what we are trying to accomplish, consider as a baseline the channel model shown in Figure 10.9. We know from Chapter 5 that the ultimate data capacity of this channel is given by

$$C = B \cdot \log_2 (\text{SNR} + 1)\,. \tag{10.45}$$

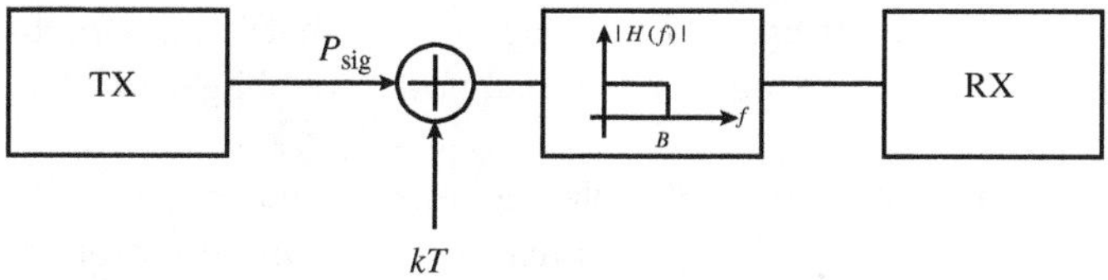

Figure 10.9 A canonical noisy channel.

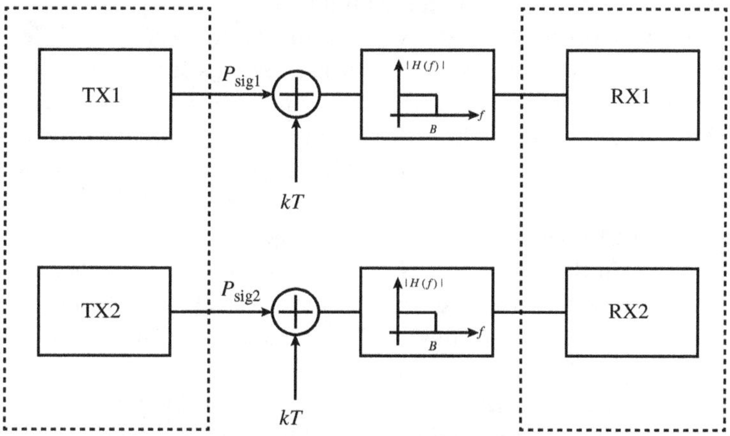

Figure 10.10 Parallel, non-interfering noisy channels.

Let's fill in some numbers, now. Let $B = 20$ MHz, and let the SNR be $\frac{P_{sig}}{kTB} = 23$ dB (which is 200 in linear units). If $T = 300$ K and the Boltzmann constant is its usual 1.38×10^{-23} J/K, then we have that the signal power is 16.6 pW. Regardless, with an SNR of 200, we have

$$\begin{aligned} C &= 20\text{ MHz} \cdot \log_2(200) \\ &= 20 \times 10^6\text{ s}^{-1} \cdot 7.64 \\ &= 153\text{ MBPS}. \end{aligned} \tag{10.46}$$

Now imagine an alternate setup in which we have two parallel communications channels that do not interfere with each other, and where we nevertheless keep the total signal power constant such that $P_{sig1} = P_{sig2} = \frac{1}{2}P_{sig}$. See Figure 10.10. In this case, the capacity of this double-barreled channel is just twice the capacity of each one in isolation, so we have

$$C = 2 \times 20\text{ MHz} \cdot \log_2(100) = 266\text{ MBPS}. \tag{10.47}$$

Keep careful track, now, of what we have just accomplished. Using the same band of spectrum (again without interfering...we will come back to that), and without transmitting any more total power, we have achieved almost a doubling of the data throughput.

This, here, is the excitement of spatial multiplexing. Energy and spectrum are both precious and finite resources, and yet the hunger for data capacity throughout in the world grows exponentially. The idea that we could somehow grow data capacity by cleverly exploiting space instead of the usual energy and spectrum...well, you can see the reason behind the tumult.

Before we get into the most complicated form of spatial multiplexing, the understanding of which requires the mighty singular value decomposition, we should quickly review *other* tricks for exploiting space to multiply capacity without consuming spectral or energy resources. The cleanest, most obvious thing we can do is make use of orthogonal polarizations of our EM waves, which we did not explicitly discuss in Chapter 7 but which is easily understood. See Figure 10.11. Recall from Chapter 8 that for a Hertzian dipole oriented in, say, the $\hat{x}$ direction, the radiated E field also points in the $\hat{x}$ direction. A receiving antenna on the other end will, it turns out, be insensitive to E fields that are orthogonal to its own orientation. In this way, we can use so-called "dual-pol"[11] antennas to achieve the dream scenario depicted in Figure 10.10.

Figure 10.12 shows another way we can do it. Fresh off our exploration of beamforming in Section 10.2, and being careful to align our transmit beams and receive antennas, we can also take advantage of space to circumvent Nature's limits on energy and spectrum.

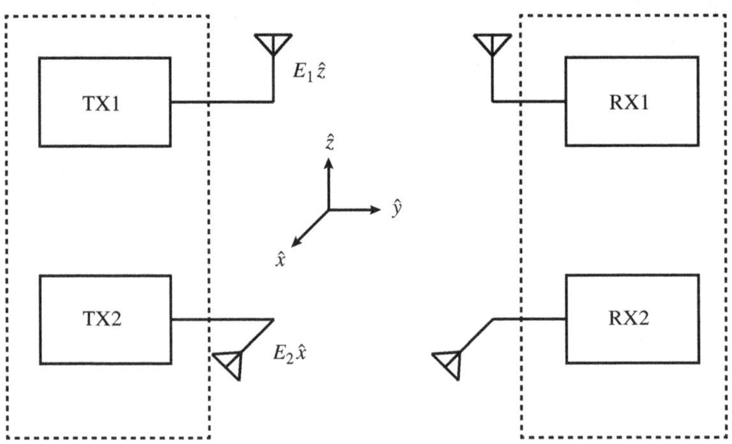

Figure 10.11 Spatial multiplexing using orthogonal polarizations.

11 Dual polarization.

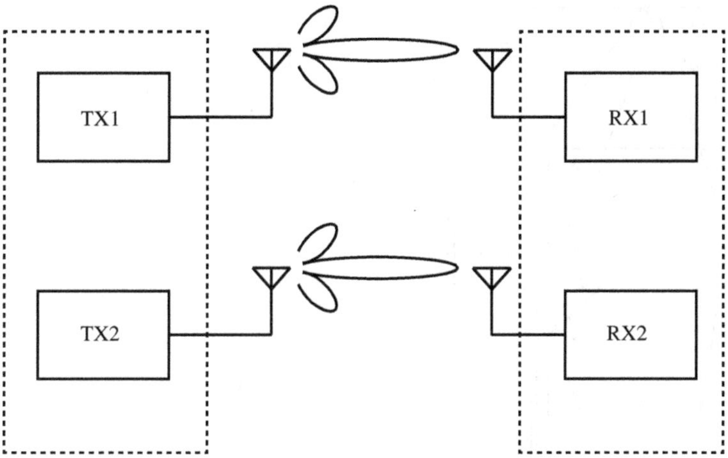

Figure 10.12 Spatial multiplexing using beamforming.

These two techniques, dual polarization and tight beamforming, are each solid, reliable, always-there techniques.[12] But we move on now to a form of spatial multiplexing that is much more obscure, in that it can only be revealed through a sophisticated application of linear algebra. Further, it relies on the propagation environment being extremely complex, rich in reflections whose analysis yields hidden parallel paths. See what you think.

We start with many transmit and many receive antennas, adjacent to highly reflective surfaces as shown in Figure 10.13. When we speak of a complex propagation environment, what we mean is that there are many, many ways for an EM wave that starts out at TX_i to wind up at RX_j. You might imagine this to be plausible the case in an indoor environment filled with walls, furniture, people walking around, etc.[13] In such an environment, the intelligence running our network would perform a "channel estimation" to determine the matrix showing how signals transmitted from antenna s_i to are received by antenna r_j. Writing it out explicitly:

$$\begin{bmatrix} a_{11} & a_{12} & \cdots & a_{1N} \\ a_{21} & a_{22} & \cdots & a_{2N} \\ \vdots & \vdots & \vdots & \vdots \\ a_{M1} & a_{M2} & \cdots & a_{MN} \end{bmatrix} \begin{bmatrix} s_1 \\ s_2 \\ \vdots \\ s_N \end{bmatrix} = \begin{bmatrix} r_1 \\ r_2 \\ \vdots \\ r_M \end{bmatrix}, \tag{10.48}$$

12 Okay, beamforming is less so if you are not operating purely line-of-sight with no reflections.
13 This is most definitely *not* the case in high-power, macrocell deployments.

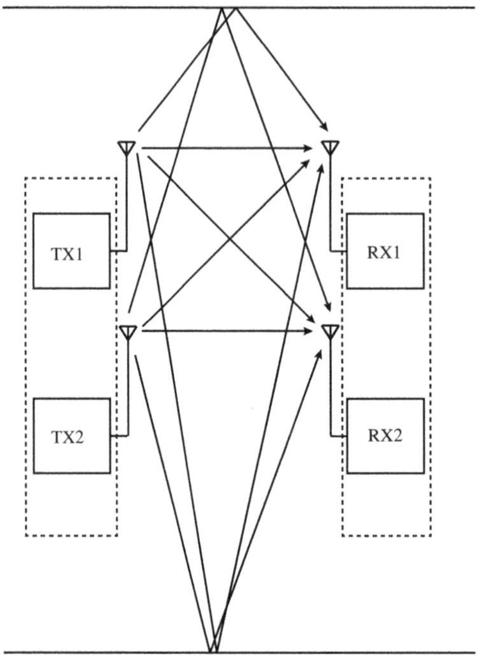

Figure 10.13 A multipath environment as the setup for MIMO.

which we then write as in the more compact

$$A\vec{s} = \vec{r}. \tag{10.49}$$

(Note that with this setup, we have officially entered the realm of MIMO systems.) The game from here is to determine, for a given set of coefficients a_{ij} in the matrix A, if there are *hidden, parallel, orthogonal channels that allow us our overall capacity along the lines of Eq. 10.47*. And the great news is that we can answer that question almost immediately by doing a singular value decomposition of A, reviewed in detail in Chapter 9. We reproduce the singular value decomposition of matrix A here:

$$A = U\Sigma V^{\mathrm{T}} = \begin{bmatrix} \uparrow & \uparrow & & & \uparrow \\ \\ \vec{u}_1 & \vec{u}_2 & \cdots & \cdots & \vec{u}_{\mathrm{R}} \\ \\ \downarrow & \downarrow & & & \downarrow \end{bmatrix} \begin{bmatrix} \sigma_1 & 0 & 0 & 0 & \cdots \\ 0 & \sigma_2 & 0 & \cdots & \cdots \\ 0 & 0 & \sigma_3 & \cdots & \cdots \\ 0 & \vdots & \vdots & \ddots & \vdots \\ 0 & \cdots & \cdots & \cdots & \sigma_{\mathrm{R}} \end{bmatrix} \begin{bmatrix} \leftarrow & \vec{v}_1 & \rightarrow \\ \leftarrow & \vec{v}_2 & \rightarrow \\ \leftarrow & \vdots & \rightarrow \\ \leftarrow & \vec{v}_{\mathrm{R}} & \rightarrow \end{bmatrix}. \tag{10.50}$$

The thrilling conclusion we drive to is that if we have more than one singular value σ_i, then we have more than one orthogonal channel! We have only to devote ourselves to seeing how we can make use of this insight.

Start by examining Figure 10.14, where we look at the possibility of exploiting a matrix A of rank 2. For now we think of the v_{ij}s as static, multiplicative weights, and the $m_i(t)$s as the time domain messages that we want to send. Mathematically, we render the sum of signals coming out of the antennas in Figure 10.14 as

$$\begin{bmatrix} v_{11} \\ v_{12} \\ \vdots \\ v_{1N} \end{bmatrix} m_1(t) + \begin{bmatrix} v_{21} \\ v_{22} \\ \vdots \\ v_{2N} \end{bmatrix} m_2(t). \tag{10.51}$$

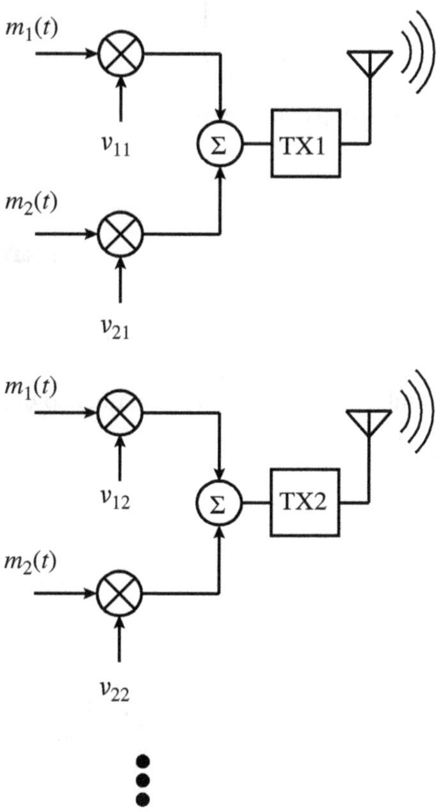

Figure 10.14 A MIMO transmitter.

Our signals leave the transmit antennas, cross our communications channel, and wind up at our receive antennas. Mathematically, we receive

$$\begin{bmatrix} r_1 \\ r_2 \\ \vdots \\ r_M \end{bmatrix} = A \cdot \left(\begin{bmatrix} v_{11} \\ v_{12} \\ \vdots \\ v_{1N} \end{bmatrix} m_1(t) + \begin{bmatrix} v_{21} \\ v_{22} \\ \vdots \\ v_{2N} \end{bmatrix} m_2(t) \right), \tag{10.52}$$

which we rewrite using the SVD. You may start to get the idea:

$$\begin{bmatrix} r_1 \\ r_2 \\ \vdots \\ r_M \end{bmatrix} = U\Sigma \begin{bmatrix} \leftarrow & \vec{v}_1 & \rightarrow \\ \leftarrow & \vec{v}_2 & \rightarrow \\ \leftarrow & \vdots & \rightarrow \\ \leftarrow & \vec{v}_R & \rightarrow \end{bmatrix} \cdot \left(\begin{bmatrix} v_{11} \\ v_{12} \\ \vdots \\ v_{1N} \end{bmatrix} m_1(t) + \begin{bmatrix} v_{21} \\ v_{22} \\ \vdots \\ v_{2N} \end{bmatrix} m_2(t) \right). \tag{10.53}$$

Recalling that the $\vec{v}_i$s are orthonormal, our next step comes easily:

$$\begin{bmatrix} r_1 \\ r_2 \\ \vdots \\ r_M \end{bmatrix} = U\Sigma \begin{bmatrix} m_1(t) \\ m_2(t) \\ 0 \\ \vdots \end{bmatrix}. \tag{10.54}$$

We said that A had a rank of 2, so only σ_1 and σ_2 are nonzero. We can thus do the next matrix multiplication as follows

$$\begin{bmatrix} r_1 \\ r_2 \\ \vdots \\ r_M \end{bmatrix} = U \begin{bmatrix} \sigma_1 m_1(t) \\ \sigma_2 m_2(t) \\ 0 \\ \vdots \end{bmatrix}. \tag{10.55}$$

Let's fill in, now, the matrix U as the collection of orthonormal column vectors that we expect it to be:

$$\begin{bmatrix} r_1 \\ r_2 \\ \vdots \\ r_M \end{bmatrix} = \begin{bmatrix} \uparrow & \uparrow & & & \uparrow \\ \vec{u}_1 & \vec{u}_2 & \cdots & \cdots & \vec{u}_R \\ \downarrow & \downarrow & & & \downarrow \end{bmatrix} \begin{bmatrix} \sigma_1 m_1(t) \\ \sigma_2 m_2(t) \\ 0 \\ \vdots \end{bmatrix}. \tag{10.56}$$

Recall that our rules for matrix multiplication allow us to write

$$\begin{bmatrix} r_1 \\ r_2 \\ \vdots \\ r_M \end{bmatrix} = \begin{bmatrix} \uparrow \\ \vec{u}_1 \\ \downarrow \end{bmatrix} \sigma_1 m_1(t) + \begin{bmatrix} \uparrow \\ \vec{u}_2 \\ \downarrow \end{bmatrix} \sigma_2 m_2(t). \tag{10.57}$$

The only thing remaining is to bridge the gap between the abstract mathematics and how we think about implementing this strategy in an electronic machine. We have arrived at the receiver. Recall that the u_{ij}s came from the SVD of the matrix A, which itself came from a channel estimator that told us the complex gains connecting each transmit antenna to each receiving antenna. We put these u_{ij}s to work here according to Figure 10.15. Figure 10.15 is nothing less than the mechanization of the mathematical expression

$$\begin{bmatrix} r'_1 \\ r'_2 \\ \vdots \\ r'_M \end{bmatrix} = \begin{bmatrix} \leftarrow & \vec{u}_1 & \rightarrow \\ \leftarrow & \vec{u}_2 & \rightarrow \\ \leftarrow & \vdots & \rightarrow \\ \leftarrow & \vec{u}_M & \rightarrow \end{bmatrix} \cdot \left(\begin{bmatrix} \uparrow \\ \vec{u}_1 \\ \downarrow \end{bmatrix} \sigma_1 m_1(t) + \begin{bmatrix} \uparrow \\ \vec{u}_2 \\ \downarrow \end{bmatrix} \sigma_2 m_2(t) \right). \tag{10.58}$$

One final step:

$$\begin{bmatrix} r'_1 \\ r'_2 \\ \vdots \\ r'_M \end{bmatrix} = \begin{bmatrix} \sigma_1 m_1(t) \\ \sigma_2 m_2(t) \\ 0 \\ \vdots \end{bmatrix}. \tag{10.59}$$

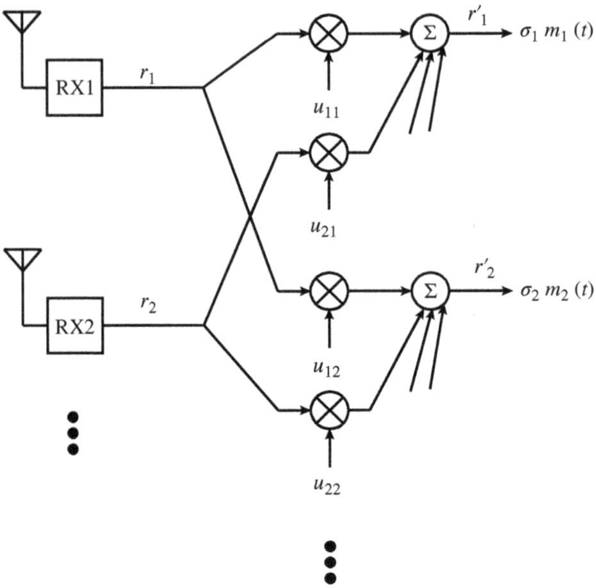

Figure 10.15 A MIMO receiver.

That is to say, our two messages reached our destination without interfering with each other.

Magic.

It bears emphasizing, one last time, that spatial multiplexing is largely opportunistic. That is to say, we do not engineer our propagation environment to ensure that multiple orthogonal channels exist. Rather, we install our communications equipment in a complicated environment, switch it all on, and see what happens. If we are fortunate, we wind up with multiple σ_is *of comparable magnitude,* and we can live out the dream of spatial multiplexing as we originally conceived. Note that in the field, in live deployments, having more than four nonzero singular values is exceedingly rare. Don't be fooled by lab bench channel emulators that tell you otherwise.

This spatial multiplexing is the third trick for multiple antennas, following spatial diversity to combat fading and beamforming. In practice, always keep your eye on the achieved MIMO rank that your channel estimator gives you.

11

Transmitter Structures

In a radio link, it is the transmitter (labeled TX in many diagrams up to this point) that dominates the power consumption. It is here that our bits of information are imposed as modulation on a sinusoidal carrier. This modulated carrier is then used to accelerate charges in the antenna, which radiates Poynting flux out into the world. Radio transmitters are rare among electronic systems in that they process both information and energy at the same time.[1]

A start to understanding the energy fundamentals of radio is found in the Shannon limit, which tells us that information capacity goes up logarithmically with transmit power. Our first refinement to that rudimentary understanding comes from applying the Friis equation, which brings in the impact of the carrier frequency, antenna gain, and distance on the energy consumption (see Section 8.5.) The reader should understand *this* chapter as supplying a new and messier piece to the puzzle, which is the energy impact that we feel because of circuit implementation choices. This aspect of the problem is supremely uninteresting to the theorists, in love as they are with Nature's limits. How, they ask, do we know that our implementations are the best that Nature allows? The answer is that we don't always know. As a practical matter, though, we cannot sit around and be paralyzed by this uncertainty. When pure theory does not give us the answers, we forge ahead with what we know from experience. In this chapter, we plunge into the problems of amplification and filtering that so dominate

1 Switching power converters are examples of electronic systems that process energy but no information. Microprocessors are electronic systems that process information but not energy if we can help it, the CV^2f power dissipation being a byproduct of parasitic and very undesirable node capacitances.

Energy Fundamentals of Radio, First Edition. Joel L. Dawson.

the energy considerations of our TX chains, taking a detour into frequency synthesis for the sake of overall literacy in the radio design problem. We wrap up with a primer on heat sinks, where the reader can start to form an understanding of the real-world implications of energy waste.

11.1 The Miller Effect in Amplifiers

You will sometimes hear it said that amplifying and processing signals at high frequencies and/or broad bandwidths is "hard" or "difficult." This is common but imprecise language. In engineering, if we have a big enough energy budget, we can make almost anything happen. Calling something difficult almost always means that we have a hard power constraint, and that our performance limits are in the vicinity of Nature's fundamental power requirements to accomplish a given Task.[2] *The smaller the gap between Nature's limit and our permitted power budget, the more clever, creative, and careful we have to be in our designs.*

Fine. What makes circuits "harder" to design for high frequencies? We saw part of the answer in Chapter 3, specifically in Section 3.1. We saw that nature resists our attempts to move voltages on capacitors quickly. That B^3 dependence is bad enough, of course. But in signal chains where we must amplify signals, the problem actually gets worse.

To begin to understand why, we consider the setup shown in Figure 11.1. We ask: what is the transfer function connecting v_{in} to v_1 and then to v_2? We have

$$\frac{v_{in} - v_1}{R} = \frac{v_1 - v_2}{1/j\omega C} = \frac{v_1(1+A)}{1/j\omega C}. \tag{11.1}$$

The insight that we seek will come from solving for v_1:

$$\begin{aligned} \frac{v_{in}}{R} &= v_1 \left(\frac{1}{R} + (1+A)j\omega C\right) \\ &= v_1 \left(\frac{1 + (1+A)j\omega RC}{R}\right) \\ v_1 &= \left(\frac{1}{1 + (1+A)j\omega RC}\right) v_{in}. \end{aligned} \tag{11.2}$$

We see that the negative gain $-A$ has given us an equivalent capacitor load for our source that is very large. For analytical comparison, we look at Eq. 11.2 alongside Eq. 3.4. We could in fact redraw Figure 11.1 as Figure 11.2. This phenomenon is called the Miller effect, and we say that the capacitance C is "Miller multiplied" by factor $(1 + A)$ from the standpoint of the input. Our study from Chapter 3 makes clear that from an energy perspective, this development is

2 See the discussion of the key Task in electronics at the beginning of Chapter 3.

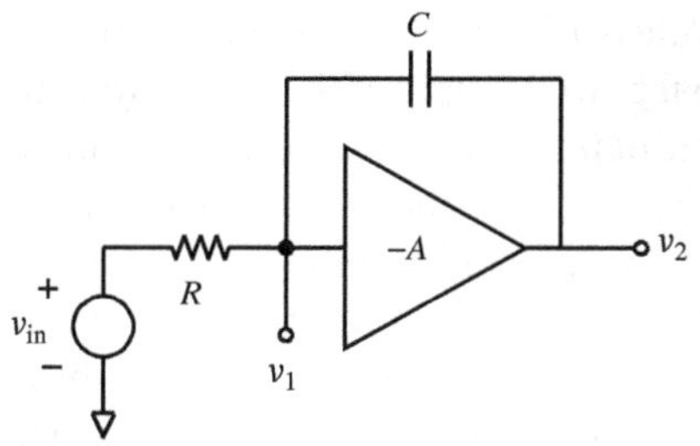

Figure 11.1 A common amplifier structure that results in Miller multiplication.

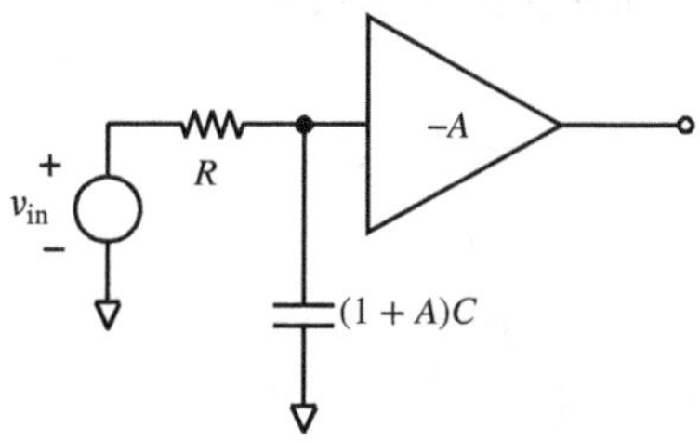

Figure 11.2 A common and useful redrawing of Figure 11.1. The focus is on the added difficulty of driving this stage due to the multiplied input capacitance.

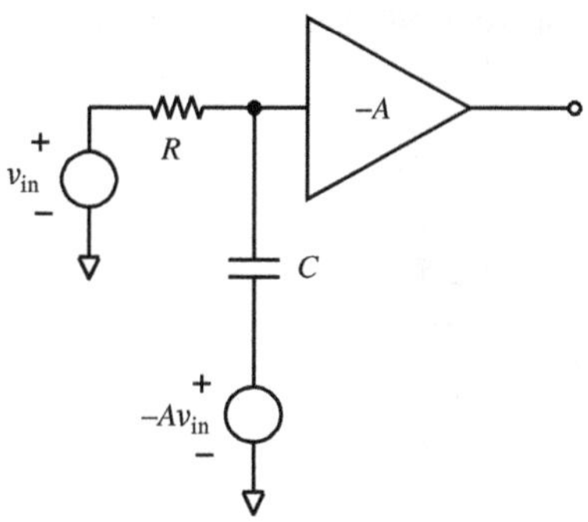

Figure 11.3 A less common but still useful redrawing of Figure 11.1. Here we see what energy sources are driving the charging and discharging of *C*.

unfortunate. While Figure 11.2 captures well the current required from voltage source v_{in}, what is less appreciated is the energy cost of Miller multiplication. To this end, we redraw Figure 11.1 once again as Figure 11.3. This rendering allows us to consider the energy demands of the totality of all voltage sources. We see that the bridging capacitor experiences a total voltage swing that exceeds the voltage swing at the output node. The source voltage does some of the required work, while the amplifier itself does the rest.[3] Regardless, *all* the energy required to move the bridging capacitor traces back to the global power supply of the whole system.

This phenomenon of Miller multiplication happens a lot. Consider the most basic amplifying structure in analog electronics, the common emitter (or common source) amplifier, as shown in Figure 11.4 (biasing details omitted). There is an entire world of analog design centered on the problem of amplification, as this function is central to most signal chains. Here we maintain a tight focus on these basic amplifying structures, *and* on the fact that every known amplifying device has a parasitic capacitance bridging the input and output terminals as shown in Figure 11.5. It bears emphasis that detailed analysis is not quite as simple as that implied by Figures 11.2 and 11.3, even if they get across the essential spirit.[4]

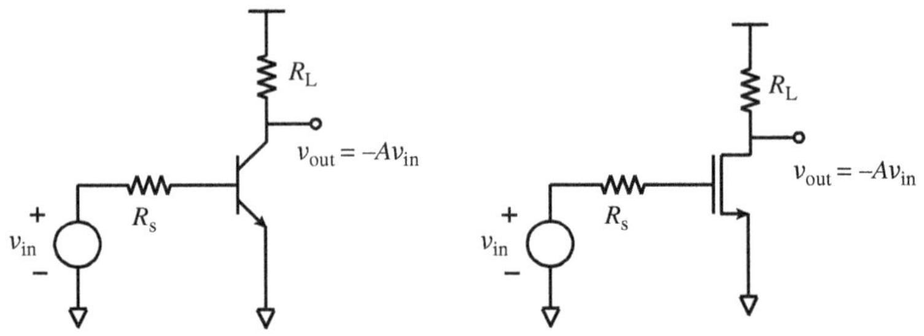

Figure 11.4 Basic amplifying structures. A bipolar transistor (NPN) is on the left, and a MOSFET (NMOS) is on the right.

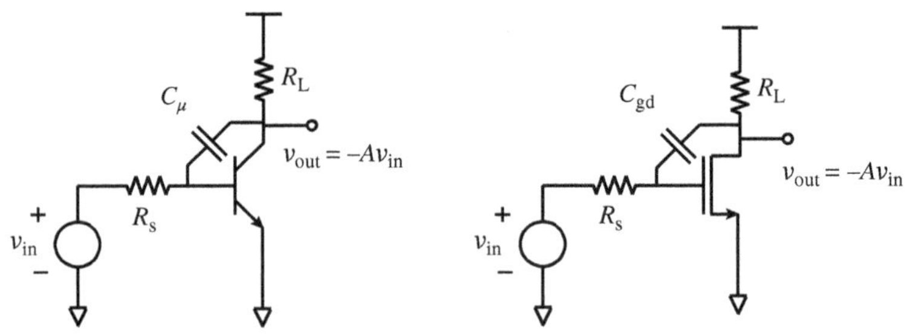

Figure 11.5 Basic amplifying structures, this time with the bridging capacitances made explicit.

3 And don't sleep on the fact that most of these amplifiers are going to be extremely low efficiency, Class A type stages. That is to say, for every joule they pour into the charging the capacitor, they take many more from the supply. We will have much more to say about Class A operation in Section 11.4.1.

4 We have used an ideal gain of A with zero output impedance, for starters. Also, the input of the amplifier is assumed to have infinite input impedance.

When it comes to our original question of why amplifying at high frequencies is difficult, we now have our answer. Parasitic capacitance is everywhere in electronic circuits, including bridging the input and output nodes of our amplifying elements. With most active devices, the gain between these nodes is *inverting*, or negative. The result is that, from a dynamical perspective, these bridging capacitances are multiplied by a factor of $\approx A$.

Tremendous effort has been expended over the years to mitigate this problem. A circuit technique known as the cascode topology is a favorite, for example.[5] And when it comes to transistor fabrication, every effort is made to minimize the size of the bridging capacitance between input and output nodes (drain-to-gate in field effect transistors, collector-to-base in bipolar transistors). This is effort well spent. The fact remains, however, that once you have minimized all your bridging capacitances, Miller multiplication makes them bigger in proportion to how much you need to amplify your signal. Also, because of where they are placed in the circuit, it is not usually realistic to "resonate out" these pesky capacitances with cleverly placed inductors. We therefore suffer energy penalties that scale as the square of the carrier frequency (f^2) and the cube of the bandwidth (B^3), in accordance with Section 3.1.3. This is a major reason power amplifier designers experience high frequency, high gain, broadband designs as "difficult," and why the lineup efficiencies of these amplifiers are lower than for lower frequency, lower gain, narrower band amplifiers.

11.2 Filters

Filter design is a huge and important topic for all sorts of communications systems. We touched their importance in Section 2.3.1, and it would be good to review that section now. In particular, Eq. 2.57 tells us that in general, we want to focus our attention on bands in the frequency domain where our signal *is*, and reject energy in bands where our signal *is not*. On the transmit side, maximizing our use of spectral real estate means placing our signal strictly within the bounds of our assigned channel so as not to interfere with another user. If you go trudging through the electrical engineering literature, you will find publications without end on the nuances of filtering. Filters are that important.

So let's get a little bit more detailed about what we mean by a "filter." Figure 11.6 shows some key characteristics of a frequency-selective filter. Region #1 is called

5 P. R. Gray and R. G. Meyer, *Analysis and Design of Analog Integrated Circuits, 3rd Edition.* USA: John Wiley & Sons, Inc., 1992.

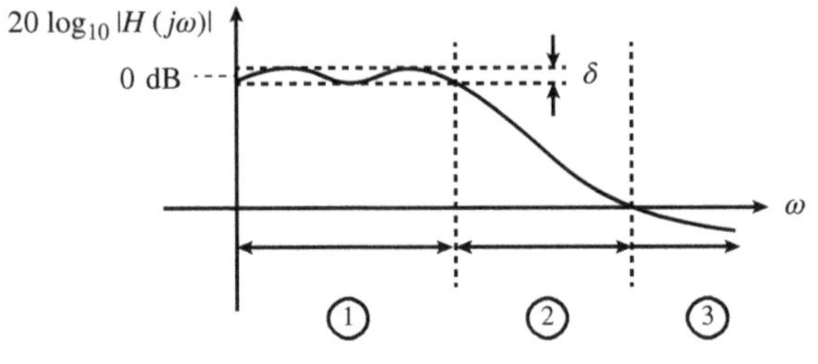

Figure 11.6 The features of a frequency-domain specification for a filter.

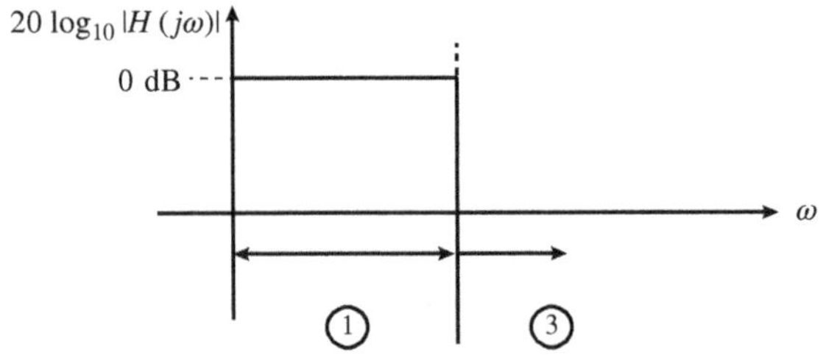

Figure 11.7 An example of an ideal filter.

the passband and is indicative of how we treat the frequencies that we *want*. The ideal is that they come through with a uniform gain and no variation in the phase response. When we invariably fall short of this ideal, we say that our filter has a "ripple in the passband," characterized by the parameter δ in Figure 11.6.

Region #3 shows how we treat frequencies that we don't want: we really don't want them! We would take infinite attenuation in this region if we could get it. This region is called the stopband.

We would like to make region #2, the transition between regions #1 and #3, extremely abrupt. That is to say, the swath of frequencies occupied by this transition should be as narrow as possible. Good, sharp filters allow us to divide up our spectral real estate with the precision of a surgeon. Figure 11.6 is a representation of filters in the real world. Figure 11.7 is how filters look in our dreams. If you have any experience at all in engineering, that there is a gap between ideal and real filters will hardly surprise you. The really interesting question is, what prevents us from getting ideal filters in the real world? Or put another way, what is the implementation cost of building filters closer and closer to the ideal?

The short answer that in practical implementations, the cost of getting closer to the ideal filter is hardware that is bigger and more complex. And bigger and more complex sometimes means that we pay an energy penalty. Let's get into it.

11.2.1 The Impulse Response

The first step to understanding the complexity trade-off is taken by examining something from linear system theory called the "impulse response" of a filter.[6] The impulse response is a time domain characterization of a filter according to its response to an impulse of the form introduced in Section 3.3.1. We almost universally label the impulse response of a filter $h(t)$.

When we speak of an impulse in time, what we mean is a pulse of unit area that is so short in time that the system we apply it to is insensitive to the pulse's exact shape. Only the area matters. It is longstanding tradition to saddle the impulse with the duty to be of "infinite height" and "zero width," but in the classroom this description produces only eye-watering confusion. Better, in the author's experience, to tie our definition of an impulse to the context of the system under study. It is then useful to have a *symbol* for our firmly-in-context impulse. The conventional symbol is shown in Figure 11.8. For a concrete example, we turn to a parallel *RC* circuit driven by a current source as shown in Figure 11.9. It is a great exercise, if you are able, to work out the response of the *RC* circuit to these two stimuli and verify that both result in substantially the same response. The physical understanding is that the resistor of 1 MΩ limits the current bleeding off the capacitor to levels far below the charging current. For example, when the capacitor voltage is near 10 V, the resistor current is only about 10 μA, which is tiny compared to the currents of 100 or 20 mA as shown in the figure.

Anyway, the point is that a very common and valuable way to characterize a filter is through its impulse response, as opposed to its frequency response $H(j\omega)$ (which we saw for the first time in Eq. 3.6). It is a fact of linear, time invariant systems that if we know the impulse response $h(t)$, we can determine the response to an arbitrary input $x(t)$ through something called a convolution integral:[7]

$$y(t) = x(t) * h(t) = \int_{-\infty}^{\infty} x(\tau)h(t-\tau)d\tau = \int_{-\infty}^{\infty} h(\tau)x(t-\tau)d\tau. \tag{11.3}$$

There's more! If you take the Fourier transform of the impulse response $h(t)$, you get nothing less than the transfer function $H(j\omega)$, an example of which we found useful in Eq. 3.7.

There is a very important takeaway here, and so we repeat: the impulse response $h(t)$ and the transfer function $H(j\omega)$ are Fourier transform pairs.

6 Traditionally, this topic would have been covered alongside other highlights of linear system theory in Chapter 2. We treat it here instead by way of reducing, as much as possible, the preliminaries for the reader before getting into the heart of radio systems. We can put it off no longer.

7 Z. Z. Karu, *Signals and Systems Made Ridiculously Simple*. Huntsville, AL: ZiZi Press, 1995.

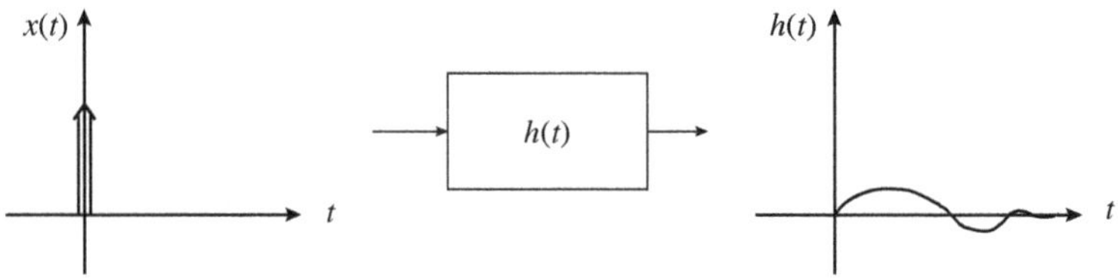

Figure 11.8 The symbol for an impulse (left), and an example response.

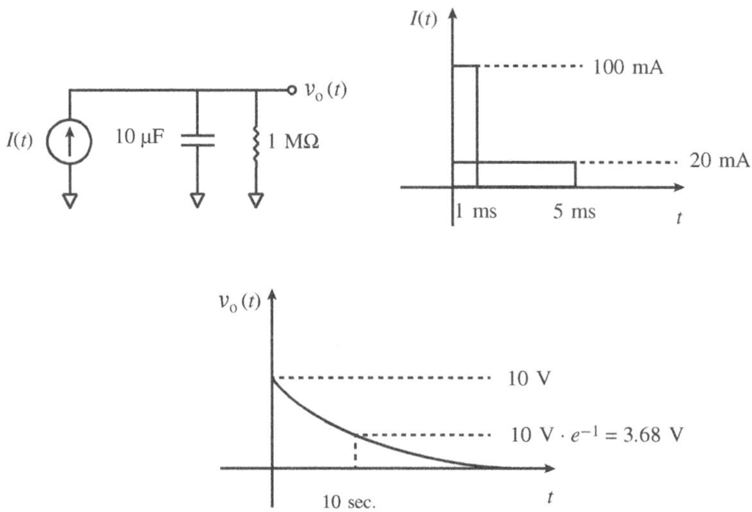

Figure 11.9 Response of a parallel *RC* to an impulse of current.

This means that when we specify a filter response in the frequency domain, we simultaneously and uniquely specify its time domain response to an impulse.

Finally, and because it will be useful later, we'll add that what holds in continuous time also holds in discrete time. Only here, instead of a convolution integral we have a convolution sum according to:

$$y[n] = x[n] * h[n] = \sum_k x[k]h[n-k] = \sum_k h[k]x[n-k]. \tag{11.4}$$

11.2.2 Finite Impulse Response Filters

Remember that we are looking at filters with a purpose. We want some understanding of how they work, of course, but we move quickly to the bigger question of the implementation cost of higher and higher performance filters. This is a difficult question at the outset, because before you take even the first analytical step you are confronted with a dazzling array of implementation choices. How can we come to a unified understanding?

Sometimes in situations like these, the best approach is to pick *one* implementation style and learn everything you can about that. Many times it happens that the insights you gain this way are then applicable to other implementations. In this case, we choose the venerable finite impulse response (FIR) filter, not the least because they are the consensus favorites in advanced systems at the time of writing. And while you can implement FIR filters in continuous time form (the surface acoustic wave (SAW) style is a good example), we are going to start with discrete time FIR implementations because they are so common today.

In discrete time, we needn't tear our hair out over the definition of the basic impulse. Its definition is straightforward, as shown in Figure 11.10. If we feed our impulse into an FIR filter with five taps, we might get an output as shown in Figure 11.11. Notice, first of all, that we get five nonzero samples out in response to our impulse and nothing more. If you come from the world of, say, *LC* filters, this behavior is new to you. In CT, the time domain responses are characterized by exponentially fading transients that in principle go on forever. Not so with an FIR filter. Here, the impulse response abruptly ends.

A look at the structure of an FIR filter will help you to understand this behavior. Consider Figure 11.12. The blocks labeled *D* in the figure are unit delays, which

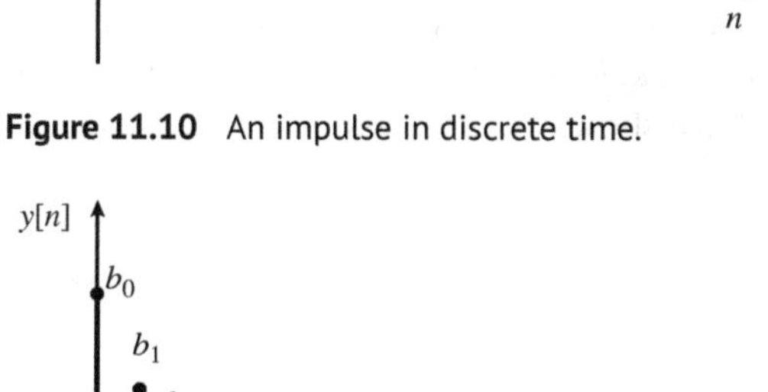

Figure 11.10 An impulse in discrete time.

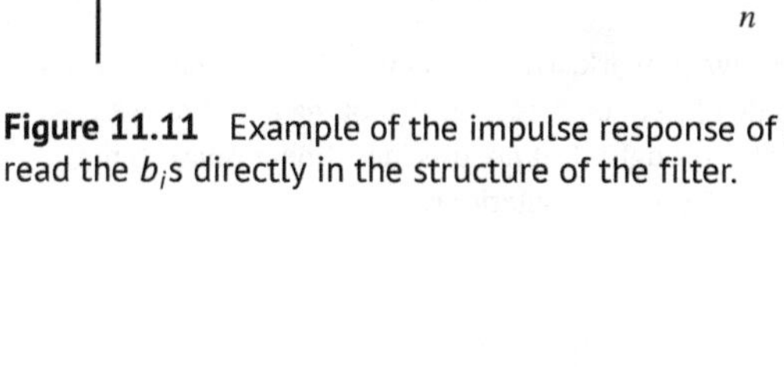

Figure 11.11 Example of the impulse response of an FIR filter. We will see that we can read the b_is directly in the structure of the filter.

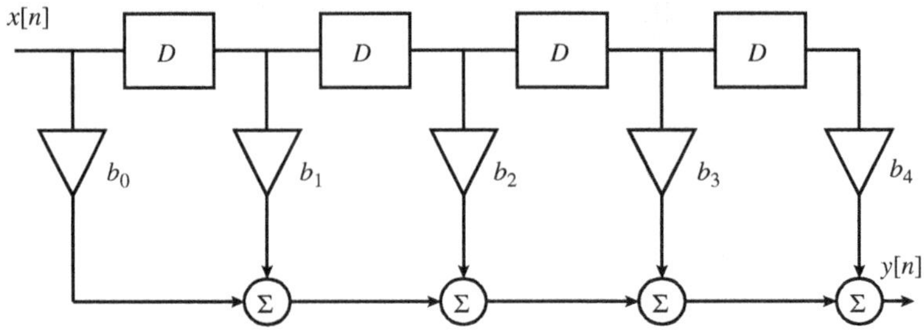

Figure 11.12 An example FIR filter structure.

pass the input unchanged to the output only delayed by a single clock cycle. Such delay blocks are very easy to realize in clocked, digital circuits. To get a basic idea of how this structure works, see if you can reconcile Figure 11.11 with Figure 11.12.

We *love* FIR filters in communications systems. One reason for this is that if we choose the impulse response to have certain symmetry properties, we can arrange for the filter to be "linear phase," which means that they introduce no phase distortion.[8] That is to say, the filter shapes the relative amplitudes of different frequencies but leaves their phase relationships unchanged.[9]

Now we get to the really important part of our development, which is discovering what makes a filter specification "demanding," and how and why that translates into a large number of taps. It turns out that, for fundamental reasons, we require a large number of taps to achieve either or both of the following in a filter response:

- We want the filter to be extremely narrowband.
- We want the filter to have sharp transitions between the passband and stopband.

To see the reason behind the first bullet, we need look no further than the Uncertainty Principle from Section 2.5. If we look, for example, at the Fourier transform pair for an ideal low-pass filter, we have shown that the widths of the two functions are governed by the inequality

$$\Delta T \Delta f \geq \frac{1}{\pi}. \tag{11.5}$$

8 A. V. Oppenheim and R. W. Schafer, *Discrete-Time Signal Processing*. Englewood Cliffs, NJ: Prentice Hall, 1989.

9 This is probably how you thought filters always work. The reason you thought that is because we are in the habit of plotting the magnitude of frequency response, looking at the phase response only dutifully and usually as an afterthought. But no, this ideal linear phase is not usually easily achievable when we don't use FIR implementations.

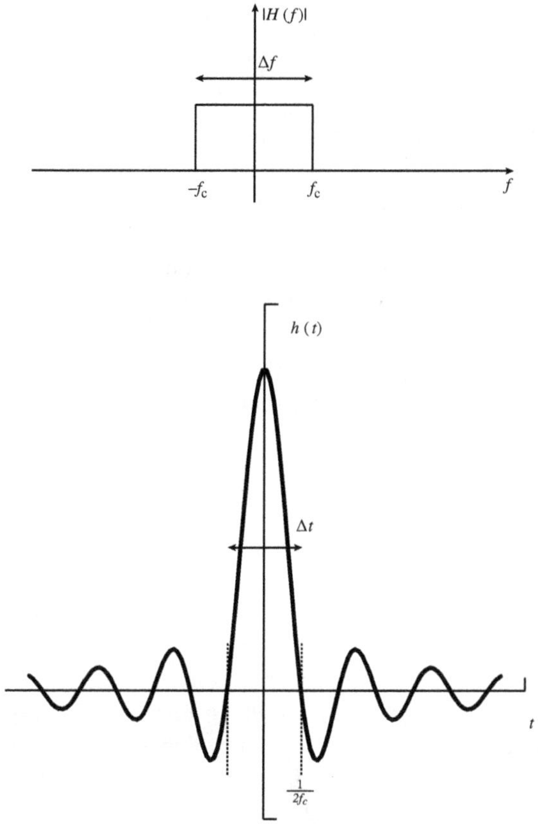

Figure 11.13 The uncertainty principle tells us that there is an inverse relation between the width in time of the impulse response and the width in frequency of the frequency response.

(Actually, in this special case, we can do better than an inequality and show that if we use the definitions of ΔT and Δf shown in Figure 11.13, then $\Delta T \cdot \Delta f = 2$.) With the Uncertainty Principle in hand, explaining the first bullet is therefore quite easy. A more narrowband filter implies a smaller Δf and therefore a larger ΔT. And a larger ΔT means a longer impulse response, which means more taps for our FIR filter implementation. It is that simple! So if, for example, you find that you are interested in selecting out a 10-Hz wide band of spectrum with your filter, you can write down immediately, without doing any design, that the time scale of the impulse response of that filter will be in the neighborhood of

$$
\begin{aligned}
\Delta T \cdot (10\ \text{Hz}) &\geq \frac{1}{\pi} \\
\Delta T &\geq 31.8\ \text{ms}.
\end{aligned}
\tag{11.6}
$$

In a discrete-time implementation with a sample period of, say, 1 ns, you can see that implies a lot of taps indeed!

The second bullet point requires a slightly more complicated development. We start by pointing out that an extremely common and practical way to design an FIR filter is to start with the ideal impulse response of the filter you want to synthesize. This response is usually of infinite extent, and so we take a finite swath of time domain samples of that ideal response as our FIR filter impulse response. The expectation is that the longer our FIR approximation, the better. In the filtering vernacular, this practice is called "windowing." We say that the impulse response for our FIR filter is given by

$$
h[n] = h_{\text{ideal}}[n] \cdot w[n]. \tag{11.7}
$$

If $w[n]$ is what we call a rectangular windowing function, $w[n] = 1$ for $0 \leq n \leq M$,[10] and $w[n] = 0$ otherwise.

It turns out we need one more piece of trivia from undergraduate linear system theory. When we multiply time sequences together, as we do in Eq. 11.7, their respective Fourier transforms get *convolved* in the frequency domain. Figure 11.14 shows the discrete-time Fourier transform (samples of which form the DFT) of a rectangular window of length $M + 1$. Now in the frequency domain, when we apply this windowing function, we convolve the two frequency responses shown in Figure 11.15. Formally, the convolution is given by

$$
H(e^{j\Omega}) * W(e^{j\Omega}) = \frac{1}{2\pi} \int_{-\pi}^{\pi} H(e^{j\theta})W(e^{j(\Omega-\theta)})d\theta. \tag{11.8}
$$

This convolution results in the smearing of the ideal low-pass filter function as sketched in Figure 11.16.

The grand conclusion is that if we want really sharp transitions between a passband and a stopband, we require M to be very large for our windowing function, and the larger the better. And M, of course, tells us how many taps are in our FIR filter, which is our measure of complexity.

---

10 Here we default to real-time signal processing, in which filters are constrained to be causal. Our impulse responses therefore cannot be nonzero for $n < 0$, despite what we show in the ideal low pass filter response. When striving to approximate an ideal low pass filter in the real world, then, we not only window as described here, we also tend to introduce a *delay* to shift the impulse response into causality.

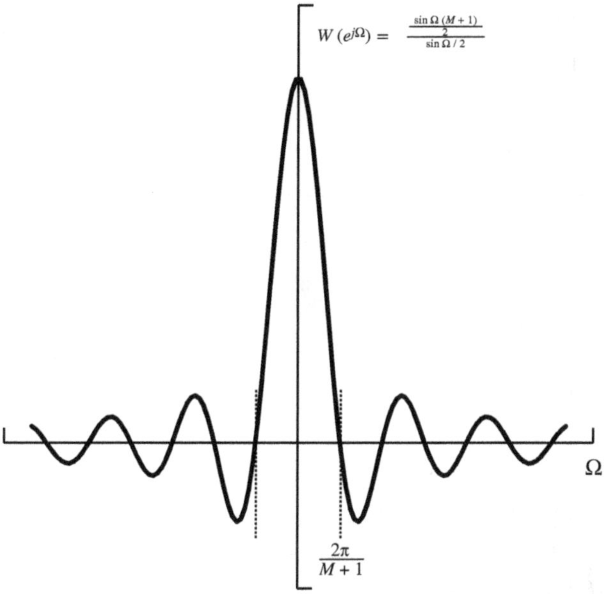

Figure 11.14 The discrete-time Fourier transform of a rectangular window of $M + 1$ samples.

This is our deep plunge into one implementation style of filters. The *generalizable* takeaway is that highly selective, narrowband filters with sharp transitions between passband and stopband require impulse responses $h(t)$ that are *long in time*. We'll now apply this conceptual insight to understanding other types of filters.

11.2.3 Other Filter Implementations

It turns out that a wide variety of filter realizations rely on a combination of resonant structures, with *LC* filters and cavity filters being outstanding examples. Recall that for an *RLC* tank circuit, in the underdamped case its impulse response goes as

$$v_0(t) = e^{-t/2RC}\left(A_1 \cos \omega_d t + A_2 \sin \omega_d t\right). \tag{11.9}$$

It's a detail, but the definition of ω_d was

$$\omega_d = \sqrt{\omega_0^2 - \left(\frac{1}{2RC}\right)^2}, \tag{11.10}$$

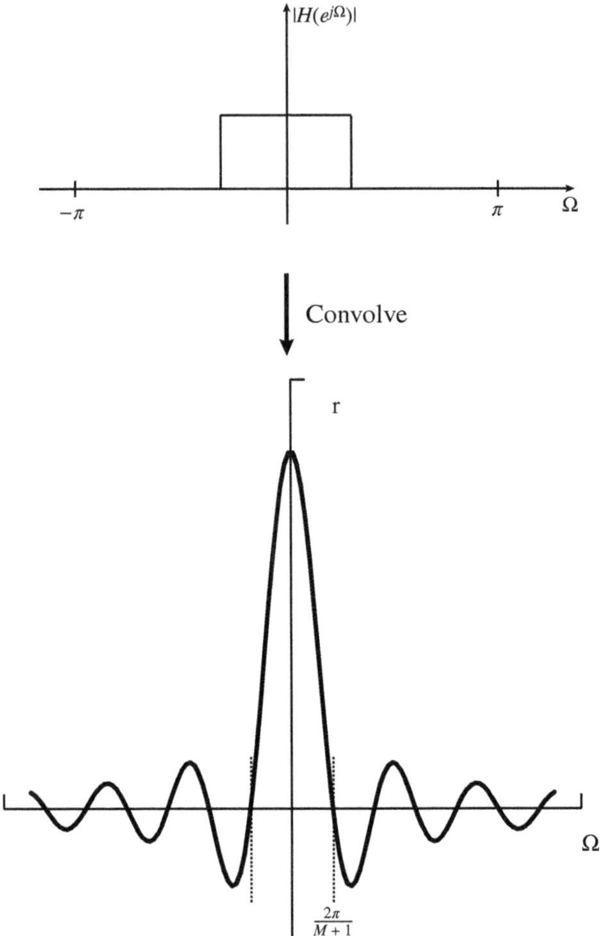

Figure 11.15 When we multiply by a windowing function in the time domain, we convolve the Fourier transforms.

Also, A_1 and A_2 are set by initial conditions. The original discussion is in Section 3.3.1. While we're at it, the definitions of Q are given by

$$Q = \frac{R}{\omega_0 L} = \frac{R}{\sqrt{L/C}} = \omega_0 RC. \tag{11.11}$$

In this case, the key to having an impulse response that lasts a long time is to have the RC product be large in some sense.

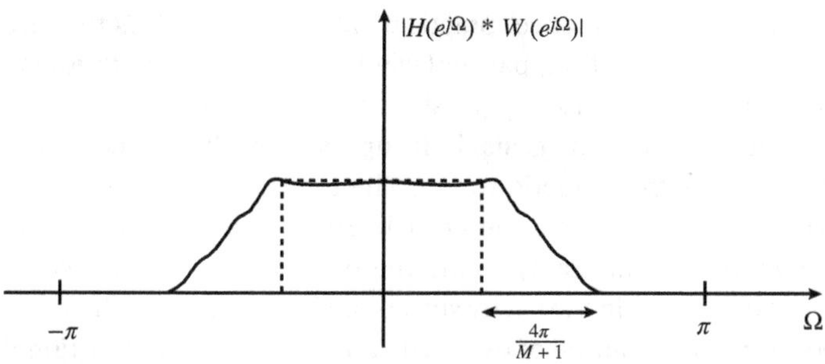

Figure 11.16 As a result of the convolution, our sharp filter edges are widened according to the main lobe width of the Fourier transform of our windowing function. The longer our window (in time), the thinner this lobe, and therefore the sharper our filter transitions.

But the key to generalizing to other filter structures built on resonances is to tie this requirement of a "large" *RC* product to *Q* of the resonance. This is easy enough to do, reaching for, in this case, the most convenient form of *Q* for the point we are trying to make:

$$Q = \omega_0 RC. \tag{11.12}$$

So what we *really* mean by a large *RC* product is that we require that the *Q* of our resonant structures be high.

It helps from this point to tie the *Q* discussion to the discussion of windowing for FIR filters. Recall that for our windowing discussion, we stuck with the rather brutal, straight-ahead rectangular window that abruptly cut off in time what would otherwise be an infinitely long impulse response. When it comes to resonant structures, one may think of the finite *Q* of our resonators as imposing a *window with a soft, exponential taper* on an underlying, ideal impulse response. We can still talk rigorously about a ΔT width for this windowing function.[11] From here, the last step is to tie the necessity of long impulse responses to the implications on the size and complexity of various implementations. To wit:

- A SAW filter is a type of FIR filter that exploits piezoelectricity to perform the filtering function. Very fun. What we take from the development of this section, though, is that a demanding filter specification calls for a SAW implementation with many taps. This leads to a physically large filter with a painful amount of insertion loss.
- For *LC* filters, we see that the demands of high *Q* drive us to very high ratios between *L* and *C*. (This is a good time, actually, to ponder the tragedy of the

11 For an example of the mathematics of deriving widths of a distribution, see Section 2.5.1.

square root in Eq. 11.11). In practice, practical values of *L*s and *C*s that are good for high frequency operation, particularly for *L*s, are limited in range. This basic fact greatly compromises aggressive *LC* filter designs.

- Cavity filters are, as the name suggests, built up using multiple microwave cavity resonators for electromagnetic waves. The total stored energy in the cavity is determined by integrating the electric and magnetic field density throughout the volume of the cavity, while the primary loss mechanism is joule heating due to currents in the cavity surfaces. The Q of a cavity therefore scales in proportion to the volume of the cavity and is inversely proportional to the surface area. It is no contest: cavity volumes scale as the cube of the key physical length dimension, while the surface area scales as the square. Available cavity filters bear this out, as the more demanding the filter specification, the bigger and heavier the cavity filter is.
- Even IIR digital filters (infinite impulse response filters, which allow feedback and therefore can have long, exponentially decaying tails to the impulse response) are not immune. Long, slowly decaying impulse responses require extremely high precision, floating point digital representations to match, otherwise those exquisitely decaying tails get lost in the quantization noise. If we are serious about our filter specifications, it turns out we need to meet that challenge with suitably complex computation structures.

And so it goes. If you are a system architect, any plans that you have that depend on extremely selective filters will bring with them a high implementation, and possibly high energy cost. It is what Nature requires.

11.3 A Note on Frequency Synthesis

We will tread lightly here. We present the tip of the iceberg, and then leave it to the reader's curiosity to pursue further if desired. As is *always* the case, you can do a better job with frequency synthesis if you spend more energy on the Task. That is actually okay, as there are far bigger energy consumers in our radio systems. Our energy analysis here can be safely regarded as secondary. But the problem of coordinating carrier frequencies between transmitter and receiver is so central to the basic functioning of a radio link that we could not leave the topic untouched.

So the purpose of this section is to remind readers of the need for frequency synthesis, and then give some idea as to how we pull this off in the real world.

11.3.1 The Problem of Matching Carrier Frequencies

Throughout this entire book, we have taken it as a given that the transmitter and the receiver would agree absolutely on f_0, the carrier frequency. But that is an

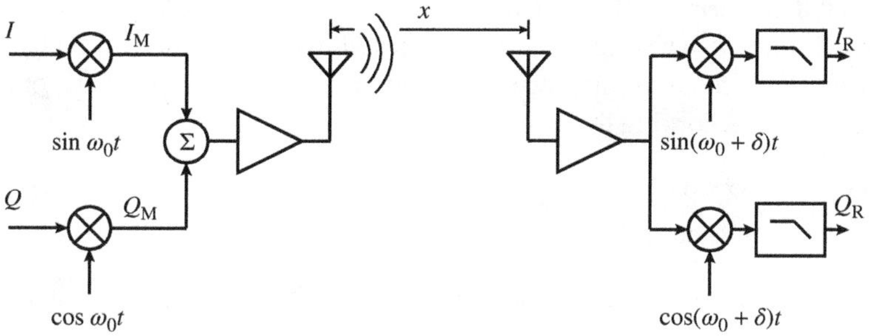

Figure 11.17 A transmitter and receiver, showing the inevitable carrier frequency mismatch.

extraordinary demand. If our transmitter local oscillator is at 1.000000000 GHz, and our receiver is merely 1 Hz away at 1.000000001 GHz, we would all agree that that is pretty good. But still, let's look at what a slight carrier frequency mismatch between transmitter and receiver means for our communications link.

We start with the setup in Figure 11.17. Up until this point, it has been axiomatic that ω_0 on the transmit side is equal to ω_0 on the receive side. If you have any experience in engineering at all, once you thought about it you would be skeptical of such a claim. And so in Figure 11.17 we make our doubt explicit by notating a frequency offset of δ.

What is the consequence of this frequency offset? Well, it is straightforward to derive the received signals I' and Q'. A helpful reminder of our trigonometric identities:

$$\begin{aligned} \sin(\omega_0 + \delta)t &= \sin \omega_0 t \cos \delta t + \cos \omega_0 t \sin \delta t \\ \cos(\omega_0 + \delta)t &= \cos \omega_0 t \cos \delta t - \sin \omega_0 t \sin \delta t. \end{aligned} \tag{11.13}$$

From here,

$$\begin{aligned} I' &= (I \sin \omega_0 t + Q \cos \omega_0 t)[\sin \omega_0 t \cos \delta t + \cos \omega_0 t \sin \delta t] \\ Q' &= (I \sin \omega_0 t + Q \cos \omega_0 t)[\cos \omega_0 t \cos \delta t - \sin \omega_0 t \sin \delta t]. \end{aligned} \tag{11.14}$$

Yet more trigonometric identities prove themselves handy:

$$\begin{aligned} \sin^2\theta &= \frac{1}{2} - \frac{1}{2}\cos 2\theta \\ \cos^2\theta &= \frac{1}{2} + \frac{1}{2}\cos 2\theta \\ \sin\theta\cos\theta &= \frac{1}{2}\sin 2\theta. \end{aligned} \tag{11.15}$$

We drop, as usual, all terms in $\sin 2\omega_0 t$ and $\cos 2\omega_0 t$ as they are filtered out before baseband processing. We'll also spare ourselves the tedium of carrying around factors of $\frac{1}{2}$ since we are focused on conceptual insight. If we do this, we have our demodulated symbol according to

$$I' = I\cos\delta t + Q\sin\delta t \tag{11.16}$$
$$Q' = -I\sin\delta t + Q\cos\delta t.$$

The sharp-eyed reader may recognize a rotation matrix here similar to what we saw in Eq. 9.13:

$$\begin{bmatrix} I' \\ Q' \end{bmatrix} = \begin{bmatrix} \cos\delta t & \sin\delta t \\ -\sin\delta t & \cos\delta t \end{bmatrix} \begin{bmatrix} I \\ Q \end{bmatrix}. \tag{11.17}$$

The picture to have in your mind is that if we transmit a static symbol (I, Q), what we receive is a symbol that rotates with angular speed δ radians per second. See Figure 11.18. Now it is true that, to a certain extent, a frequency mismatch will always be with us. Fortunately, digital signal processing rides to the rescue with common algorithms for detecting carrier frequency mismatches and compensating for them.

From here, there are two more things to know. First, despite the aforementioned nod to digital signal processing, we *do* seek to minimize the associated computational overhead by matching frequencies as well as we can in hardware. The principal method of minimizing frequency offsets is to tie all internal frequency generation to high quality crystal references. Such references are spectacularly stable over temperature, and it is not unusual to have their center frequencies specified to within 30 ppm (!). Then, the second thing to know is that there are typically many frequency references needed in a modern transceivers. We would therefore be in a bit of trouble if we needed one extremely accurate (and therefore high Q, and *therefore* bulky and expensive) crystal for every frequency reference in the system.

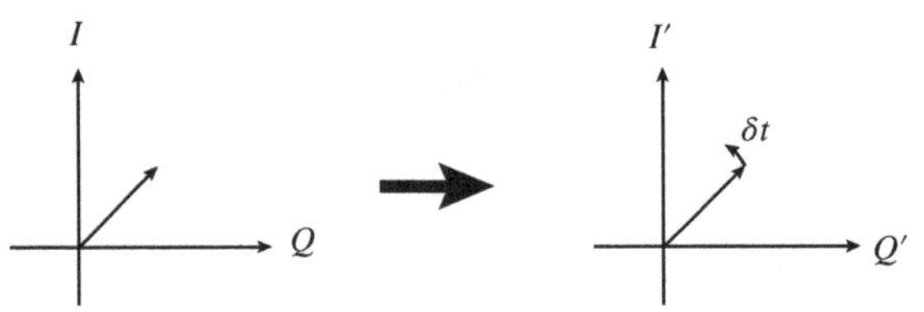

Figure 11.18 The impact of frequency mismatch between a transmitter and a receiver. If we transmit a static symbol, it seems to rotate in the receiver with an angular speed of δ radians per second.

What we need is a way to *derive* all or most of the frequency references that we need from a single, stable crystal. This gets us into the art and science of frequency synthesis, the key to which is an understanding of the basic phase-locked loop. We will get into the basics of this fundamental building block in the next section, but as you study, know two things: (1) we use high quality frequency references because they allow us to maximize the use of our spectral real estate; (2) as with all electronics, the higher "quality" of frequency reference we require, usually the greater its power consumption. This second point is somewhat less serious than the power consumption of, say, power amplifiers, just because it is unusual for the frequency synthesis function to figure prominently in the system power budget. But this basic fact of design is important to keep in mind.

11.3.2 Phase-locked Loops and Frequency Synthesis

A phase-locked loop (PLL) is a foundational functional block for communications systems, and PLLs are a world unto their own. They are, in their own right, a coming together of multiple electrical engineering subdisciplines. What is important and relevant for this book is a basic grasp of this critical function. This will strengthen your literacy in the overall radio problem.

The beginning of understanding PLLs is to get used to keeping track of the *phase* of a sinusoid as the *variable of interest*. The less said the better from here—we should jump in right away to the example of a phase detector.

Consider Figure 11.19. We see an analog multiplier, which produces at its output a voltage proportional to the arithmetic product of the two input voltages. Our contention is that if we look at the output of the multiplier and apply a low-pass filter, we get an output voltage v_0 that is responsive only to the phase difference between the two input sinusoids.

This is easy enough to check. Recalling that $\cos(\omega_0 t + \phi) = \cos\omega_0 t \cos\phi - \sin\omega_0 t \sin\phi$, we have the output of the multiplier as

$$\begin{aligned}
&\cos\omega_0 t \cdot [\cos\omega_0 t \cos\phi - \sin\omega_0 t \sin\phi] \\
&\quad = \cos^2\omega_0 t \cos\phi - \cos\omega_0 t \sin\omega_0 t \sin\phi \\
&\quad = \left(\frac{1}{2} + \frac{1}{2}\cos 2\omega_0 t\right)\cos\phi - \left(\frac{1}{2}\sin 2\omega_0 t\right)\sin\phi. \qquad (11.18)
\end{aligned}$$

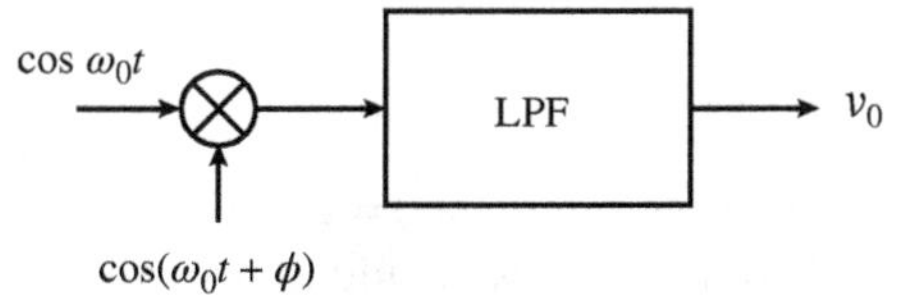

Figure 11.19 Using an analog multiplier and a low-pass filter as a phase detector.

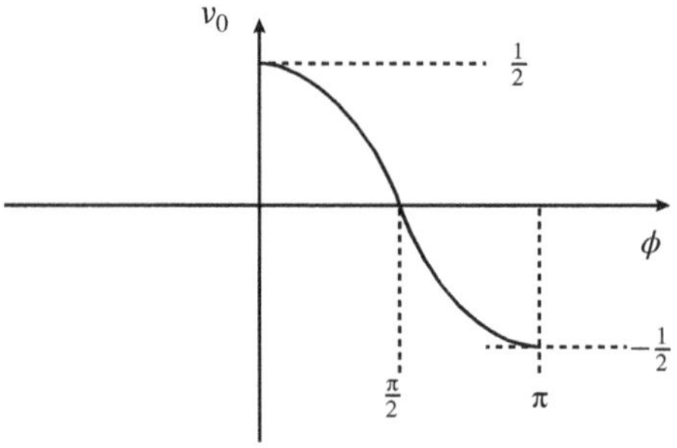

Figure 11.20 After a low-pass filter, this is the output voltage versus phase offset that we get from an analog multiplier as a phase detector.

Now we see where the low-pass filter comes in. If we eliminate the $2\omega_0$ terms, we are left with

$$v_0 = \frac{1}{2}\cos\phi. \tag{11.19}$$

A graph is particularly helpful here; see Figure 11.20. We see that we can look at the output of what we will now call a phase detector, and determine the phase offset between the two incident sinusoids. If we see $+\frac{1}{2}$ V steady at the output, we must be in phase; if the output is 0 V, the two inputs must be in quadrature; and if the output is at $-\frac{1}{2}$ V, then the inputs must be completely out of phase. And if we are determined enough to invert the $\cos(\cdot)$ nonlinearity in detail, we can successively read off all phase offset values in between.

The phase detector is the *whole game* when it comes to understanding PLLs. We have to get used to looking at otherwise familiar circuit and system blocks and reading their function through their effect on the phase of the sinusoids that we present to their inputs.

There are a few more conceptual pieces to solidify. Normally when we think of a phase, we have in mind a throwaway parameter included for completeness at the tail end of the argument of a sin or cos:

$$v(t) = \cos(\omega_0 t + \phi_0). \tag{11.20}$$

That is, we think of ϕ_0 as the "phase" and ω_0 as the (angular) frequency ($= 2\pi f_0$). A key shift to make for a PLL frame of mind is *to think of the entire argument of the cosine as the "phase."* And why not? Don't $\omega_0 t$ and ϕ_0 have the same units, after all? So the point now is that we get used to thinking of sinusoids with generalized phase:

$$\cos(\phi(t)), \tag{11.21}$$

where very often, $\phi(t) = \omega_0 t + \phi_0$ like you are used to. But now that we have freed our minds, we could in principle make $\phi(t)$ more complicated. Consider, for example,

$$\phi(t) = \omega_2 t^2 + \omega_1 t + \phi_0. \tag{11.22}$$

How would we now talk about $\cos(\omega_2 t^2 + \omega_1 t + \phi_0)$ in terms of our old concept of frequency of a sinusoid? The answer is that we also generalize our definition of frequency, and define it as the *instantaneous rate of change of the generalized phase.* That is,

$$\omega \triangleq \frac{d\phi(t)}{dt}. \tag{11.23}$$

Applying this definition to the most common case of $\phi(t) = \omega_0 t + \phi_0$, we see that if we take the first time derivative $\frac{d\phi(t)}{dt}$ the instantaneous frequency is ω_0 just like we are used to. Now let's examine the instantaneous frequency for the phase as described in Eq. 11.22.

$$\begin{aligned} \omega(t) &= \frac{d\phi(t)}{dt} \\ &= \frac{d}{dt}[\omega_2 t^2 + \omega_1 t + \phi_0] \\ &= 2\omega_2 t + \omega_1. \end{aligned} \tag{11.24}$$

This is referred to in signal processing circles as a "linear chirp."

Finally, we complete the circle by pointing out that if the instantaneous frequency is the time derivative of the phase, then the phase must be the integral of the instantaneous frequency:

$$\phi(t) = \int_{-\infty}^{t} \omega(t)dt. \tag{11.25}$$

If generalized phase is new to you, and you are curious to understand more about PLLs, then don't rush this section. Getting really comfortable with generalized phase is a key sticking point for many who struggle to understand PLLs.

The next step is to look at an actual PLL and examine it block by block. Figure 11.21 is an example of a generic PLL. Remember that the essence of a properly functioning feedback loop is that the "error," V_e in this case, is *driven* and *held* to zero. We will come back to this.

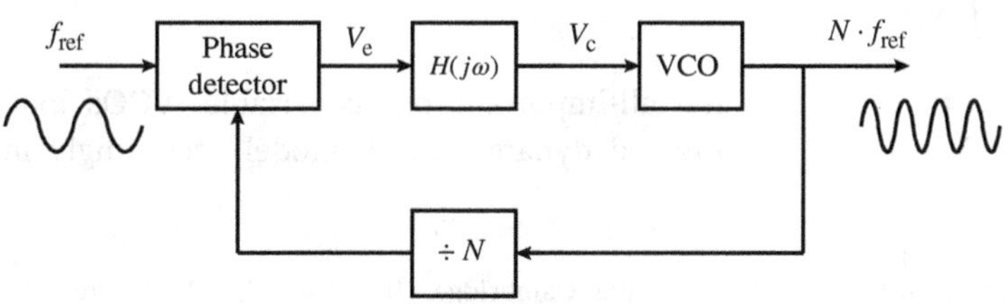

Figure 11.21 Overview of a phase-locked loop.

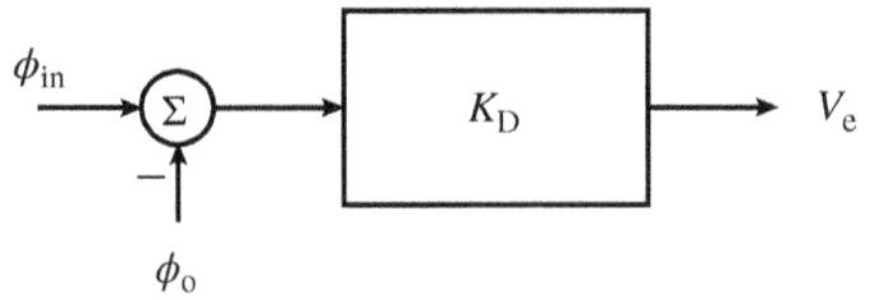

Figure 11.22 Redrawing our phase detector for system analysis.

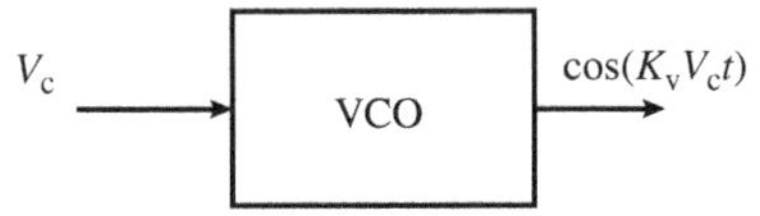

Figure 11.23 The voltage controlled oscillator.

In our block-by-block examination, we start with the phase detector. We've already looked at this a little bit and can probably agree on a sanitized representation as shown in Figure 11.22. In practice, we sometimes have to be a little flexible. If we go with the implementation concept represented by Eq. 11.19, for example, our $V_e = 0$ condition would actually correspond to $\phi_{in} - \phi_0 = \pi/2$. Sometimes this is okay, as in some applications all that matters is that the input and output waveforms have a static phase offset, whatever that offset happens to be. (When an offset is not okay, we have to use a different trick for our phase detector.[12]) Whatever we decide, we begin transitioning our diagram from one focused on implementation details to one that facilitates dynamic analysis with the phase detector as shown in Figure 11.22.

The VCO block is the "voltage controlled oscillator." This circuit is just like a normal oscillator, except the frequency of its output sinusoid is roughly proportional to an input control voltage. See Figure 11.23. Mathematically, we are interested in relating the control voltage V_c to our all-important phase variable. We have, in short order:

$$\begin{aligned} \omega_0 &= k_v V_c \\ \frac{d\phi_0}{dt} &= k_v V_c \\ \phi_0 &= k_v \int V_c(t)dt. \end{aligned} \tag{11.26}$$

So apparently, with respect to the all-important phase variable, VCOs look like integrators. We draw the sanitized dynamic block model accordingly in Figure 11.24.

12 B. Razavi, *Design of CMOS Phase-Locked Loops.* Cambridge, UK: Cambridge University Press, 2020.

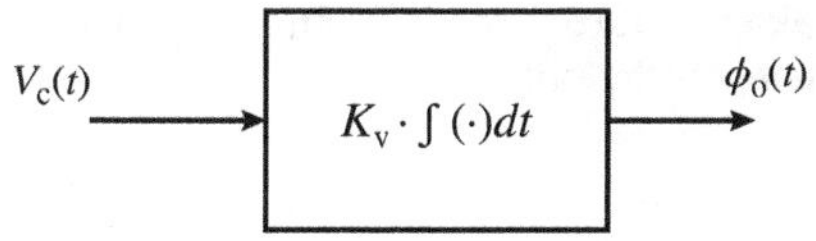

Figure 11.24 The voltage controlled oscillator for system analysis, with phase as the variable of interest.

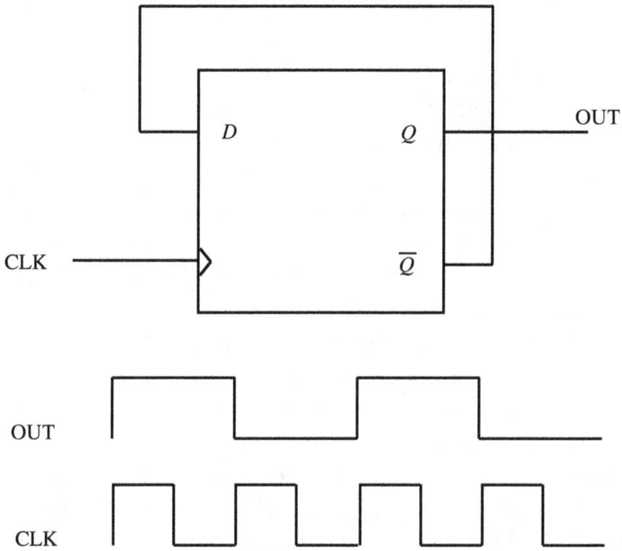

Figure 11.25 Using a D flip-flop to implement a divide-by-2 frequency divider.

Frequency dividers tend to sound more exotic than they are. A big thing to remember is that at the circuit level we tend to blur the line between analog circuits and digital circuits. We make sure that our sinusoids are of a large enough amplitude that they can be processed meaningfully by digital gates. Once you have made this leap, frequency dividers become nothing more than re-labeled digital counters. Figure 11.25 is an easy example of a divide-by-2 using a single digital flip-flop.

So how does a frequency divider affect the phase variable? We start with how it affects frequency:

$$\omega_{\text{out}} = \frac{1}{N}\omega_{\text{in}}$$
$$\frac{d\phi_{\text{out}}}{dt} = \frac{1}{N}\frac{d\phi_{\text{in}}}{dt}. \tag{11.27}$$

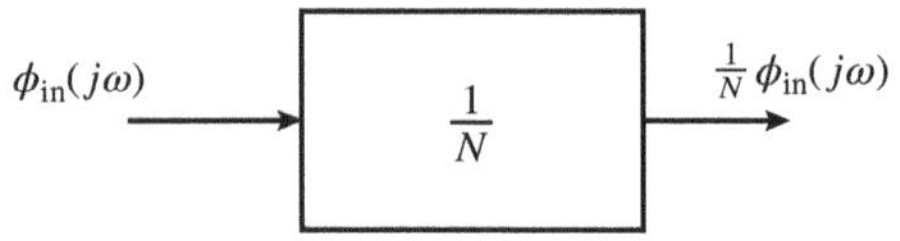

Figure 11.26 How a frequency divider acts on the phase variable.

The derivative is a linear operator, so...

$$\phi_{out} = \frac{1}{N}\phi_{in}. \tag{11.28}$$

We duly illustrate this as the system block in Figure 11.26.

The final bit from Figure 11.21 to address is the $H(j\omega)$ block, variously called the "loop filter" or the "loop compensation." This block is where we address the heart of the feedback system design problem, deciding such things as the speed of response, the size of the steady-state error, and stability margins. Feedback theory is its own jungle: the author has attempted to make navigating that terrain as short and straightforward as possible in a previous book.[13] For specifics on feedback compensators for PLLs, Razavi's book is an excellent reference.[14]

For our purposes, the right thing is to assume that $H(j\omega)$ is well-chosen according to all relevant metrics so that the feedback loop behaves well. The essence of a well-behaving loop is that $V_e = 0$, as we said before. What we'll add now is that feedback systems can be unstable if we do not design them properly, and so a big part of $H(j\omega)$'s job is to manage in that department. That settled, we now redraw Figure 11.21 using sanitized system blocks, and give ourselves one more frequency divider at the input to the whole thing. Figure 11.27. If we accept that $V_e = 0$, then what we have is that

$$\frac{\phi_{in}}{M} = \frac{\phi_o}{N} \tag{11.29}$$

$$\phi_o = \frac{N}{M}\phi_{in},$$

and therefore

$$\omega_o = \frac{N}{M}\omega_{in}. \tag{11.30}$$

This is the conceptual heart of what is known as an "integer-N frequency synthesizer." We use as a frequency reference a single sinusoid derived from a high accuracy crystal (this is where ϕ_{in} comes from), and then digitally select N and M to give us a wide array of clock frequencies for all of our circuit timing and carrier frequency needs.

---

13 Joel L. Dawson, *A Guide to Feedback Theory*. Cambridge, UK: Cambridge University Press, 2021.

14 Ibid.

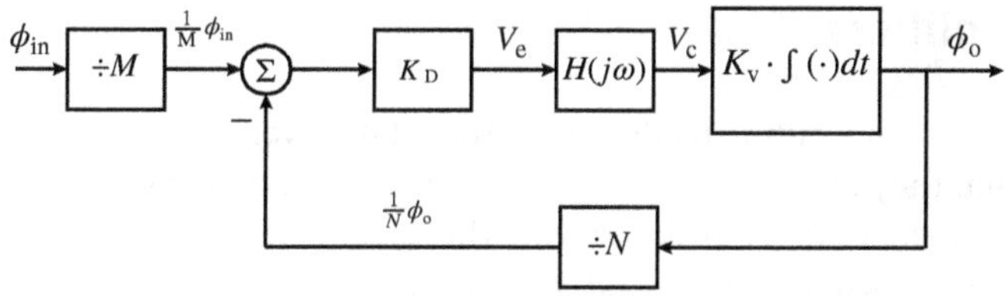

Figure 11.27 A redrawing of the basic PLL for system analysis.

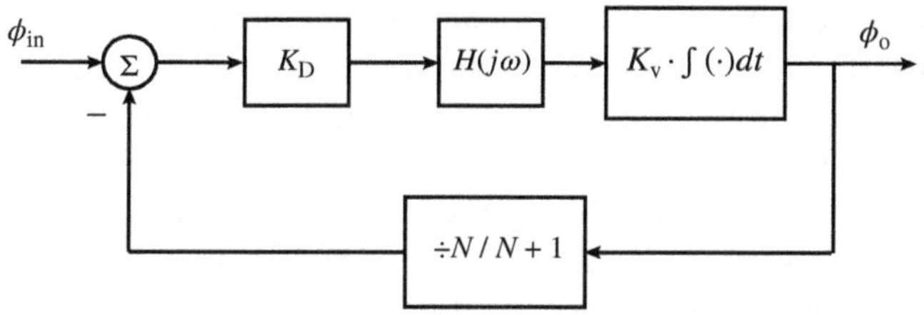

Figure 11.28 A fractional-N synthesizer. The basic principle is that the feedback divide ratio is dithered between N and $N + 1$ at a rate that is fast compared to the dynamics of the loop. The PLL acts essentially as an interpolation filter, and the system realizes a frequency multiplier of somewhere between N and $N + 1$.

Finally, we can go all the way down the rabbit hole of frequency synthesis and look at "fractional-*N*" synthesizers as shown in Figure 11.28. In this scheme, the divider in the feedback path is toggled back and forth between N and $N + 1$, spending a deliberately chosen fraction of time in each position. The dynamics of the overall loop, set by $H(j\omega)$, are chosen to make the loop respond slowly compared to the time scale of dithering between the two divider settings. The net result is continuous interpolation between $N \cdot \omega_{in}$ and $(N + 1)\omega_{in}$.

In terms of power consumption for the frequency synthesizer, a major consumer is the VCO block itself. There are many ways to realize a VCO, but an extremely common method centers a parallel *RLC* tank circuit at its heart. Knowing this allows us to make use of Eq. 3.56 to understand the energy costs of moving around the performance space. Specifically, tanks with higher Q allow us to lower energy consumption; it is more expensive to require a higher voltage swing (A^2); and, all else being equal, we can expect higher energy consumption when we operate at higher frequencies.

Frequency synthesis is, again, a whole field unto itself. Our purpose in discussing it here is to help the reader appreciate how modern transceivers can be frequency agile without having to have banks of heavy, expensive crystals. Additionally, the hope is to demystify even a little bit how frequency synthesis is done in practice.

11.4 Power Amplifiers

The power amplifiers (PAs) in a radio link dominate the total power consumption. They are unusual in that they must to process energy and information at the same time. Examples of circuit blocks that *don't* have to do this include microprocessors, which are designed to process information and only consume energy because parasitic capacitors are everywhere in the world, and power converters, which need to deliver energy but very often need only hold a given voltage indefinitely. Example circuit blocks in the same unfortunate boat as the power amplifier include the driver amplifiers for electrical motors when speed and acceleration must be tightly controlled, or the amplifiers for audio speakers, where energy is put out in the form of acoustic waves and where signal fidelity is at a premium. For power amplifiers and other circuits in its class, part of the power consumption is due to the fact that its output really is power and not, say voltage. And then usually the *bigger* part is due to the inefficiencies we incur when we also try to have extremely high signal fidelity. Figure 11.29 serves as a reminder of the place of the power amplifier in our link. Power amplifiers are in the power business because they supply the Poynting flux that flows out of the transmit antenna and into space. They are in the information business because all of our Shannon limit hopes are pinned on the detailed modulation of our carrier. If our power amplifier distorts its signal, we are messing with our bits!

The purpose of this section is to clarify the energy trade-offs associated with the power amplifier. Historically there has been a tendency to deemphasize mastery of these trade-offs because of the false belief that we can prioritize data throughput (about which there is no hesitation) and then worry about energy later. The truth is that infrastructure scale wireless networks are *always* energy constrained. Mastering the energy trade-offs, therefore, is the key to maximizing the data capacity of the network.[15]

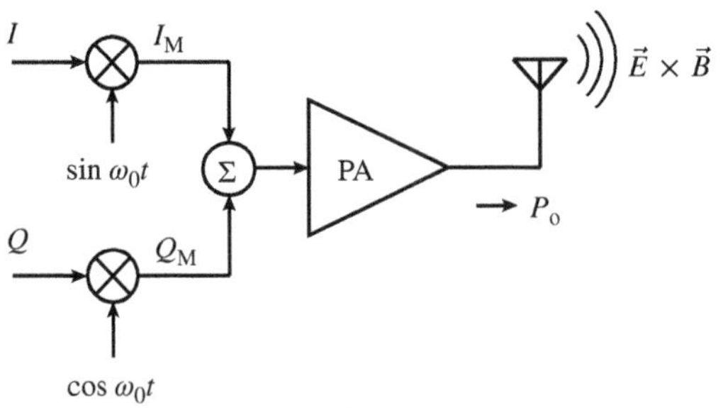

Figure 11.29 The place of the power amplifier (PA) in a transmitter. Not shown are common sources of loss between the PA output and the antenna, such as duplex filters, circulators, directional couplers, etc.

15 At the time of writing, network data traffic was compounding at 20–30% *per year*, and along with it, roughly 1:1, the energy demands of the network. As human beings, we don't really

Anyway, as the largest power consumers in our radio links, there is a lot at stake when it comes to the energy trade-offs. Let's get into it.

11.4.1 Power Amplifier Preliminaries

To get some feel for the energy trade-offs at the heart of power amplification, it will help to get a little beyond The Least You Need to Know About Analog Electronics from Chapter 3. For those readers with an analog circuit background, a careful review of key, representative circuit topologies here will be stimulating and helpful. For those readers without that background, the important stuff with regard to energy is a detailed understanding of the current and voltage waveforms at the output terminal (drain, collector) of the amplifying transistor. If readers will take this importance on faith, the author will attempt to lay out the implications of those waveforms clearly.

Figure 11.30 shows the conceptual heart of the generic PA structure. The idea is that v_{RF} is a modulated sinusoid centered at the resonance frequency of the parallel *RLC* tank. Meantime, L_{big} is effectively an open circuit for frequencies at or near the carrier frequency, while C_{big} is a short circuit for the same. It is instructive to look at the waveforms for i_D, V_A, and V_B when the PA is putting out the absolute maximum output power. See Figure 11.31. If you have a background in analog electronics but have not designed much with inductors, you may be surprised to see the drain voltage rising *above* the positive supply voltage rail. It will help keep your bearings with inductors to remember that the average voltage $\langle V_{DD} - V_A \rangle$

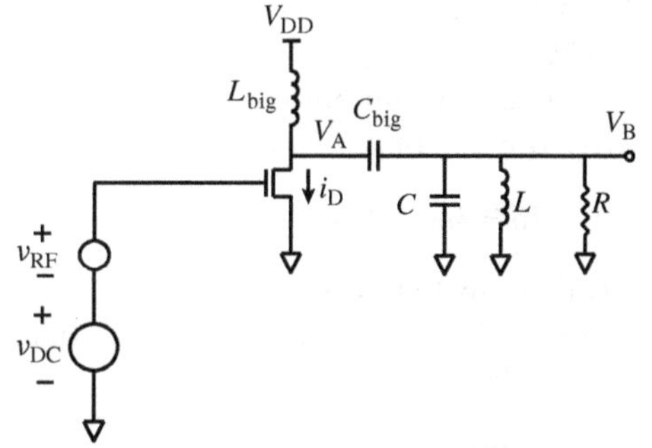

Figure 11.30 A generic power amplifier structure.

experience exponential rises as a continuous process. Rather, whatever is compounding goes from trivially small to astoundingly large seemingly overnight. This is the "magic" of compound interest, for example, or the astonishment that is Moore's Law scaling. We have now definitely crossed this threshold with the energy consumption of our communications infrastructure.

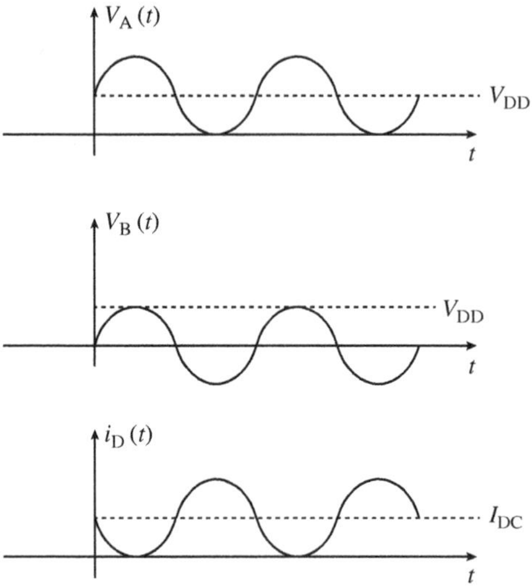

Figure 11.31 Voltage and current waveforms for the generic power amplifier structure at maximum output power.

must equal zero. If that were not true, $\langle \frac{dI_L}{dt} \rangle$ would be nonzero, implying an inductor current that rises or falls indefinitely. Power converter designers use this little morsel all the time to gain insight in analysis, together with the fact that the average current through any capacitor must be zero. The governing constitutive relations are $L\frac{dI_L}{dt} = V_L$ for inductors and $I_C = C\frac{dV_C}{dt}$ for capacitors. Taking the average of both sides of each equation, we get $L\langle \frac{dI_L}{dt} \rangle = \langle V_L \rangle$ and $\langle I_C \rangle = C\langle \frac{dV_C}{dt} \rangle$. It must be that $\langle \frac{dI_L}{dt} \rangle$ and $\langle \frac{dV_C}{dt} \rangle$ are zero, and so therefore are $\langle V_L \rangle$ and $\langle I_C \rangle$.

The inductor L_{big} is what is known as an "RF choke." It passes DC and low-frequency current but presents an extremely high impedance at RF frequencies. We write the total current through the inductor as a sum:

$$i_D = I_{DC} + i_{rf} \sin \omega t. \tag{11.31}$$

We then say that it is I_{DC} that flows through the inductor, while $i_{rf} \sin \omega t$ goes out through C_{big} to the load. This means that the average power dissipated in the load is given by

$$P(t) = (i_{rf} \sin \omega t)^2 \cdot R \tag{11.32}$$
$$\langle P(t) \rangle = \frac{1}{2} i_{rf}^2 R.$$

Now suppose that we make the RF signal current as big as it could possibly be for a given bias drain current. That is to say, we raise the amplitude of the RF current right up to the point where the transistor starts to "cut off," or where the *total* drain current vanishes (which happens to be at the peak of the voltage waveform). Looking at Eq. 11.31, we meet the cut off condition when

$$i_{\text{rf}} = i_{\text{peak}} = I_{\text{DC}}. \tag{11.33}$$

The total power consumption of this amplifying stage is

$$P_{\text{DC}} = I_{\text{DC}} V_{\text{DD}} = i_{\text{peak}} V_{\text{DD}}. \tag{11.34}$$

What all of this means is that the efficiency of the stage η, defined as *the ratio of power delivered to the load to the power drawn from the supply*, is given by

$$\eta = \frac{\frac{i^2_{\text{peak}} R}{2}}{I_{\text{DC}} V_{\text{DD}}} = \frac{\frac{i^2_{\text{peak}} R}{2}}{i_{\text{peak}} V_{\text{DD}}} = \frac{i_{\text{peak}} R}{2 V_{\text{DD}}}. \tag{11.35}$$

If we look at Figure 11.31 specifically to see that the drain voltage is pulled all the way down to zero, this corresponds to the absolute maximum signal current we could transmit.[16] Pulling the drain voltage all the way down to zero means that $i_{\text{peak}} R = V_{\text{DD}}$. We can therefore write the efficiency η at maximum possible signal power as

$$\eta_{\max} = \frac{V_{\text{DD}}}{2 V_{\text{DD}}} = 50\%. \tag{11.36}$$

This is an extremely well-known result for power amplifiers. You will often hear it said that the "maximum efficiency for a Class A power amplifier is 50%." What we have shown here is the analysis behind that fun fact.

It turns out that you can do better than 50% if you play around with bias points, and actually have the transistor on for only part of the cycle. If, for example, you only have the transistor conducting for 50% of the period of the sinusoidal carrier, you have what is called a "Class B" amplifier, and you can show that your efficiency gets as high as 78.5% for maximum signal power in these conditions. Anything between 100% conduction and 50% conduction is called, logically enough, "Class AB" operation, and yields an efficiency suitably between the extremes of Class A and Class B. And this is all before the myriad games you can play with output networks that are more exotic than a parallel *RLC*.[17]

There is much more to explore, but those additional details are mostly a distraction from the system-level knowledge that we seek. The point for now is that power amplifiers are not 100% efficient, even when we are passing a bland, information-free, unmodulated carrier. We will see that the news gets substantially worse when we modulate the carrier, as we must to send information.

---

16 It is at this level of input drive that we say the amplifier is "saturated." Stronger input drive will no longer cause an increase in output power.

17 Cripps, Steve C. *RF Power Amplifiers for Wireless Communications, Second Edition.* Norwood, MA: Artech House, 2006.

You may recall from Chapter 6 that when we send information on a carrier sinusoid, we modulate both its amplitude and phase. And a major insight from Section 5.3 is that wider range of amplitudes that we allow ourselves, the more bits we can encode in a given time sample. This in turn means more bits per second for a given allowed bandwidth. It turns out, however, that the amplitude modulation that is so desirable from the standpoint of *spectral* efficiency forces quite a degradation of the *energy* efficiency of our power amplifiers, and therefore of our whole radio link. We will get into the details shortly.

In the meantime, when it comes to amplitude modulation there is an outstanding characteristic of OFDM signals (see Section 6.2.5) that deserves special mention. A reasonable model to carry around in one's head about OFDM signals is that it is a tightly grouped set of N sinusoids that are roughly even in amplitude, and let's call that amplitude A. The *average signal power* of a group of N identical amplitude sinusoids is actually $N \cdot \frac{1}{2}A^2$. However, the instantaneous peak power that could occur if all the sinusoids were suddenly, exactly in phase is N^2A^2. This means that the peak-to-average-power ratio, or PAPR, goes as $2N$.[18] This means that for tens, hundreds, or more subcarriers, the PAPR in principle can be quite large. In practice, what saves us is the extreme rarity of having all of those carriers lining up exactly in phase. If our hardware happens to "clip," or fail to reproduce, the highest theoretical peaks, it's okay. But in general, OFDM signals have high PAPRs, and we'll see what this means for the energy efficiency of our power amplifiers and therefore our cellular network.

We look back now at our Class A PA, which we'll promote as a surrogate for any PA whose principal means of modulating its output power is via the amplitude of its input drive. Clearly the PA must be sized and biased to accommodate the peaks if we are to transmit with no distortion. If the PAPR is given to us in dB, then the average power is

$$\text{PAPR} = 10^{-\frac{\text{PAPR}}{10}} \cdot P_{\text{peak}}. \tag{11.37}$$

Now we go back to our efficiency calculation of Eq. 11.35, where the efficiency at peak output power was given by

$$\eta = \frac{\frac{i_{\text{peak}}^2 R}{2}}{I_{\text{DC}} V_{\text{DD}}}. \tag{11.38}$$

18 In dB, $10 \cdot \log_{10}(2N)$.

Table 11.1 Theoretical efficiencies for a Class A power amplifier in backoff. Reminder: this is the efficiency you get with a perfect, ideal power amplifier. Real world amplifiers are worse.

PAPR	Class A η
0 dB	50%
3 dB	25%
6 dB	12.5%
7 dB	10%
8 dB	8%
10 dB	5%

To figure out things at backoff, we scale the numerator by the backoff factor

$$\eta_{\text{backoff}} = 10^{-\frac{\text{PAPR}}{10}} \cdot \left(\frac{\frac{i_{\text{peak}}^2 R}{2}}{I_{\text{DC}} V_{\text{DD}}} \right). \tag{11.39}$$

We already worked out that the part in parentheses is 50%. So our efficiency at backoff is simply

$$\eta_{\text{backoff}} = 10^{-\frac{\text{PAPR}}{10}} \cdot 50\%. \tag{11.40}$$

We've been pretty big fans of equations in this book, but there is no better way to capture the horror of backoff efficiency than with a table. Behold Table 11.1.

And to really drive our point home, let's focus on that last row, for which the PAPR is 10 dB (not an outlandish number by any means). Here, 5% efficiency *means* that we draw 20 W from our power supply for every watt we deliver as Poynting flux out of the antenna. This means, in turn, that for every watt we successfully transmit, a staggering 19 W of heat is dissipated to our heat sinks. And now we start to do the math: a middle-of-the-road cell tower today might transmit 40 W of average output power per antenna, *after* losing 2–3 dB to post-PA loss (cavity filters, antenna cables, circulators, etc.). An aggressive deployment might have eight antennas per sector, times three sectors on a tower...times 10,000 towers for a medium size network. You get the idea. At the time of this writing, cellular networks consume 3–5% of the world's electricity, and that is before we've made much headway on the rollout of 5G service.

This energy stuff is not just an academic concern.

As a final wrap-up to this section, you can sometimes do a little better than this backoff efficiency. An ideal Class B amplifier, for example, has an efficiency degradation that goes roughly as the square root of the backoff rather than linearly with the backoff (i.e. it goes as $10^{\frac{1}{2}\left(-\frac{\text{PAPR}}{10}\right)}$), and then for Class AB it goes somewhere between linearly and as the square root.[19] If you are trying to figure out, for example, how a Doherty amplifier's efficiency falls with backoff from its efficiency peak, you won't go too far wrong if you assume a Class B-ish rolloff.

All of that said, what we have outlined in this section might be reasonably called "the efficiency challenge of power amplifiers." When spectral efficiency is at a premium, it forces us into modulation strategies that have a nonzero PAPR. If we do nothing special in the hardware, we bleed energy efficiency at a painful rate.

We will look at what we can do to fight for energy efficiency in the coming sections. Before that, though, we should also examine the linearity challenge posed by the power amplifier. Put simply, nonlinearity in the PA costs us spectral efficiency, which is...bad. We will see how in a Class A PA linearity and efficiency are totally at odds with each other. The deep message to the system architect is that choices such as modulation strategy have a profound impact on the energy efficiency of the network, and these impacts are due to Nature herself. The detailed mathematical form of the trade-off is highly dependent on our engineering creativity, and we will get into some of that. But the reader would be well served by getting into the mental habit of connecting architectural choices with energy trade-offs that are independent of implementation details.

11.4.2 Nonlinearity in Signal Chains

We kicked things off in the book with a discussion of LTI systems (Chapter 2) because we *love* LTI systems. The reasons for our ardor are threefold: (1) the math works out beautifully for LTI systems; (2) the math works out beautifully for LTI systems; and (3) Nature allows us to get away with linear approximations to obviously nonlinear systems a jaw-droppingly high percentage of the time. If you ever take yourself on a walking tour of nonlinear system theory, you will find a large number of highly specialized results that were so difficult to conceive that their originators still have their names attached to them. But in wild contrast to LTI systems, with nonlinear systems generalizable insight is hard to come by.

Sometimes, though, Nature does not allow us to get away with linear approximations. Communications links are a great example of this type of system, especially

---

19 In Class A amplifiers, the transistors conduct current for all of each carrier cycle; in Class B amplifiers they conduct current for *half* of each carrier cycle; and Class AB amplifiers are somewhere in between. There are also Class C amplifiers, wherein the transistor conducts for less than half of each carrier cycle. Class C amplifiers can get you high efficiency in exchange for difficulty delivering any power.

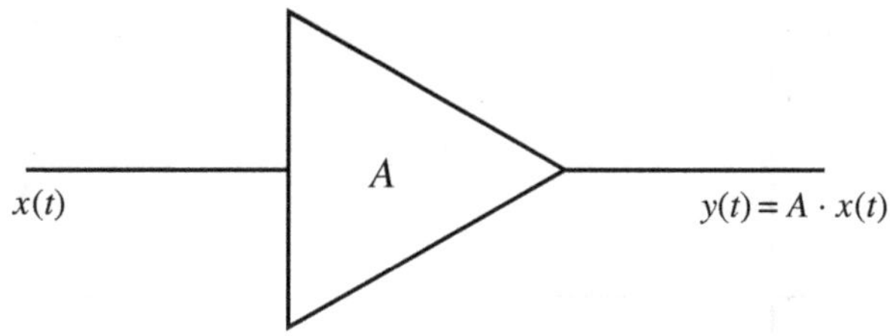

Figure 11.32 A perfectly linear amplifier.

when, as we must, we get detailed about trade-offs between energy and spectral efficiency.

We begin our analysis by reminding ourselves of what a purely linear signal block is and of how it behaves. In particular, when we say that an amplifier is linear and memoryless, we mean what we show in Figure 11.32. The output at any given time is simply a scaled version of the input.[20]

Now suppose instead that we have an explicitly nonlinear amplifier. We have the input/output characteristic as follows:

$$y(t) = c_0 + c_1 \cdot x(t) + c_2 \cdot [x(t)]^2. \tag{11.41}$$

Suppose further that the input is given by

$$x(t) = A\cos\omega_1 t + A\cos\omega_2 t. \tag{11.42}$$

We go slowly now, and make use of our trusty trigonometric identities like $(\cos x) \cdot (\cos y) = \frac{1}{2}\cos(x+y) + \frac{1}{2}\cos(x-y)$, to give us

$$\begin{aligned} y(t) = c_0 & \\ & + c_1[A\cos\omega_1 t + A\cos\omega_2 t] + c_2A^2 + \frac{1}{2}c_2A^2\cos 2\omega_1 t \\ & + \frac{1}{2}c_2A^2\cos 2\omega_2 t + c_2A^2\cos(\omega_1+\omega_2)t + c_2A^2\cos(\omega_1-\omega_2)t. \end{aligned} \tag{11.43}$$

Very interesting. What you find is that nonlinearities, such as arrive via $c_2 \neq 0$, tend to "splatter" the signal in the frequency domain. With purely linear systems this never happens. The sinusoids present at the input may get amplitude scaled and phase shifted, but they do *not* spontaneously give rise to new spectral party

20 The "memoryless" part is that the output at any time does not depend in any way on past inputs. The impulse response is itself an impulse; the amplifier does not filter the signal in the way that real amplifiers do and must. We will stay with memoryless systems in our analysis in the interest of favoring broadly applicable insight for the system level architect. If and when you are tasked with implementing distortion reduction strategies for power amplifiers, particularly for broadband applications, memory effects must be included in the model of the nonlinearity.

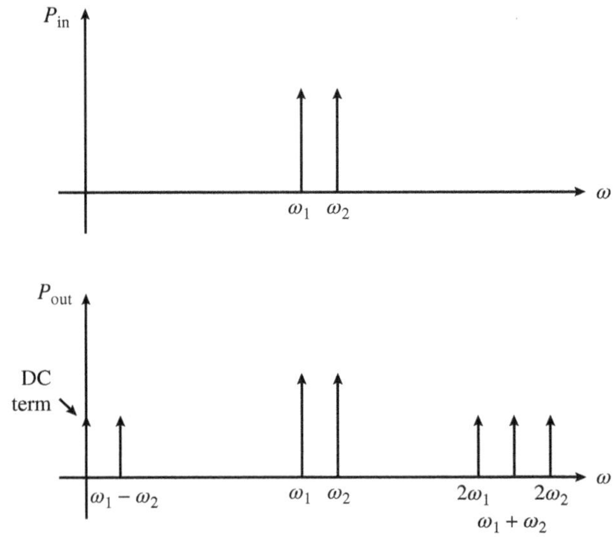

Figure 11.33 Passing two tones through a second-order nonlinearity.

guests. Let's look graphically at the experiment we just tried, a so-called two-tone test on a second order nonlinearity. See Figure 11.33.

Another way to describe what just happened is that as a result of the nonlinearity we get an *expansion in bandwidth*. That is to say, after passing a signal through a block with strong nonlinearity, we find ourselves occupying more spectral real estate *without a corresponding increase in the information we are sending*. In the Shannon driven world of communications, where

$$C = B \cdot \log_2 (\mathrm{SNR} + 1), \tag{11.44}$$

this waste of bandwidth is a cardinal sin.

Now as it happens, even-order nonlinearities tend to be less serious in many radio communications systems. If we look at what happened in Eq. 11.43, for example, in addition to the desired products at ω_1 and ω_2, we have artifacts that are either near DC or at roughly twice our center frequency. Either way, these artifacts are usually easy to filter out. But now let's take things up a notch and consider a nonlinear transfer function with terms up to third-order:

$$y(t) = c_0 + c_1 \cdot x(t) + c_2 \cdot [x(t)]^2 + c_3[x(t)]^3. \tag{11.45}$$

The linear term is what we want, of course, and now we know that artifacts arising from the quadratic (squaring) term may be annoying but can be filtered out. (The DC term, attributed to nonzero c_0, is also easily dispatched.) But the third-order term, and indeed all odd-order terms, are a different beast. The problem is that they generate "cross-modulation" artifacts that fall either uncomfortably close to or entirely within the band of interest. You cannot filter them out.[21] One way or another, you have to make sure that they never arise too strongly in the first place.

Let's look at what the third-order term introduces in response to our two-tone input $x(t) = A\cos\omega_1 t + A\cos\omega_2 t$. If we work our trigonometric identities what results from the cubic term in Eq. 11.45 is

$$\begin{aligned} c_3[x(t)]^3 &= \frac{c_3A^3}{4}[\cos 3\omega_1 t + \cos 3\omega_2 t] \\ &\quad \frac{3c_3A^3}{4}[\cos(\omega_1 + 2\omega_2)t + \cos(\omega_1 - 2\omega_2)t \\ &\quad + \cos(2\omega_1 + \omega_2)t + \cos(2\omega_1 - \omega_2)t]. \end{aligned} \tag{11.46}$$

It is clarifying to make a substitution:

$$\begin{aligned} \omega_1 &= \omega_0 - \frac{B}{2} \\ \omega_2 &= \omega_0 + \frac{B}{2}. \end{aligned} \tag{11.47}$$

Pictorially, we draw our two tones now as in Figure 11.34. Insight follows naturally as a byproduct of tedious accounting:

$$\begin{aligned} 3\omega_1 &\to 3(\omega_0 - B/2) = 3\omega_0 - 3B/2 \to \text{Easily filtered} \\ 3\omega_2 &\to 3(\omega_0 + B/2) = 3\omega_0 + 3B/2 \to \text{Easily filtered} \\ \omega_1 + 2\omega_2 &\to 3\omega_0 + B/2 \to \text{Easily filtered} \\ 2\omega_1 + \omega_2 &\to 3\omega_0 - B/2 \to \text{Easily filtered} \\ \omega_1 - 2\omega_2 &\to -\omega_0 - 3B/2 \to \omega_0 + 3B/2 \to \text{Problem} \\ 2\omega_1 - \omega_2 &\to \omega_0 - 3B/2 \to \text{Problem} \end{aligned} \tag{11.48}$$

If you look again at the picture in Figure 11.35, what we see is that the third-order distortion products do indeed fall close to our band of interest. This is a huge headache.

As you think about it, you can start to see an outline of what we mean when we talk about "spectral spreading." This was only a two-tone experiment, of course, but one could reasonably speak of a "bandwidth" as the separation between the two tones. After going through the third-order nonlinearity, and assuming a

21 The selectivity of such filters would make them impractically large and complex; see Section 11.2.

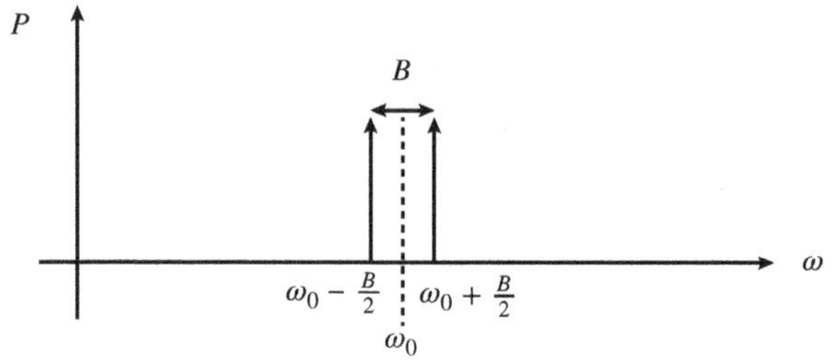

Figure 11.34 Using our two tones to define a signal bandwidth.

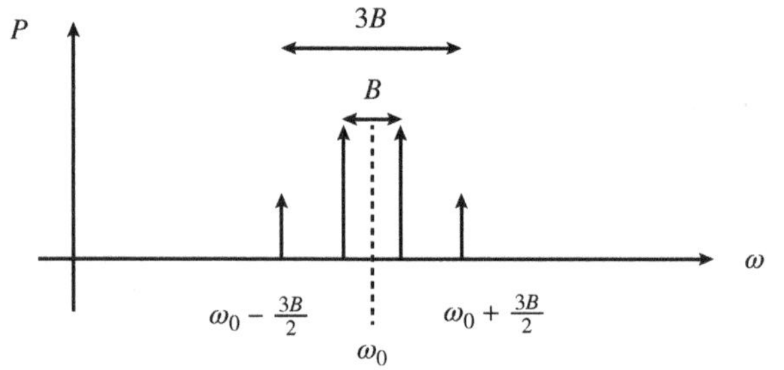

Figure 11.35 A third-order nonlinearity expands our bandwidth by a factor of three.

filtering of the super high frequency artifacts (e.g. 3ω), we are left with a set of tones that spans three times the original bandwidth.

So the last jump that we have to make is from a two-tone experiment to a more general, narrowband signal. What can we say specifically about a continuous frequency distribution as shown in Figure 11.36, once it is passed through our third-order nonlinearity? It helps to remember first that we can write our signal using the Fourier transform as

$$x(t) = \frac{1}{2\pi} \int_{-\infty}^{\infty} X(j\omega)e^{j\omega t}d\omega, \tag{11.49}$$

and recall that this integral is mere shorthand for

$$x(t) = \frac{1}{2\pi} \lim_{\Delta\omega \to 0} \sum_{k=-\infty}^{\infty} X(jk \cdot \Delta\omega)e^{j(k \cdot \Delta\omega)t}\Delta\omega. \tag{11.50}$$

From here, it is helpful to step out of undergraduate analysis mode and step into hard-nosed, numerical computation mode. In this mode, we toss the limit symbol and just set $\Delta\omega$ as small as we need:

$$x(t) \approx \frac{1}{2\pi} \sum_{k} X(jk \cdot \Delta\omega)e^{j(k \cdot \Delta\omega)t}\Delta\omega. \tag{11.51}$$

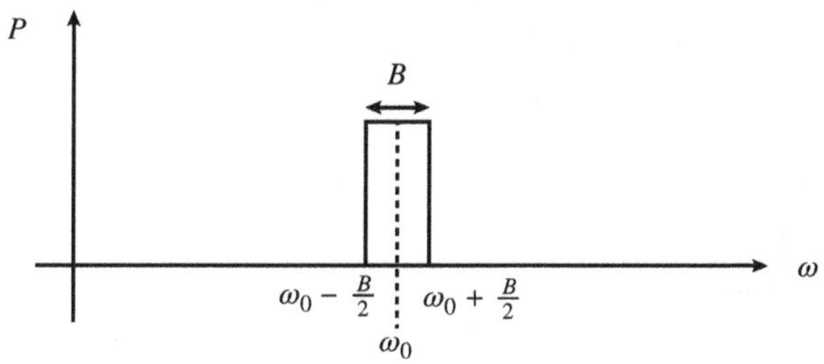

Figure 11.36 A signal with total bandwidth of *B*.

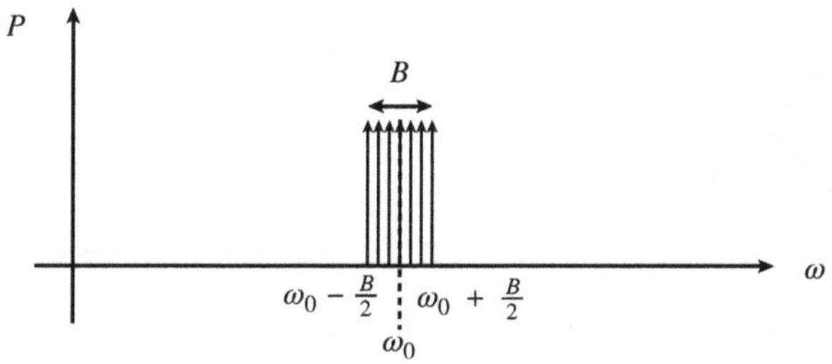

Figure 11.37 Discretizing our frequency domain representation *B*.

In this mindset, our formerly continuous frequency distribution looks more like Figure 11.37.

At last, we see that the key to jumping from the two-tone experiment to a more general narrowband signal lies in the following two realizations:

- The outer limits of the distorted frequency spectrum will be defined by mixing products involving the sinusoids at $\omega_0 - B/2$ and $\omega_0 + B/2$.
- All other mixing products necessarily fall between these two extremes.

It is worth taking your time and internalizing these assertions. Their simplicity is actually analytically freeing. Notice we do *not* demand of ourselves to know the exact amplitudes (and phases) of all of our distortion products, as a computation engine is better suited to that task. For hand analysis, we are content to see that there is a 3× bandwidth expansion, and maybe, as a flourish of insight, the tones on the very outskirts will be smaller than those in the middle. We may therefore draw a sketch or a cartoon as shown in Figure 11.38. This 3× bandwidth expansion is what third-order distortion is famous for. Generally what you see on paper here is what you see in the laboratory, and you see similar spectral spreading with all odd-order nonlinearities. As a practical matter, what we find in most real power amplifiers is that the third- and fifth-order nonlinearities affect us most strongly. We will have more to say about this, but this is why a rule of thumb for digital

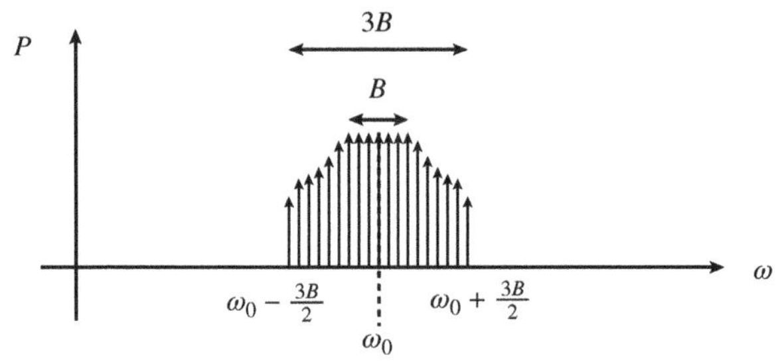

Figure 11.38 Another look at 3× bandwidth expansion.

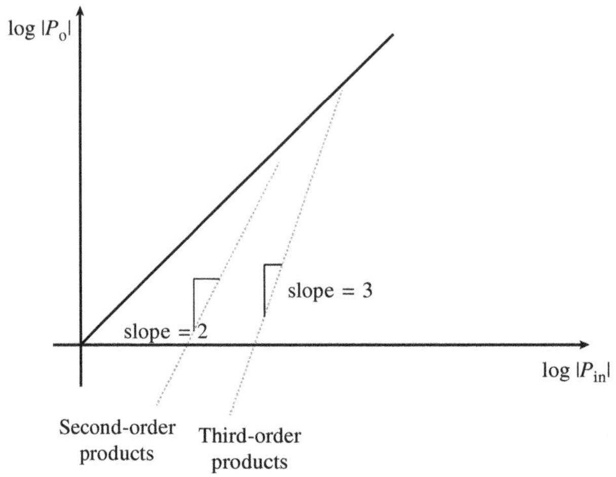

Figure 11.39 A comparative look at how linear and distortion products scale in the output with increasing input signal power.

compensation of nonlinear power amplifiers requires that we have the processing power for 3–5× the bandwidth of our desired signal.

We close out this section on nonlinearity by expanding on a particularly important fact. Notice from Eq. 11.43 that the strength of the second-order products went as the square of the original signal amplitude. Notice similarly that all the third-order products in Eq. 11.46 rise and fall as the cube of the input signal amplitude. It is instructive to examine on a log–log plot, as shown in Figure 11.39. This plot actually points to a strategy for dealing with nonlinearity. If we want to decrease the power of our distortion products relative to the linear output, one possibility is to *back off our input drive relative to the level that saturates the amplifier.*[22]

22 If you push hard enough, the amplifier "saturates" or "clips," which is where further increase of the input drive fails to increase the output power.

The further we back off, the greater the ratio of linear products to distortion products. (The other thing to know is that this ratio is typically how we characterize "how linear" an amplifier is.) Notice that for every dB we back off the drive, linear products decrease by 1 dB, the second-order products decrease by 2 dB, and the third-order products decrease by 3 dB.

Of course, we recall from Section 11.4.1 that the further we backoff from saturation, the lower the efficiency of the stage. We thus arrive at the conceptual heart of a hugely critical trade-off in electronic communications systems. Now we turn our attention rather urgently to this trade-off, and see what we can do to mitigate it.

11.4.3 The Linearity/Efficiency Trade-off

Having stumbled into the famous linearity/efficiency trade-off, it is instructive to look at some numbers. As a start, let us say that the average power for our transmitted signal is 40 W, and that our PAPR is 6 dB. Looking at Table 11.1, we see that for a Class A amplifier the best efficiency we could ever hope for is 12.5%. This means that our power consumption is $\frac{1}{0.125} \cdot 40\text{ W} = 320\text{ W}$.

But the news gets worse. We look at the output from our PA on a spectrum analyzer, and see what we depict in Figure 11.40. We see that outside of our assigned channel, our distortion products spill energy into the adjacent channel. We'll suppose that the signal power in the adjacent channel is due to third-order products, and what we measure is that the power in the adjacent channel is only 30 dB down from the power in the main channel. An extremely common specification at the time of writing is that we want these products to be 45 dB down. What to do?

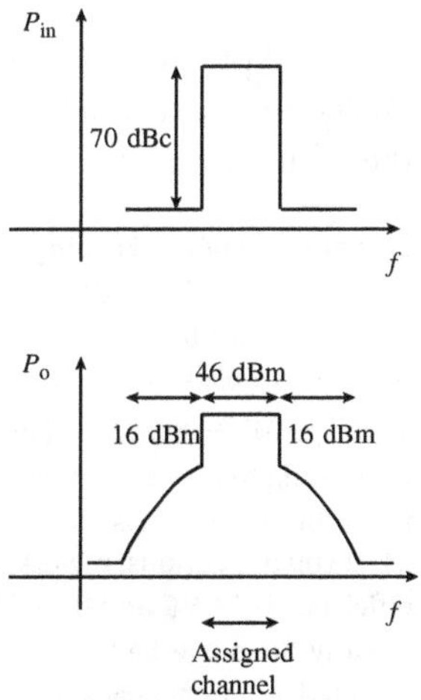

Figure 11.40 A look at a signal spectrum at the input and then at the output of a distorting power amplifier. Using the lower axes, we would say that the "adjacent channel power ratio," or ACPR, is 30 dB.

Well, the analysis from Section 11.4.2 gives us one path to explore. If these distortion products do indeed arise from third-order nonlinearity, then we know that for every dB of backoff in output power, we drop 3 dB in power for the third-order distortion products and therefore gain 2 dB of separation between the signal power and the power in the adjacent channel. To pick up the 15 dB of separation that we need, then, we must build a *bigger* PA, such that 46 dBm of output power represents an additional 7.5 dB of backoff from the peak output power.[23] Having committed to this design choice, we can re-engage with Eq. 11.40 to calculate the efficiency of our new PA:

$$\eta = 10^{-\frac{13.5}{10}} \cdot 50\% = 2.2\%. \tag{11.52}$$

We can see fully what it has cost us to meet our linearity target, which we strive for because we are trying to be responsible users of this finite spectrum resource. We started with a power consumption of 320 W; we are now at an astounding $40\ \text{W} \cdot \left(\frac{1}{0.022}\right) = 1800\ \text{W}$.[24]

Serious business.

This is why the linearity/efficiency trade-off is such a big deal, and why so much effort has been expended in research and development to do better than backoff as a strategy. Our industry has made a great deal of progress along essentially two paths. The first path is to use *digital predistortion* to compensate for nonlinearity in the PA, so that we can get better linearity without having to resort to backoff. In combination with predistortion, we have expended a great deal of effort and creativity in power amplifier structures that are not so rigid in their load impedance and/or bias conditions, the basic idea being that we can *dynamically size the PA in response to the signal envelope fluctuations.* This will make more sense when we get into examples. Meantime, we turn our attention first to digital predistortion, or DPD.

A good picture to have in your mind for DPD is Figure 11.41. The idea behind DPD is that we precede the PA with a nonlinearity that is the functional inverse of the PA nonlinearity. The resulting cascade is perfectly linear.

---

23 For those of you keeping track, our original design delivered 40 W of average output power assuming a PAPR of 6 dB, which means that the PA was designed for a peak power of 160 W. We are now asking for a design that can still deliver 40 W of average power, but now 40 W represents a back off of 6 dB + 7.5 dB = 13.5 dB. The peak power of our new PA is therefore $40\ \text{W} \cdot 10^{\frac{13.5}{10}} = 895\ \text{W}$. It is worth gaining proficiency in these shocking little calculations, translating frequently between power in logarithmic units (dB) and linear units (W). If you stay in dB for too long, it is too easy to lose track of the energy implications of our design choices.

24 If you are curious, rework this problem assuming Class B characteristics. There, the peak efficiency is 78.5%, and the efficiency falls off as the square root of the back off rather than linearly. So, not as bad. But the linearity that you start with is worse than with Class A, so the required additional back off will be greater.

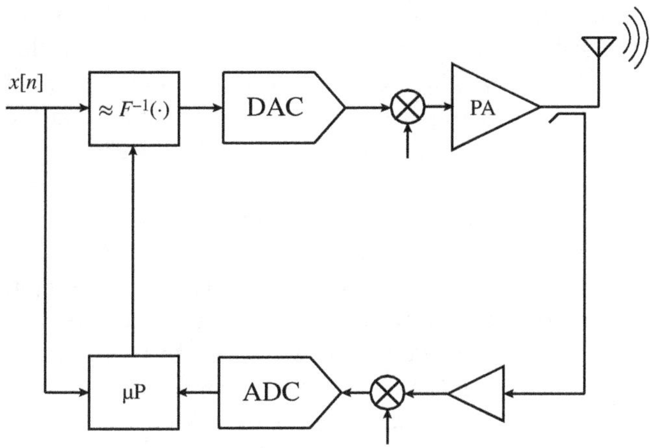

Figure 11.41 Adaptive digital predistortion. The inverse function $F^{-1}(\cdot)$ is computed such that the cascade of it and the nonlinear PA is, end-to-end, a linear function that therefore does not produce distortion products.

As you might expect, there is a whole art and science to doing the adaptive nonlinear modeling that gets us $F^{-1}(\cdot)$. One of the tools that we have is the least squares method of Section 9.7, which would allow us to construct a nonlinear model of the PA. Resist the temptation to analytically invert that model once we have it, as you can use that nonlinear model to construct a lookup table instead. That is to say, suppose that you have a model for $f(x)$. You first construct a finely spaced table of values of x and values of $f(x)$. Then, whenever you desire output $f(x_0)$, you simply go to your table and locate the corresponding x_0.[25] The overall ambition behind this technique is that we get our desired linearity *without* having to do any backoff. We see in our numerical example, using a Class A amplifier, that this would save us nearly 1.5 kW!

Now the benefits of DPD in this case are obvious, so much so that we have brushed aside any power penalty associated with the digital signal processing power itself. Moore's law scaling being the miracle it has been, we are right to assume that any digital power would be negligible. But this leads us to the question of situations where the trouble of DPD is "not worth it." Specifically, what if our output power was only 0.1 W (100 mW) instead of 40 W?

Running the same numbers as before, relying only on back off blows up our power consumption from $\frac{1}{0.125} \cdot 100\text{ mW} = 800\text{ mW}$ to $100\text{ mW} \cdot \frac{1}{0.022} = 4.5\text{ W}$.

25 It can get more complicated, especially when there are memory effects to compensate for, but this simple model conveys the right spirit and will serve for system level understanding.

At the time of writing, that delta of 3.7 W would buy you a *lot* of floating-point operations per second (FLOPS), far more than would be necessary for a good DPD implementation. And this observation remains true at least for another order of magnitude, e.g. dialing the output power down to a measly 10 mW. From the admittedly narrow perspective of saving overall system power, there are few situations where DPD is "not worth it." What people usually mean when they say such things is that in their particular application power is free or unimportant, and/or pales in comparison to the cost of additional parts and engineering effort.

Another important nuance is that in order to successfully predistort, the input signal chain must be able to support a bandwidth consistent with the nonlinear spreading caused by the PA. So if, say, the PA is verified to have fifth-order nonlinearity, the analog signal chain driving the PA must be able to support a bandwidth of $5B$.[26] Bearing in mind the *RLC* energy costs as described in Section 3.3.3, and assuming that the analog lineup feeding the PA is built around parallel resonant structures, we may associate an analog energy cost to running DPD that goes as $(5B)^2$. Despite appearances, this is not usually a huge concern. But it does come up, for example, in mm-wave links where the ambition is to support bandwidths of 400 MHz or more.

All of this said, we should remind ourselves that we are fighting for an efficiency of 12.5% in this example. And we should further remember that 12.5% is an idealization of which we fall short in practice(!). Numerically, then, our radio transmitter is justly described as a "space heater that also transmits radio signals." We can, and must, do better.

Beyond DPD, there are two basic approaches to modulating the output power while keeping the efficiency high. What we *want* is to maintain the full voltage swing at the output allowed by the supply rail, and yet still somehow modulate the output power.

Option #1 is load modulating according to Figure 11.42. In this scenario, the idea is that we keep $i_{\text{rf}} \cdot R'_{\text{L}} = V_{\text{DD}}$ at all times. We accomplish this by magically modulating the impedance network to vary R'_{L}. Energy conservation being what it is, we have is a scheme that varies the output power in inverse proportion to R'_{L}. To see this, recall that the average power delivered to R'_{L} for a sinusoidal current

26 You can think of it as the PA produces distortion products out to $5B$; the predistorting signal chain then produces corrections that are *small enough to be linearly processed by the PA*, which then additively cancel the distortion products at the PA output. It's all approximate, but not a *bad* approximation. And a great deal more useful than the analytically rigorous route!

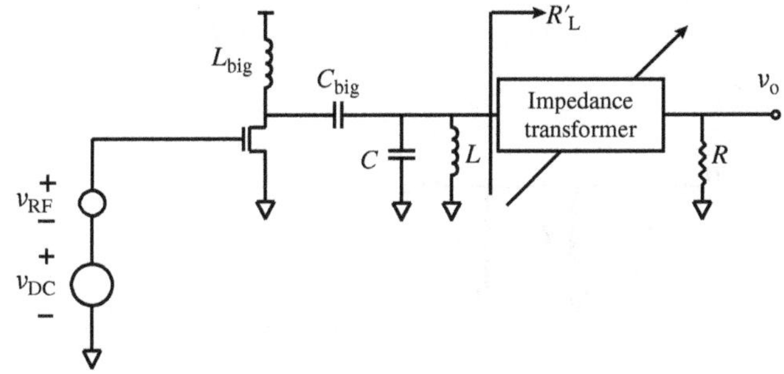

Figure 11.42 Load modulation in a generic power amplifier structure. The idea is to maintain high signal swing at the drain of the transistor, ensuring high efficiency operation. Modulating the impedance transformation then modulates the output power. (The key to understanding is to apply conservation of energy: power flowing into R'_L all winds up in R if the impedance transformer is lossless.)

drive (whose amplitude is i_{rf}) is $\frac{1}{2}i_{rf}^2 R'_L$. But we keep $i_{rf} \cdot R'_L = V_{DD}$, which means that $i_{rf} = \frac{V_{DD}}{R'_L}$. It follows, then, that

$$
\begin{aligned}
P_{out} &= \frac{1}{2} i_{rf}^2 R'_L \\
&= \frac{1}{2}\left(\frac{V_{DD}}{R'_L}\right)^2 \\
&= \frac{1}{2}\frac{V_{DD}^2}{R'_L}.
\end{aligned}
\tag{11.53}
$$

Option #2 is supply modulation, as shown in Figure 11.43. As with load modulation, our obligation is to ensure that $i_{rf} \cdot R'_L = V_{DD}$ at all times. But by varying V_{DD}, we vary the output power.

That's the list! And neither technique can be casually pulled off in practice. It is fair to say that Doherty PAs are the most successful load-modulating PAs ever deployed. When it comes to supply modulation, discrete supply modulation takes the crown, although the first field deployments are progressing as of this writing. We will discuss these two techniques in turn and digress briefly to explain why continuous supply modulation, known more commonly as "envelope tracking," always ends in tears. Finally, note that it is not usually possible to pull off either load or supply modulation without the aid of DPD to make everything work out.

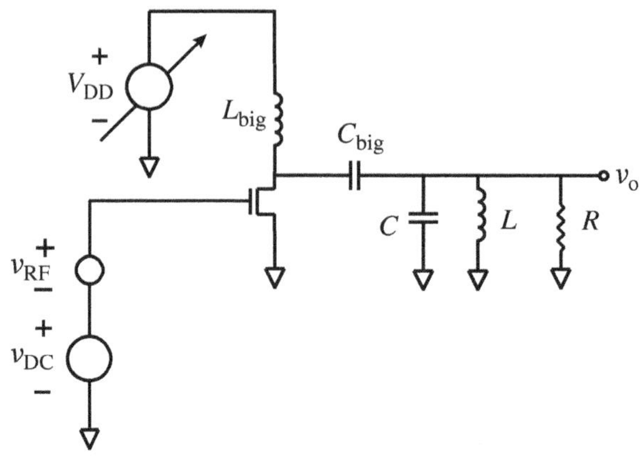

Figure 11.43 Drain modulation in a generic power amplifier structure. This time, we keep the drain bias as low as possible at all times, responsive to the signal envelope. This way, the PA is able to meet its instantaneous output power demands while staying in or near the high-efficiency, saturation region of operation.

11.4.4 Doherty: A Highly Successful Power Amplifier Technique

The Doherty amplifier is simply an ingenious idea, way ahead of its time. It was invented by William Doherty at Bell Laboratories in 1936.[27] It took until the late 1990s to explode into common application. This is because the nonlinearities introduced by the load modulation mechanism needed DPD for correction, and DPD became viable at this time. Doherty is an outstanding realization of the load modulation concept and today is deployed in nearly every cellular tower in the world.

Given how long Doherty has been in use, it is surprisingly difficult to find an explanation for its operation that is easy to follow. Doherty himself gave an excellent and very clear description in his original paper, and the curious reader is encouraged to dig it up and take a look. What we offer here is a shorter explanation that does not appeal directly to a knowledge of vacuum tubes. That said, the recommendation to look at the original paper stands.

Figure 11.44 shows how we typically draw a Doherty power amplifier today. The $\lambda/4$ segments are quarter wavelength transmission lines, whose function we will describe in sufficient detail in order for the reader to get the overall Doherty concept.

---

27 W. H. Doherty, "*A New High Efficiency Power Amplifier for Modulated Waves*," *Proc. of the IRE,* 24(9):1163–1182, Sept. 1936.

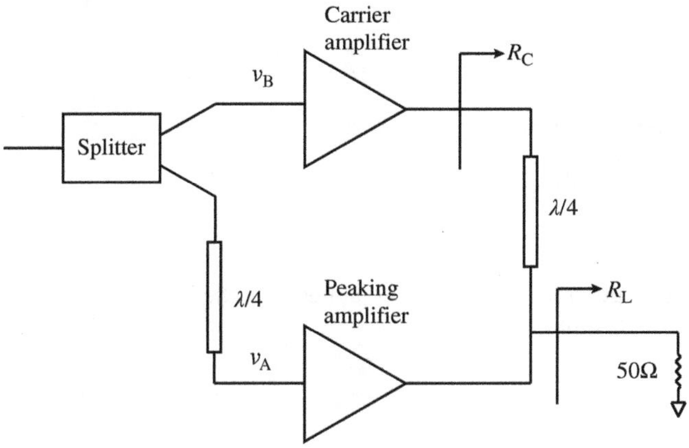

Figure 11.44 The Doherty amplifier.

We'll focus first on the most interesting region of operation, which is when the carrier amplifier is completely saturated. That is to say, its output voltage is swinging at the maximum amount as determined by its supply voltage. The power delivered in this circumstance is given by

$$P_{\text{carrier}} = \frac{1}{2}\frac{V_{\text{DD}}^2}{R_{\text{C}}}. \tag{11.54}$$

So if R_C is modulated somehow, then the power delivered by the carrier amplifier can be varied *while still staying in its high-efficiency, full voltage swing mode of operation.*

The best way to think about how we pull off the load modulation trick starts with seeing the peaking amplifier, nominally biased in Class C or thereabouts, as a large signal transconductance G_m. By G_m, here we mean

$$G_{\text{m-P}} = \frac{i_{\text{fundamental}}}{v_{\text{A}}}, \tag{11.55}$$

where v_A is the voltage amplitude of the input drive to the peaking amplifier, and $i_{\text{fundamental}}$ is the amplitude of the fundamental Fourier component of the current output, which is at the carrier frequency. If we look at the current output of the peaking amplifier's transconductance, it is something like Figure 11.45. As the overall amplifier is driven harder, and as v_A therefore increases, both I_{peak} and the conduction angle Θ increase. $i_{\text{fundamental}}$ increases accordingly. More importantly, it turns out that the magnitude of $G_{\text{m-P}}$ increases monotonically with the input drive.

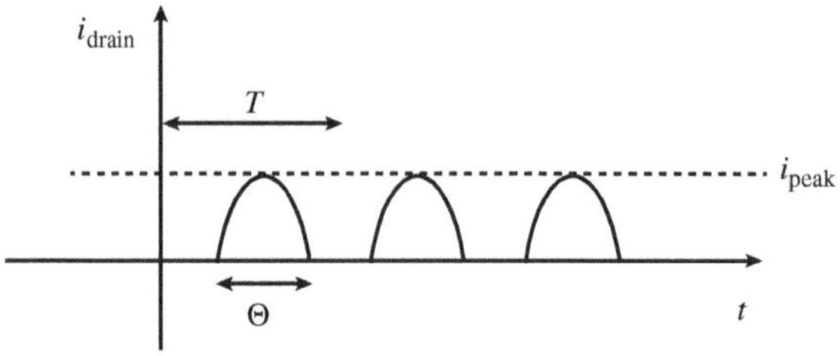

Figure 11.45 The current output from the peaking amplifier once the input drive has gotten high enough for it to activate.

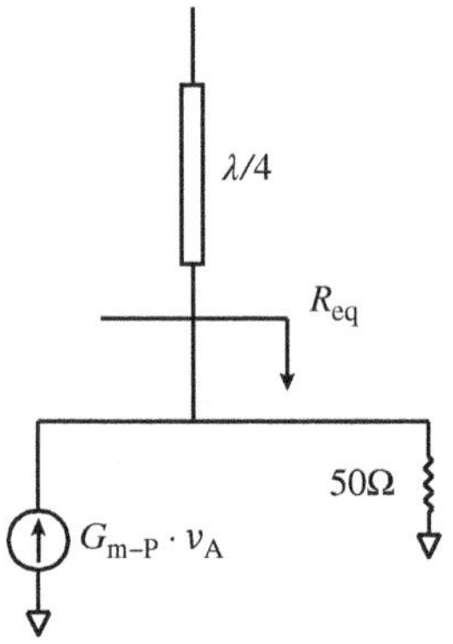

Figure 11.46 The key structure for understanding load pulling in a Doherty amplifier.

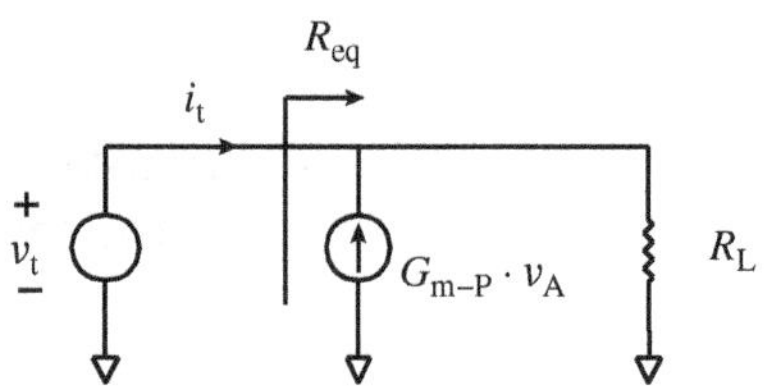

Figure 11.47 Determining the driving impedance R_{eq}.

The key to understanding Doherty comes down to understanding the impedance seen by the output $\lambda/4$ segment looking into the parallel combination of the peaking amplifier drain and R_L, as shown in Figure 11.46. We do the analysis now in the usual way, by setting up a test voltage source v_t and calculating the resulting current; R_{eq} is then v_t/i_t as shown in Figure 11.47.

But there remains one bit of cleverness, which is because the carrier amplifier and peaking amplifier are driven by the two legs of a phase matched splitter, and because we have preserved this phase matching with $\lambda/4$ transmission lines strategically placed in both the carrier and peaking paths, we have the unusual situation that v_t and v_A track each other. We capture this in our figure via proportionality

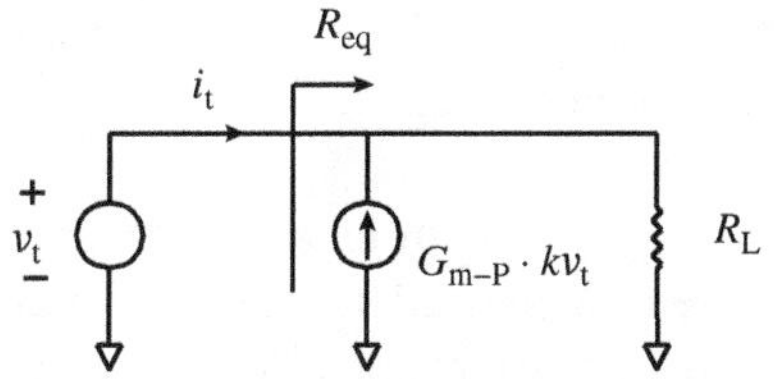

Figure 11.48 Determining the driving impedance R_{eq}, but recognizing that v_A and v_t track each other.

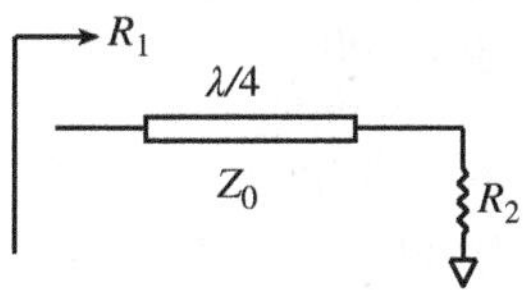

Figure 11.49 The quarter-wave transformer.

constant k as shown in Figure 11.48. Now if you work it out, what you find is that the impedance seen by the output $\lambda/4$ segment is given by

$$R_{eq} = \frac{v_t}{i_t} = \frac{R_L}{1 - G_{m-P} \cdot kR_L}. \tag{11.56}$$

It is important to remember now that G_{m-P} increases with input drive v_A as the peaking amplifier comes on more strongly. This point bears repeating, and you will see why after we supply one more piece of microwave inside baseball.

Recall that the carrier amplifier looks into this particular node through a $\lambda/4$-length piece of transmission line. Such structures are often referred to as "quarter-wave transformers." Referring to Figure 11.49, they behave in the following remarkable way. If Z_0 is the characteristic impedance of the transmission line, then we have for R_1:[28]

$$R_1 = \frac{Z_0^2}{R_2}. \tag{11.57}$$

This result allows us to go all the way back to Figure 11.44 and write for R_C

$$R_C = Z_0^2 \frac{1 - G_{m-P} \cdot kR_L}{R_L}. \tag{11.58}$$

It finally becomes clear why we have hammered on the fact that G_{m-P} increases with increasing input drive of the overall amplifier. Trusting that $|G_{m-P} \cdot kR_L| < 1$, what we see is that R_C decreases as the input drive increases, and therefore the

28 In Section 6.1, we waxed poetic about the "narrowband assumption" in RF circuits. What we see with the quarter-wave transformer is an example of a whole genre of impedance transformation tricks that are only valid over a narrow frequency band.

power delivered by the carrier amplifier *increases* despite the fact that its output voltage swing stays constant. This is the essence of load pulling.

Pretty neat.

We close this section by returning to an earlier remark about how the Doherty structure needed DPD in order to be successful. We have glossed over quite a lot of detailed biasing considerations in the rush to get to the master trick of load pulling. The truth is that the input-output gain of the overall amplifier is not constant as we go from small inputs to large inputs, and this nonuniform gain is the very definition of nonlinearity. Doherty himself was quite aware of this nonlinearity issue and, in fact, proposed correcting it by enclosing the whole amplifier within an analog feedback loop. This turns out to only be plausible for extremely low modulation bandwidths, but that was no barrier in 1936. Still, bandwidth requirements grew rapidly over the ensuing decades, leaving the Doherty amplifier somewhat orphaned since its nonlinearity could not be tamed for demanding applications. It was not until the 1990s that DPD rose to the occasion, and at that time the Doherty amplifier rightly assumed its place as a masterful sacrifice of analog linearity in the service of efficiency. With DPD to reclaim the lost linearity, a huge step was taken forward in communications technology.

11.4.5 Backoff: Doherty's Achilles Heel

Having zeroed in on the circuit mechanics that make Doherty work, it helps to step back and see what we have accomplished. The best and easiest way to appreciate Doherty in practice is to look at a plot of its drain efficiency vs. its output power, as shown in Figure 11.50. When properly matched to signal statistics, we see that in addition to high efficiency at maximum output power, we also have an efficiency "tent" that occurs a few dB below that. The best is to match the location of this tent to the signal PAPR. You can see that we do so much better than Class A if we get all our ducks in a row. Really impressive, and a great advance on the power amplifier state-of-the-art. Prior to Doherty, high-efficiency PAs were really only possible if we confined ourselves to constant envelope modulation. Here you can see that if we keep the PAPR constant, which is an *extremely light* restriction, we can do quite well.

What you can see from Figure 11.50, however, is that we have to commit ahead of time to the average power being constant. And then, in the field it is worse than this, in that we have to commit the hardware ahead of the time to the *maximum* average output power that we expect to use. And *then*, somehow, it gets even worse: deployed base stations are almost never operating beyond half maximum output power, corresponding to a high traffic condition in which half of the OFDM subcarriers are being used. The result is that...there is almost certainly not

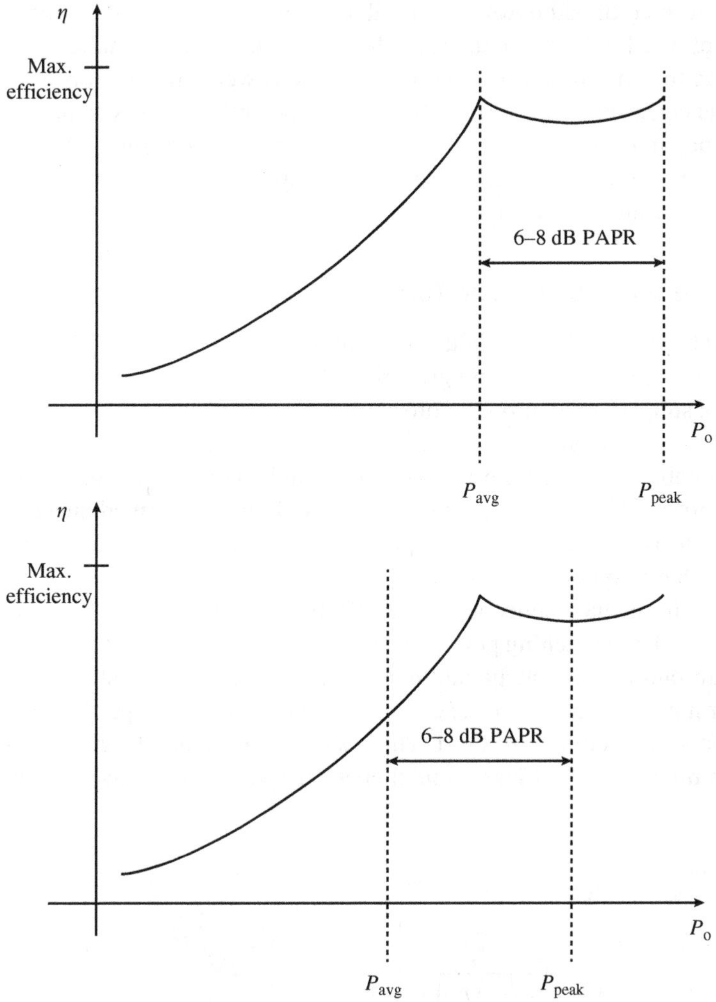

Figure 11.50 On the axes above, we have the average power lined up exactly with the famous Doherty peak, 6–8 dB below peak power. On the axes below, we have how almost all Doherty amplifier are run in the field, which is to say, never at maximum capacity. A quick and dirty estimate for the drain efficiency is the efficiency at P_{avg}.

a single Doherty amplifier on Earth, in a field deployment, that is operating in its efficiency tent.[29]

---

29 When all the subcarriers are being used, that is called "being fully allocated" and corresponds to the maximum average output power. It also, necessarily, corresponds to the link operating at absolute maximum capacity. This is a condition that network planners justly fear,

There is one more bit of bad news. If you talk to network planners, the people who actually deploy radios in the field, they will wryly observe that no matter how much you reduce the output power of their radios, the power consumption *stays the same!* This is consistent with a roughly Class A back off efficiency behavior from the efficiency tent in Figure 11.50. The power consumption, remember, is $P_{DC} = P_o/\eta$, and that for Class A, the efficiency η is proportional to output power.
We would like very much to do better.

11.4.6 Envelope Tracking: Don't Do This

Doherty is a great example of load modulation, and you will recall that at the end of Section 11.4.3 we mentioned another great branch of power amplifier solutions along the lines of supply modulation. Before getting to a technique that is proving to be successful, we will spend a little bit of time on a technique called envelope tracking. It is probably true that almost every serious student of radio communications has been intrigued by this idea. We shall see that despite its immediate conceptual appeal, efforts to extend the technique beyond all but the narrowest band implementations have no hope of success.

So what is the "immediate conceptual appeal"? It is easy to explain. We start with the state of modern switching power converters, which can convert an input DC voltage to an output DC voltage and deliver real power at astonishing efficiencies (95% or higher in some cases). Seeking to leverage this concept, the ideal envelope tracker is shown in Figure 11.51. The basic concept is that the switching power converter might be 95% efficient, and then we suppose that we have a really

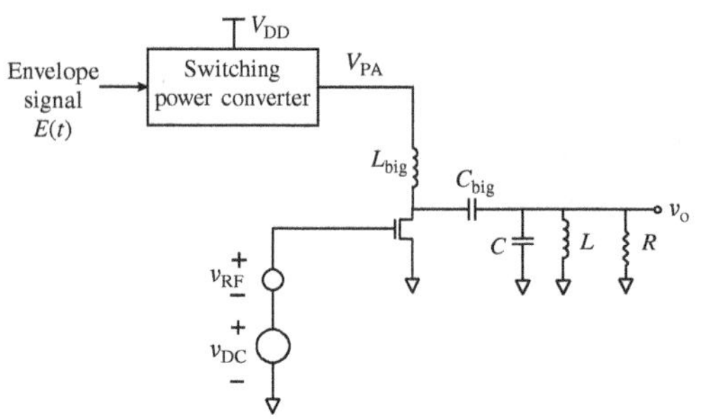

Figure 11.51 Simplified depiction of envelope tracking. The gate drive is sometimes depicted as constant-envelope, but in implementation practitioners are usually more nuanced than that. Otherwise, the beating heart of the technique is the switching power converter.

because it means that they cannot accommodate even one more user. They therefore build in a lot more capacity than they think they will ever use, all but guaranteeing that the Doherty amplifiers will be outside of their efficiency sweet spot.

well-executed switching power amplifier that gets as high as 80% efficiency (e.g. $P_{out}/P_{supply} = 0.8$). This simple arithmetic, $0.95 \times 0.80 = 0.76$...

...has launched a thousand ships. An efficiency of 76% in an RF power amplifier under real-world conditions (say, with a PAPR of 7 dB) would be astounding.

The critical error lies in thinking that the switching power converter can maintain its high efficiency when moving at the rate of a high-bandwidth signal envelope. To see why, consider the operation of the venerable buck converter shown in Figure 11.52. That capacitor, highlighted in the figure, is The Problem:

1. This is no mere parasitic capacitance. This is a decoupling capacitor, a deliberate part of the *LC* filter put there to minimize power supply ripple. This capacitor is big, and big by design.
2. Recall from Section 3.1.3 that we derived the power required to impose a low pass spectrum of bandwidth B on the capacitor of an *RC* circuit. It was terrible, in both amplitude and bandwidth, weighing in at a crushing $P \propto A^2 \cdot B^3$.

Now for item #2 you might object that what we have here is more complicated than a simple *RC*. Fair enough, this does warrant some explanation. A fuller capture of how the buck converter interacts with a power amplifier stage is shown in Figure 11.53. This is a lot to take in, but the thing to keep your eye on is that switching converters only make sense when the switching frequency is far above the highest frequency of, in this case, the envelope. The inductor L is chosen such that on the time scale of the envelope, it does not introduce significant reactance. When it comes to the energetics of charging the capacitor, then, all we really have are ohmic losses in the diode and switch, and then of course the size of the capacitor itself. It's a complicated R through which we charge that capacitor, but still an R at the end of the day. Our scaling rule holds.

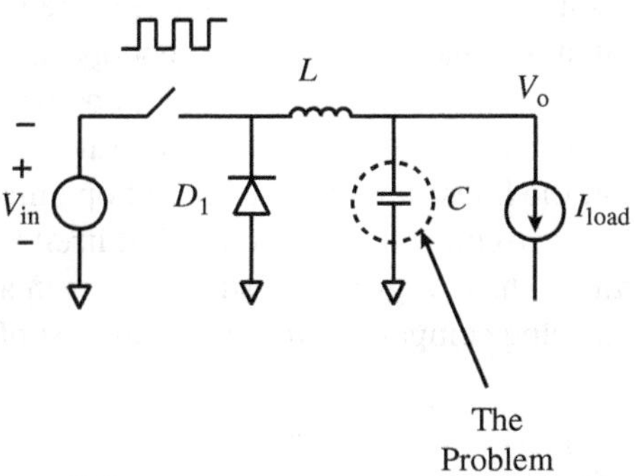

Figure 11.52 The venerable buck converter, presented here as a stand-in for DC-DC switching power converters. The essential design conflict lies in sizing the capacitor: you want it to be big, in order to minimize the supply ripple that comes from switching; you also want it to be small, to minimize the energy penalty that comes from moving V_o.

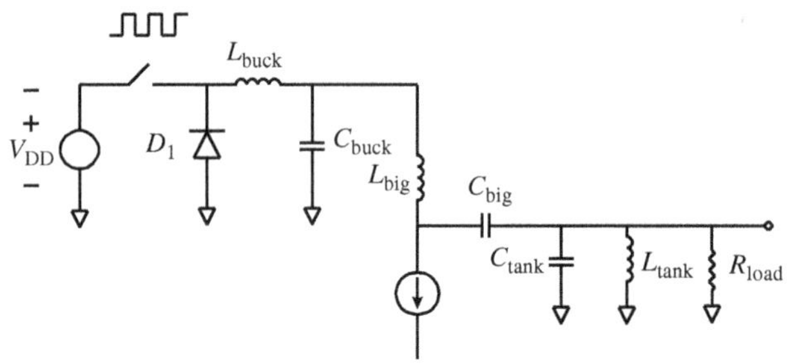

Figure 11.53 Connecting a buck converter to a generic PA structure. For analysis purposes, the key is to realize is that inductor L_{buck} presents very little reactance at frequencies relevant to the envelope. We thus have complicated but ohmic charging and discharging of capacitor C_{buck}.

There is one more aspect to the problem, which is that the envelope of a band-pass signal turns out to be 3–5× wider in bandwidth than the bandwidth of the signal itself. This is basically because of the awkwardness that befalls us due to zero crossings in $m(t)$, assuming that our modulated signal is $m(t) \sin \omega t$. It turns out that our envelope signal for the purposes of driving the power converter is $|m(t)|$, and that absolute value operation is a nonlinearity that spreads the spectrum.

There is simply no way of getting around all of this. The 3–5× expansion in bandwidth for the envelope, in addition to forcing us to move a big capacitor around over a large range very quickly, also pushes us to very high switching frequencies for the switching converter itself.[30]

The difficulty of high switching frequencies has always been clear. In response, for a long time state-of-the-art implementations relied on the switching converter to supply the low-frequency parts of the envelope current, and then add a high-power *linear* amplifier in summation to supply the rest of the envelope current. Proponents of this approach rightly pointed out that the bulk of the energy supplied is, in fact, in the lower frequencies of the envelope, and so the hope was that supplying this part with a high efficiency switching converter would save the day. What undermines this strategy is that the linear amplifier, a kind of op-amp on steroids, is a Class A or AB structure of extremely low efficiency that itself is *still* being asked to drive a large capacitance. The energy cost of supplying even a small high frequency current blows up to being comparable to the energy cost of

30 You figure that you want to have a switching frequency at least 10× the highest frequency in your envelope. For a 20 MHz communications channel, then, a 5× bandwidth expansion times 10× that for smooth supply modulation quickly gets us to switching frequencies on the order of 1 GHz. Our power converter is suddenly in the realm of RF power amplifiers in terms of operating frequency, without the benefit of the narrowband circuit tricks we would reach for in our panic.

the switching stage. No matter how you slice it, successful real-world implementations are hard to find for signal bandwidths exceeding about 5 MHz.[31]

The problems don't end there for envelope tracking. Hidden in real-world implementations is a need for extreme timing matching between the supply modulator path and the nominally constant-envelope "phase path." This problem gets worse at higher signal bandwidths, but even in the 5 MHz bandwidth range you find practitioners of this technique preoccupied with nanosecond-scale timing mismatches in order to keep the output spectrum from falling to pieces.

In short, avoid envelope tracking because the physics are all wrong. Note that *average* power tracking, where you move that big capacitor slowly in response to slow changes in average output power, is good and sensible. But it does not pay to get more ambitious than that.

11.4.7 Discrete Supply Modulation

While envelope tracking has fundamental problems, discrete supply modulation has been made to work extremely well. It is the only technique that the author has seen that gets higher efficiency than Doherty PAs, reducing the overall power consumption of the lineup by 30% at the same output power. With regard to signal bandwidth, we have seen examples reaching 200 MHz and beyond.

The author's involvement with this technique began at the invention stage in 2007–2008 with colleagues and students at MIT, and transitioned quickly to industry (Eta Devices, Inc., founded in 2010, and then continuing after a partial acquisition as Eta Wireless, Inc., founded in 2016).[32] As a result of the quick transition from academia to industry, almost all of the published literature on this technology exists in the form of patent filings. Appendix A is an up-to-date listing of published patents on the critical architecture, digital predistortion, and power conversion intellectual property behind the early implementations. We are fortunate that with the power conversion aspect, there are also a number of journal papers that have been published, and we include those as well. It is a great pleasure to acknowledge the large number of people whose talents, creativity, and supreme dedication were needed to take discrete supply modulation from idea to reality.

Figure 11.54 shows one realization of the concept. A simplified supply selection strategy is as follows: at all instants, select the *lowest* supply setting that still allows the PA to meet its instantaneous output power target. Modern, low-loss power switches (often gallium nitride FETs) allow supply state transitions on time scales

31 Beware of publications and marketing materials that shout otherwise. It is common practice at higher bandwidths to "exclude the loss of the supply modulator." There is no useful application for which such accounting is sensible.

32 In marketing materials, we tried a few names that we thought were catchy, like "ETAdvanced," and "Digital ET."

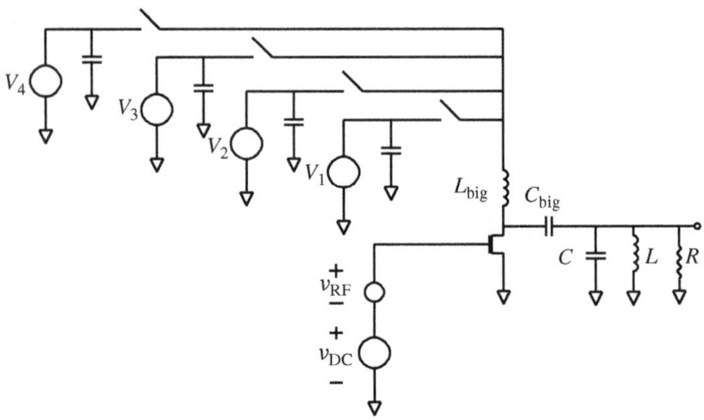

Figure 11.54 Simplified implementation of discrete supply modulation. Only one of $V_1 - V_4$ is connected to the PA at any given time.

of a few nanoseconds. Clever, switched-capacitor power converters allow four or more voltages to be generated without having to build multiple buck/boost power converters when it comes to V_1–V_4.

The thing to focus on here is all of the faults that discrete supply modulation does *not* share with envelope tracking. The decoupling capacitors at the output of each individual supply can be as large as we want them to be because they are static and/or move very slowly. The simple supply selection strategy means that we are not trying to toggle any one switch 3, 4, or 10× faster than the highest frequency of the envelope. There is no linear power stage working overtime to move a heavy capacitor that does not want to be moved. And then a *big* one: the signal error penalty for timing mismatch between the forward path and the supply modulator path is localized to small time slices surrounding each transition.[33]

A major development effort for discrete supply modulation centered around digital predistortion. Abrupt supply transitions in the drain of the PA *will* cause a lot of spectral splatter if they are uncompensated for. Compensation ultimately takes the form of "cancellation pulses" in the input signal that smooth out artifacts of abrupt supply transitions. If that sounds far-fetched and intense, that is how it sounded to the DPD architects in the early days at Eta Devices, too. But their final solution was accurate and almost shockingly robust, standing up repeatedly and reliably under intense customer demo conditions.

33 As opposed to with ET, where timing mismatch means you are making errors at all times.

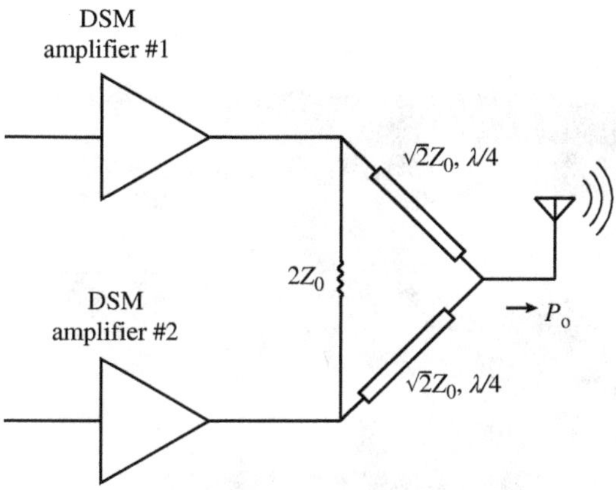

Figure 11.55 Doubling the output power with no loss in efficiency. Typically, an RF splitter is used on the input side to drive both amplifiers with the same drive. The loss of the Wikinson combiner on the output is almost exactly overcome by the increased efficiency due to more supply states. The Wilkinson combiner causes the PAs to see a constant output impedance. There is a temptation to use lossless combiners instead, but the varying impedance seen by the PAs causes no end of heartbreak. Approach with caution.

One disadvantage of Figure 11.54 compared to a Doherty is that it relies on one, large RF device to produce all of its output power, whereas with Doherty the load is wisely split between two RF power devices (one for the carrier amplifier, and one for the peaking amplifier). Ultimately, the implementation that became the workhorse for demonstrations in cellular infrastructure (average output powers of between 20 and 80 W) was what we called a "two-way architecture." It is shown in Figure 11.55. In our implementations we used straight-ahead isolating (Wilkinson) power combiners, which do have the disadvantage of dissipating power in the isolation resistor.[34] This loss can be limited if one imposes sensible limits on the allowed combinations of the two power supplies, in our case 44, 43, 33, 32, 22, 21, and 11. There is a net gain in efficiency that comes from jumping the number of supply states from four to seven, which almost perfectly balanced the loss of the combiner. The net effect was a doubling of output power compared to Figure 11.54 with no measurable loss in efficiency.

Figure 11.56 captures an example output spectrum from a two-way implementation. This is 20 W average output power with a 7 dB PAPR (so, 100 W peak) for a 20 MHz LTE signal. With all losses included (supply generator, supply modulator, and the final stage of the PA), the efficiency is 68%. Moving the bandwidth down to 5 MHz, this structure became the first in the world to achieve 70% drain

34 On the positive side, the power amplifiers feeding the combiner inputs see a constant impedance rather than a signal-dependent one.

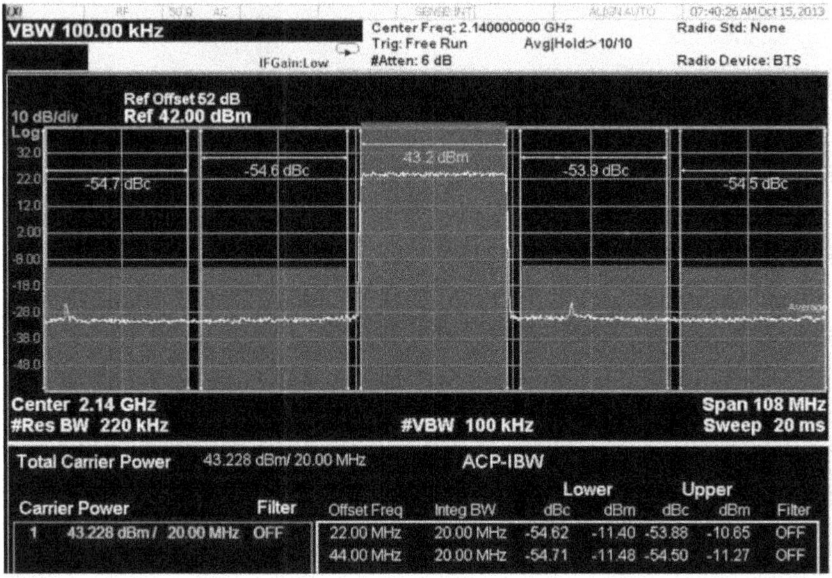

Figure 11.56 Example spectral output of for a two-way implementation of discrete supply modulation. The normal linearity requirement is that power in the adjacent channel be somewhere between −45 and −55 dBc.

efficiency under fully modulated conditions, including losses of supply generation and supply modulation.

11.4.8 Efficiency Metrics for Power Amplifiers

It turns out that Nature does not yield a high efficiency power amplifier easily when it comes to physical implementations. Losses as small as 0.1 dB can have quite a dramatic numerical effect when one is trying to demonstrate 70% efficiency.

So what exactly do we mean by “efficiency”? In the case of a supply modulated final stage, *any sensible measure of efficiency must take into account the energy required to generate and modulate the supply.* There is no way around this, since modulating the supply is the key challenge. A picture is valuable here; see Figure 11.57 for what we might call the “drain efficiency.”[35] The only meaningful drain efficiency to report here is $\eta = P_o/P_{in}$ as shown. In addition, the most important efficiency measurement is the one taken while the PA is passing a fully modulated carrier, *while* the desired linearity metrics (ACPR, EVM, etc.) are being met. You simply would not believe the amount of mischief that occurs in the real

35 The output node of the field effect transistor used in power amplifiers is called the “drain.”

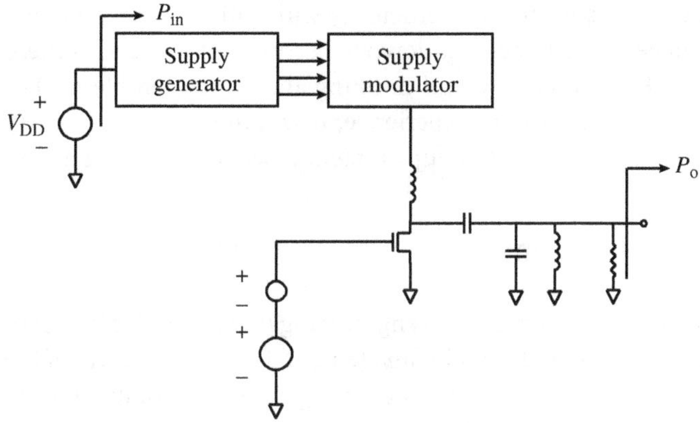

Figure 11.57 Properly defining drain efficiency.

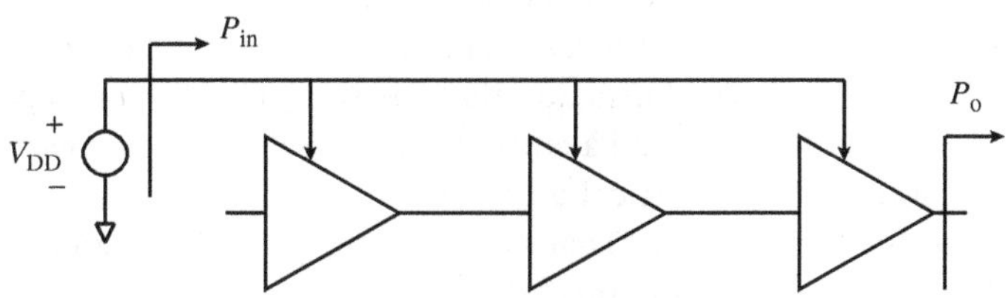

Figure 11.58 Properly defining lineup efficiency.

world around this issue. In many long, detailed presentations, this is the one measurement you cannot find. Don't fall into this trap.

All of that said, there remains one game that can be played when discussing drain efficiency even as it has been carefully laid out here. Sometimes you can get final PA stages to give you higher drain efficiencies by driving their gates with a very high voltage swing. Drain efficiency as captured here fails to penalize for the energy dissipated in order to drive the gate capacitance of the power device. Also, a final stage that has low gain may require multiple gain stages to precede it in order to enable delivering enough power to the load. The door is thus open to a false or at least less-than-useful results: eye-catching, impressive drain efficiency, but the usefulness gets undermined for real applications.

To guard against this, it is useful and important for any PA engineering effort to keep an eye on the overall lineup efficiency, as captured in Figure 11.58. To the overall power and heat management system, *this* is the metric that matters most.

Finally, readers with a background in PAs may wonder at the exclusion of something called "power added efficiency," or PAE, which, admittedly, is standard at the time of this writing. PAE is formulated to inflict a penalty for low power gain, which is good. But it does not capture any inefficiency with which the power delivered to the input is itself generated. It comes down to personal preference, as there is certainly a way to use PAE in one's way of thinking about these problems. For the author, though, after many years of experience, it was most helpful to carry around drain efficiency, stage gain, and lineup efficiency. See what works for you.

11.5 A Primer on Heat Sinks

There is perhaps no topic in communications engineering that is simpler in theory and more complex in practice than thermal management. Heat sinks get rid of the heat in our systems, moving it from our current-carrying electronic components to the surrounding environment. Without this heat removal, the components in our system would get hotter and hotter until they destroyed themselves. Very often it is the heat sinks, though, that make our equipment big, bulky, and heavy.

To understand thermal management, it helps to start out at the logical extreme of a resistor covered in a perfect thermal insulator that keeps any heat from escaping. See Figure 11.59. Suppose that we had a current I passing through the resistor. What happens? A well-trained electrical engineer will say quickly that "I^2R joules/sec, or watts, are dissipated as heat." That would normally be true, of course, but in this case we have made dissipation impossible.

What now?

Well, basically the temperature of the resistor rises until the resistor destroys itself. It does help to get analytical and quantitative about this, and to do so we will need a material property of resistor known as the "specific heat capacity." It is usually give the symbol C, and it is the amount of energy that is needed to raise one unit of mass of the material by one increment in temperature. The SI units of C are joules per Kelvin per kilogram ($\mathrm{J \cdot kg^{-1} \cdot K^{-1}}$). It helps to think, for example, of how much energy had to be added to a material sample to raise it to its current temperature. If U is the amount of heat energy added, and M is the mass, we have

$$T = \frac{1}{CM} \cdot U \tag{11.59}$$

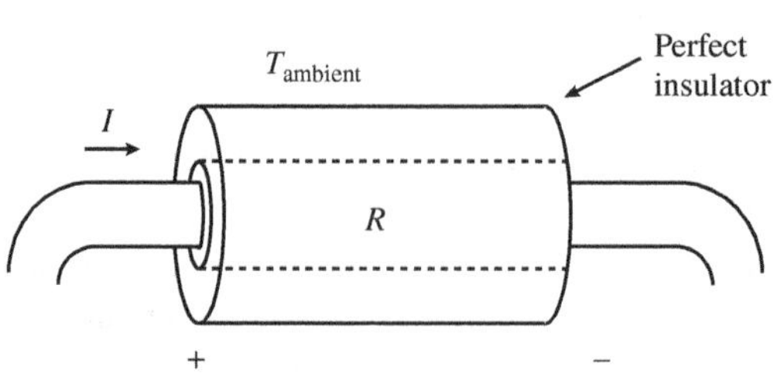

Figure 11.59 A resistor wrapped in a perfect thermal insulator.

and

$$\frac{dT}{dt} = \frac{1}{CM} \cdot \frac{dU}{dt}. \tag{11.60}$$

Now in the case of our resistor, we have that $\frac{dU}{dt} = I^2R$. When covered by a perfect insulator, its temperature must be rising at a rate of

$$\frac{dT}{dt} = \frac{1}{CM} \cdot I^2R. \tag{11.61}$$

A common 1/4 W carbon composition resistor today weighs 0.155 kg, and the best figure gleaned from the internet for the specific heat capacity is 1500 joules per kg per K. So a 100 Ohm resistor carrying 50 mA of current (and therefore trying to dissipate the full 1/4 W), would experience the temperature rising at a rate of

$$\frac{dT}{dt} = \frac{1}{1500\ \text{J} \cdot \text{kg}^{-1} \cdot \text{K}^{-1} \cdot 0.155\ \text{kg}} \cdot (1\ \text{mA})^2 \cdot 100\ \Omega = 1.1 \times 10^{-3}\ \text{K/s} \tag{11.62}$$

if, again, it is wrapped in this perfect insulator. At this rate, it would take 9.1×10^4 seconds to raise the temperature by 100 degrees (Kelvin or Celsius), which is about 25 hours. And remember that is 25 hours of cumulative operation, since the perfect insulator keeps the resistor from cooling back down when the current stops. It would not take much operation for many real world machines for this resistor to reach the point of melting down.

The problem is clearer, now. Our electrical components need to dissipate heat energy. If that heat energy has no place to go, it stays local and the temperatures of our components just rise and rise and self-destruction is inevitable. A thermal engineer would say of our present situation that the perfect insulator has "infinite thermal resistance." If we define a temperature difference between the resistor and the surrounding air, $\Delta T = T_{\text{res}} - T_{\text{air}}$, there is a very Ohm's-law-looking governing heat flow into the environment:

$$Q = \frac{\Delta T}{R_\theta}. \tag{11.63}$$

Q is heat flow (watts, or joules/sec), and R_θ is the conventional symbol for thermal resistance. Compare this equation to the actual Ohm's law:

$$I = \frac{V}{R}. \tag{11.64}$$

Keeping this relation in mind for heat flow is useful because it reminds us that we can think of combining thermal resistances the same way we combine electrical resistances in series and in parallel.

Our thermal resistance is given by

$$R_\theta = \frac{\Delta x \cdot \rho}{A}, \tag{11.65}$$

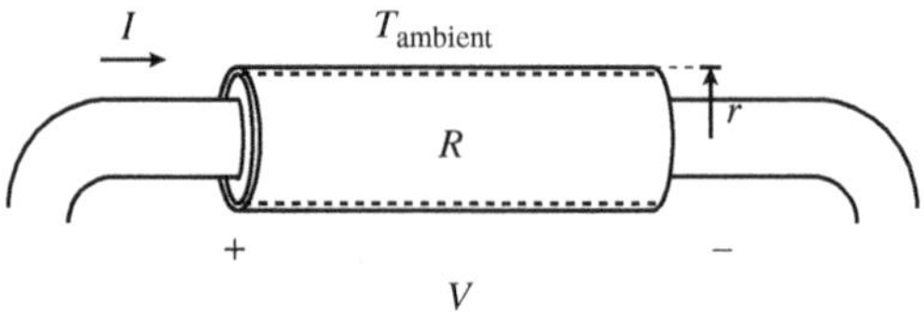

Figure 11.60 A resistor wrapped in a very thin insulator coating.

where Δx is the thickness of the material in the direction of the heat flow; ρ is the thermal resistivity of the material (units K·m/W); A is the cross-sectional area perpendicular to the flow of heat. Figure 11.59 is a little unfortunate as an example in this regard, because as one goes outward radially from the resistor the relevant surface area grows due to the geometry of the cylinder. To fix this, we'll go back and revise the figure to have the coating material be very thin as shown in Figure 11.60. We treat the cross-sectional area now as $2\pi rl$ for purposes of hand analysis.[36]

We now get to the conceptual heart of the thermal management problem:

- A given electronic component generates heat at a rate of I^2R, or V^2/R, or IV.
- The rate of heat flow from each component is given by

$$Q = \frac{\Delta T}{R_\theta} = (T - T_{\text{amb}}) \cdot \frac{A}{\Delta x \cdot \rho}. \tag{11.66}$$

- In order for the heat flow to match the rate of heat generation, Nature requires each component to settle at a temperature *higher* than the surrounding ambient temperature according to

$$T = \left(\frac{\Delta x \cdot \rho}{A}\right) \cdot Q + T_{\text{amb}}. \tag{11.67}$$

In the case of our perfect thermal insulator coating, ρ is infinite and our simple expression predicts infinite temperature, which means self-destruction. Beyond that, Eq. 11.67 shows us clearly the things that a good thermal engineer will obsess over: large surface areas through which to disperse the heat; materials with low thermal resistance; thin barriers between our component and the surrounding air. For performance reasons in electronics we tend to prefer that our electronics operate at lower temperatures rather than higher (lower thermal noise floor, for example, and transistors tend to be faster).

Bearing these principles in mind can help you to start to understand heat sinks. Figure 11.61 shows a classic structure. We love aluminum for heat sinks because

36 A thicker coating could then be treated as concentric thin shells, whose thermal resistances combine simply in series.

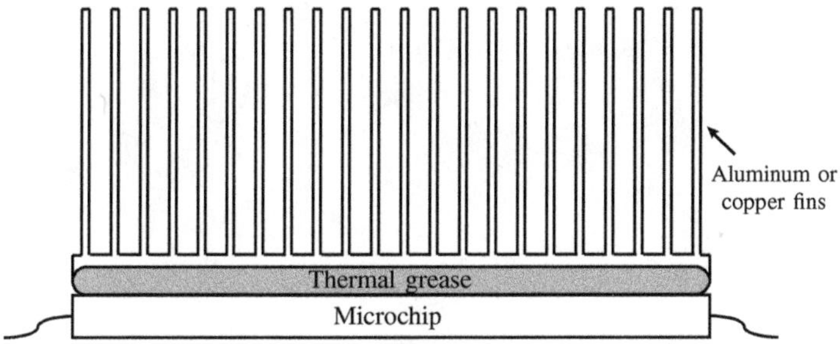

Figure 11.61 A representative heat sink structure.

it is inexpensive, has great thermal conductivity, is light relative to other metals, and has strong manufacturability properties. Sometimes when we are really in trouble though we will use copper, which while being heavier and more expensive does have higher thermal conductivity. The spiky aspect of the aluminum/copper structure that you see in the figure is what we call "fins." Use of fins is a way to increase surface area A of Eq. 11.67 while keeping our overall linear dimensions as small as we can.

The main takeaway from this section is that wasting energy as heat in electronics is a major driver of size and weight in our final products. With power amplifiers in particular, you can start to see the tragedy of being careless, say, about "2–3 dB" of post-PA loss. On the high side of this range, we are *doubling* the output power of the PA, and therefore doubling the heat that must be dissipated, which drives the physical size and weight of our heat sinks. On the other hand, increases in efficiency of PAs can dramatically reduce the heat that needs to be dissipated. Modern microprocessors are another famously difficult heat management problem, producing as they do an extraordinary amount of heat in an extraordinarily small area. Digital architects have long been big on power consumption for this reason.

12

Energy and the Shannon Limit

For much of the history of communications systems since Shannon's famous paper, the energy required to run communications networks has been a relative afterthought.[1] It is not difficult to see how this state of affairs came about. Data traffic in our communications infrastructure has been rising exponentially for decades. By this point in this book, we are wise to the fact that rising data traffic also means rising energy consumption. Only now, at the time of this writing, are the costs associated with this energy so high that they meaningfully affect the short-term profits of cellular carriers. Before this key threshold is crossed, energy is of concern only to the academic. *After* this threshold is crossed, energy-aware engineering becomes a high-stakes practice. The criticality only increases going forward: data traffic continues to compound at an annual rate of 20–30%, a veritable rocket ship.

For the modern network architect this means that all networks going forward are energy constrained. The challenge before was to deliver high data capacity given a scarcity of spectral resources; the challenge now is to deliver maximum data capacity given scarce spectral resources *and* scarce energy resources. The purpose of this chapter is to put spectrum and energy on equal footing.

12.1 An Energy Scaling Rule for Wireless Links

We are in position to complete the energy scaling rule for radio links that we started in Section 8.5. Deriving a concrete power consumption number for a link is hard to do in general, because there are so many different implementation

1 This is not the same as saying *no one has ever thought about energy.* After all, heatsinks and cooling systems have been a part of communications systems for the whole ride. The point here is rather that the net energy extracted from the Earth to run our networks has, in fact, been an afterthought.

Energy Fundamentals of Radio, First Edition. Joel L. Dawson.

choices to make. But we *can* write a scaling rule that is generally useful. In Section 8.5, we wrote

$$P_{\text{sys}} \propto kTB \cdot f^2 \cdot 2^{C/B} \cdot \frac{1}{G_t G_r} \cdot r^2. \tag{12.1}$$

At that time, though, we did not know about MIMO rank, or beamforming, or dual polarization. Whatever tricks we use, we can now modify our scaling rule with a new parameter M. This M captures the number of parallel, non-interfering data paths that are available.[2] In rough numbers, we may write an energy or power scaling rule for radio links as

$$P_{\text{sys}} \propto kTB \cdot f^2 \cdot 2^{C/B} \cdot \frac{1}{G_t G_r} \cdot r^2 \cdot \frac{1}{M}. \tag{12.2}$$

It took us a long time to get here, but we made it. A good understanding of the Shannon limit, electromagnetic theory, the Friis equation, and the tricks we play with multiple antennas ultimately lead to Eq. 12.2. You can read in this simple relation a lot of trends that play out in wireless communications. You can see, for example, why those who work on mm-wave links are obsessed with antenna gain: that f^2 dependence gives you nowhere else to run. You can also see why system planners at least get curious about the idea of small cells, which make that r^2 dependence work for you.

The utility of a scaling rule starts from having an implementation for a radio link that you believe is well-conceived. All things being equal, if we want to have the same link quality but triple the distance between our transceivers, our scaling rule tells us we need to multiply our power consumption by 9x. If we have a link running on a 1 GHz carrier and want to migrate to 60 GHz, our power bill has exploded by a factor of 3600.[3] It is worth playing a little with this relation, and testing popular trends in wireless communications networks against it. Why do carriers around the world pay so much money for spectrum? Well, if B is fixed and C needs to go up exponentially, our scaling rule predicts utter catastrophe...

Note finally that we went after a scaling rule because practical details, like energy inefficiency in our radio hardware, are hard to account for rigorously. But that does not mean we will not try! We take a crack at this tough problem in Section 12.1.3.

---

2 As we did in Section 10.3, we will assume that we are in the high SNR region of the Shannon limit. This means that we will give ourselves credit for, say, doubling our capacity even if the reduction in the SNR in each channel to keep the total power constant means that we don't *quite* double the capacity.

3 This is before, by the way, we consider the *strong* absorption of 60 GHz radiation by oxygen in the atmosphere.

12.1.1 An Example Calculation

We are at a point where it is clarifying to do a numerical, concrete calculation. Let's establish Nature's limit for the minimum transmit power to reach a data capacity of 75 Mbits/sec for a 20 MHz BW channel. We start with antennas that are isotropic radiators and refine later with antenna gains from known, real-world antennas. The setup is in Figure 12.1. Suppose that the output power from our transmitter is P_o. If it truly is an isotropic radiator, then the Poynting flux density at a distance r_0 is simply $P_o/4\pi r_0^2$. Let us suppose that each antenna has radiation resistance Z_0, and that we are at first dealing with a noiseless receiver that nevertheless presents a matched input impedance to the antenna of Z_0. Under these circumstances, the power delivered to the receive input resistance is given by

$$P_r = \frac{P_o}{4\pi r_0^2} \cdot A_{iso}. \tag{12.3}$$

The effective area of an isotropic radiator happens to be $\lambda^2/4\pi$, so

$$P_r = \frac{P_o}{4\pi r_0^2} \cdot \frac{\lambda^2}{4\pi} = \frac{P_o}{16\pi r_0^2} \frac{c^2}{f^2}. \tag{12.4}$$

The noise delivered to the (noiseless) input Z_0 of the receiver is just

$$P_N = kT_A B, \tag{12.5}$$

where T_A depends on the objects in the receive field of the antenna. For terrestrial systems, we are pretty well served by setting this at 300 K.

Believe it or not, we are pretty much there. Our SNR in this noiseless receiver is

$$\text{SNR} = \frac{P_r}{P_N} = \frac{1}{kT_A B} \cdot \frac{P_o}{16\pi r_0^2} \frac{c^2}{f^2}. \tag{12.6}$$

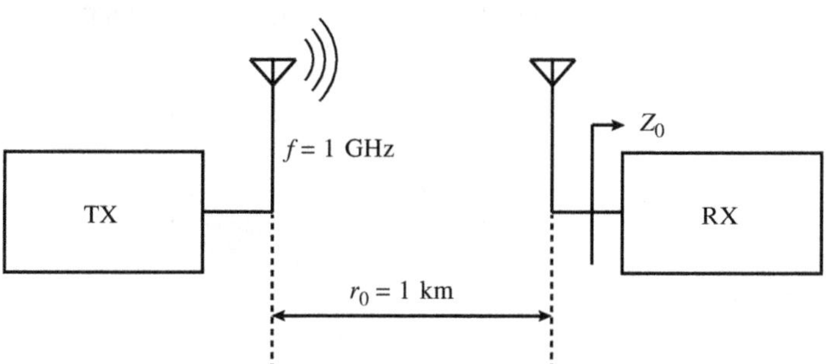

Figure 12.1 The setup for our energy calculation in a radio link.

Following a similar path to what led us to Eq. 5.10, we find our way to

$$P_o = kT_A B \cdot 16\pi^2 r_0^2 \frac{f^2}{c^2} \cdot 2^{C/B}. \tag{12.7}$$

We are knocking on the door. We have a noiseless receiver, a bandwidth of 20 MHz, a carrier frequency of 1 GHz, and a distance of 1 km, and our mandate is to get 75 Mbits/sec. The minimum transmit power that Nature requires with this setup is given by

$$\begin{aligned} P_o &= (1.38 \times 10^{-23}\,\text{J/K})(300\ \text{K})(20 \times 10^6 \text{s}^{-1}) 16\pi^2 (1000\ \text{m})^2 \frac{(10^9 \text{s}^{-1})^2}{(3.0 \times 10^8 \text{m/s})^2} \\ &\quad \times 2^{(75\times10^6\ \text{s}^{-1})/(20\times10^6\ \text{s}^{-1})} \\ &= 1.95\ \text{mW}. \end{aligned} \tag{12.8}$$

Less than you thought, perhaps? Especially with an isotropic radiator, spilling power in all sorts of directions that do not impinge on the receiving antenna? Hopefully it surprises you. And then, don't let moments of surprise in calculation go to waste. Think about what caught you off guard.

From here it is instructive to revisit our setup, make changes, and see how they impact the bottom-line number. Let's start with the distance, r_0. A real "macro" cell might be expected to have a radius as large as 30 km. Considering the quadratic dependence on distance, we see this scaling has a massive effect, moving Eq. 12.8 by nearly three orders of magnitude:

$$P_o = (\cdot) \cdot \left(\frac{30,000}{1000}\right)^2 = 1.76\ \text{W}. \tag{12.9}$$

Next we tackle our "noiseless receiver" assumption, a glaring omission if there ever was one. We draw a new figure for the receiver in Figure 12.2. We capture the effect of this new and most unwelcome noise source by modifying Eq. 12.6 like so:

$$\begin{aligned} \text{SNR} &= \frac{P_r}{P_N} \\ &= \frac{1}{kT_A B + kT_R B} \cdot \frac{P_o}{16\pi r_0^2} \frac{c^2}{f^2} \\ &= \frac{1}{k(T_A + T_R)B} \cdot \frac{P_o}{16\pi r_0^2} \frac{c^2}{f^2}. \end{aligned} \tag{12.10}$$

For the sake of comfort with the literature, it is worth noting different conventions for accounting for noise in the receiver. Again, for terrestrial links in the low GHz range, it is often appropriate to stick to an antenna temperature T_A of around 300 K. In lab work especially we take this assumption seriously. We then characterize the receive signal chain via something called the "noise factor." This is the

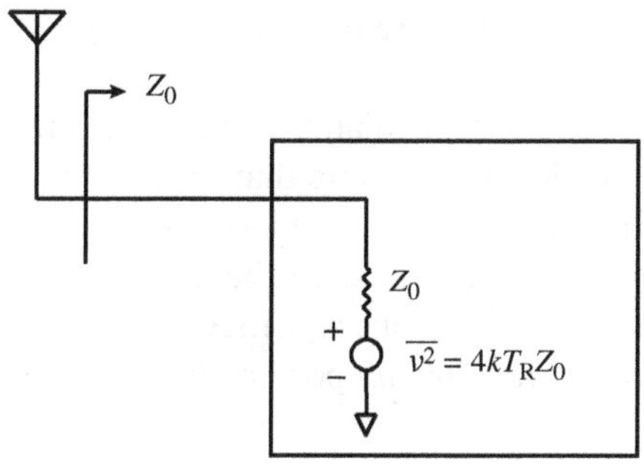

Figure 12.2 A receiver with an internal noise source.

ratio of the SNR of a noiseless receiver to the SNR of *our* receiver. Using Eq. 12.10, this looks like

$$F = \frac{\frac{1}{kT_AB}}{\frac{1}{k(T_A+T_R)B}} = 1 + \frac{T_R}{T_A}. \tag{12.11}$$

An even more commonly used metric is called the "noise figure," and it is just the noise factor but in dB:

$$\mathrm{NF} = 10 \cdot \log_{10}\left(1 + \frac{T_R}{T_A}\right). \tag{12.12}$$

When playing with numbers, it is not crazy at all for the noise figure of a complete receive chain to be 10 dB or higher. Note that this corresponds to a T_R of 9×300 K $= 2700$ K. Note further that this is *not* to say that there is a component in the receiver sitting at 2700 K. Rather, there are many contributions downstream in the signal chain, and all those contributions really add up. We can thus further scale Eq. 12.9 to include a noise factor of 10:

$$P_o = (\cdot) \cdot 10 = 17.6 \text{ W}. \tag{12.13}$$

As a final touch, suppose that we happen to know that on the transmit side, there is loss (cavity filters, circulators, cables) between the output of the PA and the antenna, and that that loss is 3 dB. (This is on the high side, but not crazy or uncommon). This factor of 2 must also be accounted for, and our PA output power swells from Eq. 12.13 to

$$P_o = (\cdot) \cdot 2 = 35.2 \text{ W}. \tag{12.14}$$

It all adds up so quickly! Note in particular that from the standpoint of transmit power, post-PA loss and noise figure have the same dB-per-dB impact. It is often true that the power consumption required in the receiver to lower its noise figure by 1 dB is quite a bit less than the power required to raise the transmit power by 1 dB. But it is always worth checking.

This section details what we mean by a "straight calculation." We'll get some practice slinging around a scaling rule in Section 12.1.2. Note that right now we are doing our calculations based on the assumption that we achieve the full, theoretical Shannon capacity in this real world link. To the extent that a real-world coding scheme falls short of the ideal, either the data capacity (here assumed to be 75 Mbit/sec) will have to be scaled down, or the transmit power will have to be scaled up.

12.1.2 When to Use a Scaling Rule Instead of Straight Calculation?

So we have a scaling rule in Eq. 12.2, and in Section 12.1.1, we dealt in absolute power. When to choose one over the other?

Well, as we saw a straight calculation can be very clarifying. And when we say that it allows you to focus on Nature's limits, you see that we also get insights into the impact of implementation nonidealities. For example, if the noise factor of state-of-the-art receive chains is known, that allows for an easy adjustment of our energy requirements. For post-PA loss, same.

Where things get trickier is in the energy cost of other parts of the system. For example, PA lineups are never anywhere close to 100% efficient. If they are 50% efficient, of course, that is a natural 2× multiplier in our straight calculation. But reliable numbers in the abstract can be very hard to come by. Same for the energy cost of frequency synthesis, the insertion loss of different filters, and the energy cost of digital signal processing.[4]

It is in this situation that a scaling rule can be very helpful. Rather than get into difficult conversations about what is possible, one only has to do the accounting on a system that already *is*. A comprehensive power consumption number for an existing link, especially one that has proven its durability and high performance in the field, is extremely valuable this way. That power consumption number has, built-in, the energy tax that we pay for all the implementation details that are so difficult to model. If you can let go of your need to be exact, the scaling rule of Eq. 12.2 can be very helpful.

12.1.3 Performance Requires Power in Electronic Machines

All of that said, there is insight to be gained about the signal chains at the heart of our communications machines. Figure 12.3 depicts the part of the system we focus

4 And, in fact, if you pin a hardware engineer down and ask for "typical" numbers, you will have, in general, a difficult time getting beyond "it depends," or a range that is so wide as to not be useful.

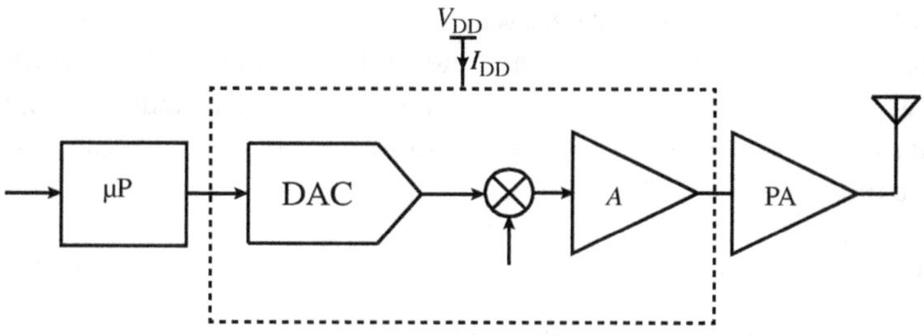

Figure 12.3 Considering the power consumption of the signal chain feeding the power amplifier.

on here. We make progress in our understanding if we get very precise about what we mean for these signal chains to be "high performance." In ordinary language, what we mean is that these signal chains have both high speed and high precision. But we know courtesy of the Uncertainty Principle (Section 2.5) that high speed means high bandwidth. And if you think about it, high SNR is an excellent way to talk about a signal chain being high precision.

For an extremely broad class of signal chains, we can safely posit the existence of a thermal noise floor. And while it is tempting (and sometimes necessary) to dive into the detailed frequency behavior of individual circuit blocks, for the most part we can zoom out and say with confidence that if the system is working, the engineers *must* have succeeded in making the signal chains flat in frequency response over the necessary bandwidth.[5] This is a powerfully simplifying insight. We may write in our minds the noise power at, say, the output of most signal chains according to

$$P_{\text{N}} = kT_{\text{N}}B, \tag{12.15}$$

for some noise temperature T_{N} and for the required bandwidth B. For a given level of precision, that is to say, a given SNR, we have that

$$P_{\text{sig}} = \text{SNR} \cdot kT_{\text{N}}B. \tag{12.16}$$

And then to the extent that the system power is proportional to the signal power, we have

$$P_{\text{sys}} \propto P_{\text{sig}} = \text{SNR} \cdot kT_{\text{N}}B. \tag{12.17}$$

5 In fact, we can go one step further and observe that the normal trick for achieving flat frequency response in RF signal chains is to introduce shunt inductance to form a parallel resonance with the output capacitance of each amplifier stage. Referencing Eq. 3.56, where we concluded that the power dissipated in such resonant structures is $\frac{1}{2}CA^2\frac{\omega_0 B}{Q}$, we see that we are on safe ground going forward if we assume that the power spent in these stages is proportional to B. And, of course, proportional to the carrier frequency. Do not lose sight of this ω_0 dependency.

Fix in your mind that for electronic systems, when we ask for high performance we are asking for speed and precision at the same time, and therefore high bandwidth and high SNR at the same time. In a nod to the fact that in many modern signal chains the "thermal" noise floor is set by quantization noise, we can recast our new scaling rule in terms of ENOB according to

$$P_{\text{sys}} \propto \left(10^{\frac{1.76+6.02\cdot\text{ENOB}}{10}}\right) \cdot B. \tag{12.18}$$

If you really want to get fancy, especially as we get close to the input of the PA, remember that there is bandwidth expansion due to digital predistortion on the order of 3–5×.

The way that this all plays out is as a general, rising anxiety as systems are asked to deal with larger and larger bandwidths. What we see is that the signal chains internal to our transmitters and receivers get hungrier and hungrier for power. If we were content with high SNR alone and slow speed, we would be fine; and if we were content with high speed but low SNR, we could also trade one against another and keep our power penalty low. But when we need both at the same time, there is a cost!

12.2 The Whole Fun of Radio Is Not Knowing Where Everybody Is

Knowledge of our scaling rule can sometimes push people to do drastic things. We have already mentioned that for mm-wave implementations, practitioners are (justly) preoccupied with high antenna gains. For another example, we have that the r^2 dependence points some people to small, targeted cells rather than a single macro cell.

Sometimes, either or both of these strategies are okay. For instance, super-narrow beam widths (corresponding to high antenna gains) can be a very useful technique if you happen to know ahead of time that your transmitters and receivers are fixed in their location. Usually the motivator in this case is that it is prohibitively expensive to lay a fiber or a cable, and for whatever reason the up-front capital is a deal breaker.[6]

For small cells, there are good ways and bad ways to act on the nudge that we get from the r^2 dependence. Small cells are a notorious shell game if the goal is to give blanket coverage to a specified *area*. Cutting the transmit radius by a factor of 10, for example, cuts the power by a factor of 100. But the area we cover has now dropped by a factor of 100 too (area $\approx \pi r^2$), so we now need 100 of these small

6 Just remember that highly directive antennas must be mechanically aligned in the field, and then one must make sure that they keep that alignment for the duration of the deployment. This is a frequently underestimated headache.

cells to cover the same area. No net benefit from a power standpoint, and quite a huge complication now in the "backhaul," or the system of fibers/cables that connect the cell towers to the network. However, if users are spread out nonuniformly across a given area, say, clumped together in distinct clusters, then small cells can be used to great power-saving effect. See Figure 12.4. The point to remember, though, is that the most fun and useful thing about wireless radio links is being able to walk around *without* having to know where all the users and towers are. In this sense, the isotropic radiator is not as wasteful as it may appear at first. We may think of it instead as spending energy for the convenience of untethered communications. One need look no further than the wireless routers that are ubiquitous

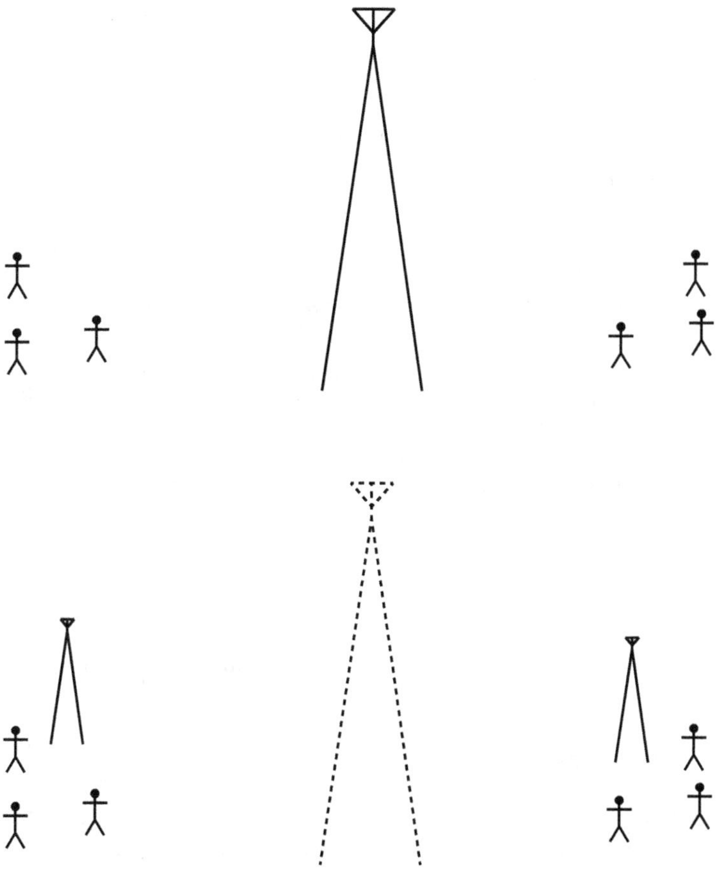

Figure 12.4 If we know ahead of time where users will cluster, small cells can be an excellent power-saving technique.

in US homes these days. No one plugs their computers into a router using an Ethernet cable anymore. This convenience is a real thing.

Don't just blindly follow the math. Keep your head up, and always be clear about what you are trying to accomplish.

12.3 Spectral Efficiency Versus Energy Efficiency

Our first order of business is to clearly define our two types of efficiency. In this section we are concerned about the use of our spectral and energy resources specifically as regards to delivering bits. For spectral efficiency, we will use

$$\eta_s = C/B, \tag{12.19}$$

which we will render in our heads as "bits per second per hertz," or "capacity per hertz."

When it comes to energy, what we want in broad strokes is a ratio with bits in the numerator and joules in the denominator. Actually, the parallel to spectral efficiency becomes better if we ask for bits/sec (capacity) in the numerator, and if that leads us to have joules/sec (watts) in the denominator that is okay. We have as a helpful example Eq. 12.7. If we ask for a ratio of capacity to watts, we only have to stick capacity in the numerator and the power number from Eq. 12.7 the denominator:

$$\eta_p = \frac{C}{kT_A B \cdot 16\pi^2 r_0^2 \frac{f^2}{c^2} \cdot 2^{C/B}}. \tag{12.20}$$

...and look what we have! η_s is already here (twice!), and so we can relate the two types of efficiency to each other:

$$\eta_p = \frac{1}{kT_A \cdot 16\pi^2 r_0^2 \frac{f^2}{c^2}} \cdot \eta_s 2^{-\eta_s}. \tag{12.21}$$

Unsightly. Let's revert to a proportionality so we can focus on the essential dependency, and rewrite slightly.

$$\eta_p \propto \frac{\eta_s}{2^{\eta_s}}; \qquad \eta_s > 1 \text{ bits per second per Hz.} \tag{12.22}$$

That last condition, $\eta_s > 1$, serves as an important reminder that all along in this chapter we have been using the Shannon limit in an approximation that we called "Region 2" from Chapter 5 (see Eq. 5.9), for which the SNR > 1. For $\eta_s > 1$, what we see is that spectral efficiency and energy efficiency are at odds when it comes to capacity. *When we ask for high capacity per hertz, we must accept lower capacity per watt, and vice versa.* It bears emphasis, however, that the exponential dependence means that modest changes in capacity per hertz lead to rather drastic changes in capacity per watt. The cost for cramming more and more data into fixed bandwidth is an exponential drop in capacity per watt, which is the same as an exponential rise in watts per capacity, which is the same as an exponential rise in joules per bit.

12.4 On the Finiteness of Spectral Real Estate

We have been designing wireless radio links based on modulated electromagnetic waves for over a century. During that time, we have found ways to use a rather wide swath of the electromagnetic spectrum. The low end features VLF, or "very low frequency" bands where we have carrier frequencies on the order of 30 kHz, whose most famous application is wireless links for submarines. On the high end, we have EHF, or "extremely high frequency" bands at carrier frequencies above 30 GHz. Here, a great deal of the excitement is due to the sheer amount of absolute bandwidth that is available, as well as the possibility of realizing many-element beamforming techniques within reasonable form factors.

It remains true that for mobile communications, there is now and always will be a sweet spot for carrier frequencies between about 500 MHz and 5 GHz. Much lower than 500 MHz, and antenna sizes get uncomfortably large. Much higher than 5 GHz, and that f^2 attenuation starts to really hammer you. In a seemingly wild coincidence, our radio astronomy friends tell us that the sky noise temperature for terrestrial links happens to have a minimum that spans almost exactly this range.[7] For fundamental reasons, 500 MHz–5 GHz is just a happy hunting ground.

This is all to emphasize that as we seek to grow the capacity of our wireless networks, we do *not* have at our disposal an infinite amount of spectrum. Avoid in particular the siren call of ever higher carrier frequencies as a way of increasing capacity through higher available bandwidth.[8] Nature requires us to be creative in other ways.

12.5 When Energy Consumption Is Not Central, It Grows

We arrive at the final section of this book and try to evaluate how well we have met the overall objective. The objective was to acquaint the reader with a genuinely helpful perspective on radio links. Radio links are, as we remarked at the outset, an extraordinarily complicated and broad topic. It is all well and good to say, "well,

---

7 J. D. Kraus and M. E. Tiuri, *Radio Astronomy,* NY, NY: McGraw-Hill Company, 1966. The key figure here is Figure 7.1 on p. 237.

8 Recall from Chapter 5 that in order for that bandwidth to help us, we have to have SNRs greater than unity first. This simple condition gets hard to ensure (to the tune of a square law) as the carrier frequency goes up.

we're just going to focus on the absolutely *key* things and treat them thoroughly." But how do you choose what to discuss?

Well, if you have read this book in anything like written order, or indeed, if you have skimmed the table of contents, you know the author's opinion. A focus on *energy consumption* imposes a marvelous hierarchy on the world of ideas that radio links inhabit. And the timing of this book is not a coincidence. Here at the beginning of 2025, it is broadly accepted in the industry that data traffic is compounding at a rate of 20–30% annually. Understanding now that we pay a 1:1 energy cost for data traffic, it is no surprise that the energy to run our wireless networks now has a huge, first-order impact on the capital costs to run communications infrastructure. Energy consumption, for so long justly an afterthought in the science of communications, has now arrived at the forefront. It would necessarily stay at the forefront even if the rocket ship of 20–30% compounding were to abruptly stop tomorrow.

There are many ways to approach finite resources in communications. Going forward, what meaningful treatments will have in common is an elevated focus on energy. And we need look no further than the hunt for spectral efficiency to see what the focus of a field of creative professionals can achieve: high-order QAM constellations, OFDM, MIMO, beamforming, error correction codes, and the like. We can look back on the publication of Shannon's famous capacity limit, and particularly on the *focus* on achieving that limit, as a force moving a whole field right up to Nature's limit. If we put energy consumption as a focus, there is every reason to expect similar miracles.

A final observation on the power of focus. Over many years in research and product development, the author has *never* checked the power consumption of one of his systems and been pleasantly surprised by how low it was. Usually, and especially when power consumption is the last item one checks or thinks about, it comes out way too high. In those instances, the crisis of missing this specification triggered the focus necessary to bring the energy consumption down.

We love fossil fuels because they carry so much energy per unit volume (or mass) that our machines can be 90% wasteful and still work. "Alternative" energy sources are not like this, despite the amazing achievements of countless creative and hard-working people to develop wind, solar, geothermal, and energy storage technologies. In the author's lifetime, alternative energy technologies have gone from exotic, expensive, and niche to practical, cheap, and scalable. What is missing in many areas, however, is an equal relentlessness about reducing the energy consumption of our machines. It always helps if we start by identifying the energy that Nature requires to do a Task, keeping a careful eye on whether or not we have the right Task. Once we have that conviction, we must be relentless about reducing inefficiencies and implementation losses, getting as close to Nature's limit as possible. In an energy-constrained world, this approach unlocks maximum performance in our machines.

A

References for Discrete Supply Modulation

The purpose of this appendix is to point to existing literature on discrete supply modulation. Most of what follows are patents issued to Eta Devices and Eta Wireless, plus some early papers and patents from our work at MIT. Generally speaking, the later the publication date, the more we knew about what bears up under the rigors of commercial application.

Figures A.1 and A.2 show system architectures that have proven themselves to be high-performance and reliable. Successful execution of discrete supply modulation requires tight integration of power amplifier design, power management (supply generator and supply modulator), and digital predistortion. The references below are organized into key engineering categories, plus references on the overall architecture.

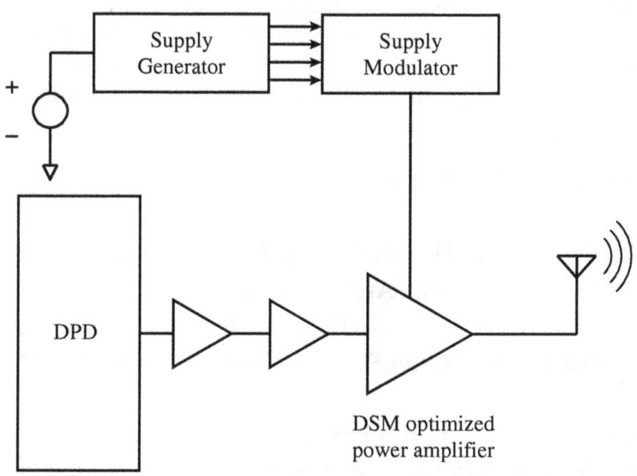

Figure A.1 "One-way" implementation of discrete supply modulation.

Energy Fundamentals of Radio, First Edition. Joel L. Dawson.

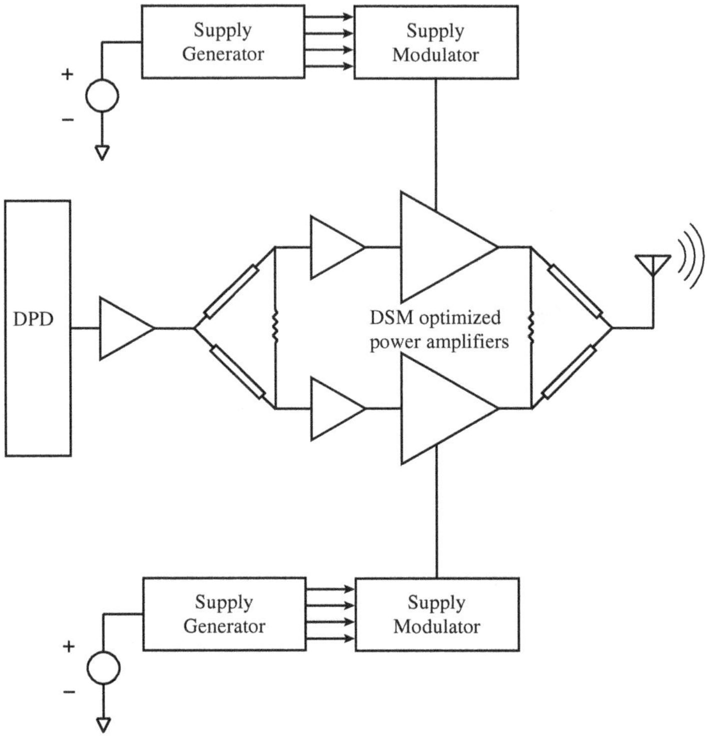

Figure A.2 "Two-way" implementation of discrete supply modulation. We used this approach for higher average output powers, intended for use on cell towers. Note in particular that by this point, we had discarded outphasing, which figures prominently in our early research. The bandwidth expansion needed to generate the two (or more) outphasing signals quickly becomes unmanageable for even modest signal bandwidths.

Appendix A.1 Overall Architecture

- J. R. Hoversten, Y. A. Tkachenko, S. H. Pakala, and J. Garrett, *"Multi-Stage Pulse Shaping Network,"* US-10992265-B2, Apr. 27, 2021.
- M. A. Briffa, J. L. Dawson, J. E. DeRoo, K. Komoni, D. J. Perreault, and O. Uyar, *"RF Amplifier Having A Transition Shaping Filter,"* US-10038461-B2, Jul. 31, 2018.
- M. A. Briffa, J. L. Dawson, J. E. DeRoo, K. Komoni, D. J. Perreault, and O. Uyar, *"Asymmetric Multilevel Backoff Amplifier with Radio-Frequency Splitter,"* US-9768732-B2, Sept. 19, 2017.
- M. A. Briffa, J. L. Dawson, J. E. DeRoo, K. Komoni, D. J. Perreault, and O. Uyar, *"Asymmetric Multilevel Backoff Amplifier with Radio-Frequency Splitter,"* US-20170070194-A1, Mar. 9, 2017.

- M. A. Briffa, J. L. Dawson, J. E. DeRoo, K. Komoni, D. J. Perreault, and O. Uyar, *"RF Amplifier Having A Transition Shaping Filter,"* US-20170063407-A1, Mar. 2, 2017.
- M. A. Briffa, J. L. Dawson, J. E. DeRoo, K. Komoni, D. J. Perreault, and O. Uyar, *"Asymmetric Multilevel Backoff Amplifier with Radio-Frequency Splitter,"* US-9537456-B2, Jan. 3, 2017.
- M. A. Briffa, J. L. Dawson, J. E. DeRoo, K. Komoni, D. J. Perreault, and O. Uyar, *"RF Amplifier Having A Transition Shaping Filter,"* US-9490752-B2, Nov. 8, 2016.
- M. A. Briffa, J. L. Dawson, J. E. DeRoo, K. Komoni, D. J. Perreault, and O. Uyar, *"RF Amplifier Having A Transition Shaping Filter,"* US-20160065138-A1, Mar. 3, 2016.
- M. A. Briffa, J. L. Dawson, J. E. DeRoo, K. Komoni, D. J. Perreault, and O. Uyar, *"RF Amplifier Having Discrete Supply Voltages,"* US-9209758-B2, Dec. 8, 2015.
- M. A. Briffa, J. L. Dawson, J. E. DeRoo, K. Komoni, D. J. Perreault, and O. Uyar, *"Transmitter Architecture and Related Methods,"* US-9166536-B2, Oct. 20, 2015.
- M. A. Briffa, J. L. Dawson, J. E. DeRoo, K. Komoni, D. J. Perreault, and O. Uyar, *"RF Amplifier Architecture and Related Techniques,"* US-9020453-B2, Apr. 28, 2015.
- M. A. Briffa, J. L. Dawson, J. E. DeRoo, K. Komoni, D. J. Perreault, and O. Uyar, *"RF Amplifier Architecture and Related Techniques,"* US-8824978-B2, Sept. 2, 2014.
- M. A. Briffa, J. L. Dawson, J. E. DeRoo, K. Komoni, D. J. Perreault, and O. Uyar, *"RF Amplifier Architecture and Related Techniques,"* US-20140118063-A1, May 1, 2014.
- M. A. Briffa, J. L. Dawson, J. E. DeRoo, K. Komoni, D. J. Perreault, and O. Uyar, *"Method and Apparatus for Multilevel Power Amplification,"* US-20140118072-A1, May 1, 2014.
- M. A. Briffa, J. L. Dawson, J. E. DeRoo, K. Komoni, D. J. Perreault, and O. Uyar, *"Transmitter Architecture and Related Methods,"* US-20140120854-A1, May 1, 2014.

Appendix A.2 Power Management

- J. Garrett, S. H. Pakala, B. Metzner, I. Duzevik, D. J. Perreault, J. R. Hoversten, and Y. A. Tkachenko, *"Multi-output Supply Generator for RF Power Amplifiers with Differential Capacitive Energy Transfer,"* US-20220149725-A1, May 12, 2022.

- J. Garrett, S. H. Pakala, B. Metzner, I. Duzevik, D. J. Perreault, J. R. Hoversten, and Y. A. Tkachenko, *"Multi-output Supply Generator for RF Power Amplifiers with Differential Capacitive Energy Transfer,"* US-11245367-B2, Feb. 8, 2022.
- J. R. Hoversten, Y. A. Tkachenko, B. Bekirov, S. H. Pakala, and J. Garrett, *"Power Management Control Over a Transmission Line for Millimeter Wave Chip Sets for Cellular Radios,"* US-11191028-B2, Nov. 30, 2021.
- D. J. Perreault, W. Li, and Y. A. Tkachenko, *"Integrated Power Supply and Modulator for Radio Frequency Power Amplifiers,"* US-10840805-B2, Nov. 17, 2020.
- D. J. Perreault, W. Li, Y. A. Tkachenko, and J. L. Dawson, *"Integrated Power Supply and Modulator for Radio Frequency Power Amplifiers,"* US-9755672-B2, Sept. 5, 2017.
- D. J. Perreault, *"Inverter/Power Amplifier With Capacitive Energy Transfer and Related Techniques,"* US-9634577, April 25, 2017.
- D. J. Perreault, W. Li, and Y. A. Tkachenko, *"Integrated Power Supply and Modulator for Radio Frequency Power Amplifiers,"* US-20150084701-A1, Mar. 26, 2015.

Appendix A.3 Digital Predistortion

- M. A. Briffa, J. L. Dawson, J. E. DeRoo, K. Komoni, D. J. Perreault, and O. Uyar, *"Linearization Circuits and Methods for Multilevel Power Amplifier Systems,"* US-10658981-B2, May 19, 2020.
- M. A. Briffa, J. L. Dawson, J. E. DeRoo, K. Komoni, D. J. Perreault, and O. Uyar, *"Linearization Circuits and Methods for Multilevel Power Amplifier Systems,"* US-20190074797-A1, Mar. 7, 2019.
- M. A. Briffa, J. L. Dawson, J. E. DeRoo, K. Komoni, D. J. Perreault, and O. Uyar, *"Linearization Circuits and Methods for Multilevel Power Amplifier Systems,"* US-10164577-B2, Dec. 25, 2018.
- M. Astrom, M. A. Briffa, J. L. Dawson, J. Hoverston, P. B. Normark, Y. A. Tkachenko, *"Digital Pre-Distortion (DPD) Training and Calibration System and Related Techniques,"* US-9979421-B2, May 22, 2018.
- D. J. Perreault, J. L. Dawson, W. Li, Y. A. Tkachenko, B. Lakshminarayanan, and J. Hoversten, *"Linearity and Noise Improvement for Multilevel Power Amplifier Systems Using Multi-Pulse Drain Transitions,"* US-9768731-B2, Sept. 19, 2017.
- D. J. Perreault, J. L. Dawson, W. Li, Y. A. Tkachenko, B. Lakshminarayanan, and J. Hoversten, *"Linearity and Noise Improvement for Multilevel Power Amplifier Systems Using Multi-Pulse Drain Transitions,"* US-20160099686-A1, Apr. 7, 2016.

- M. A. Briffa, J. L. Dawson, J. E. DeRoo, K. Komoni, D. J. Perreault, and O. Uyar, *"Linearization Circuits and Methods for Multilevel Power Amplifier Systems,"* US-20150372646-A1, Dec. 24, 2015.
- M. A. Briffa, J. L. Dawson, J. E. DeRoo, K. Komoni, D. J. Perreault, and O. Uyar, *"Linearization Circuits and Methods for Multilevel Power Amplifier Systems,"* US-9160287-B2, Oct. 13, 2015.
- M. A. Briffa, J. L. Dawson, J. E. DeRoo, K. Komoni, D. J. Perreault, and O. Uyar, *"Linearization Circuits and Methods for Multilevel Power Amplifier Systems,"* US-8829993-B2, Sept. 9, 2014.
- M. A. Briffa, J. L. Dawson, J. E. DeRoo, K. Komoni, D. J. Perreault, and O. Uyar, *"Linearization Circuits and Methods for Multilevel Power Amplifier Systems,"* US-20140132354-A1, May 15, 2014.
- M. A. Briffa, J. L. Dawson, J. E. DeRoo, K. Komoni, D. J. Perreault, and O. Uyar, *"Linearization Circuits and Methods for Multilevel Power Amplifier Systems,"* US-20140118065-A1, May 1, 2014.

Appendix A.4 Early Papers and Patents (MIT)

- J. L. Dawson, D. J. Perreault, S. Chung, P. A. Godoy, and E. W. Huang, *"Asymmetric Multilevel Outphasing Architecture for RF Amplifiers,"* US-8957727, Feb. 17, 2015.
- J. L. Dawson, D. J. Perreault, E. W. Huang, S. Chung, and P. A. Godoy, *"Asymmetric Multilevel Outphasing Architecture for RF Amplifiers,"* US-8659353-B2, Feb. 25, 2014.
- P. A. Godoy, S. Chung, T. W. Barton, D. J. Perreault, and J. L. Dawson, "*A 2.4-GHz, 27-dBm Asymmetric Multilevel Outphasing Power Amplifier in 65-nm CMOS,*" *IEEE Journal of Solid-State Circuits,* Vol. 47, No. 10, pp. 2372–2384, 2012.
- J. L. Dawson, D. J. Perreault, S. Chung, P. A. Godoy, and E. W. Huang, *"Asymmetric Multilevel Outphasing Architecture for RF Amplifiers,"* US-8164384, Apr. 24, 2012.
- J. L. Dawson, D. J. Perreault, E. W. Huang, S. Chung, and P. A. Godoy, *"Asymmetric Multilevel Outphasing Architecture for RF Amplifiers,"* US-8026763-B2, Sept. 27, 2011.
- S. Chung, P. A. Godoy, T. W. Barton, E. W. Huang, D. J. Perreault, and J. L. Dawson, *"Asymmetric Multilevel Outphasing Architecture for Multi-Standard Transmitters," 2009 IEEE Radio Frequency Integrated Circuits Symposium,* pp. 237–240, 2009.

Index

Note: Page numbers in *italics* and **bold** and n refers to figures and tables and notes.

Energy Fundamentals of Radio, First Edition. Joel L. Dawson.

www.ingramcontent.com/pod-product-compliance
Lightning Source LLC
Chambersburg PA
CBHW062139040426
41189CB00029B/18
* 9 7 8 1 1 1 9 8 9 7 7 6 7 *